QUALITY CONTROL IN FRUIT AND VEGETABLE PROCESSING

Methods and Strategies

Innovations in Agricultural and Biological Engineering

QUALITY CONTROL IN FRUIT AND VEGETABLE PROCESSING

Methods and Strategies

Edited by

Megh R. Goyal, PhD, PE

Faizan Ahmad, MTech

AAP | APPLE ACADEMIC PRESS

First edition published 2023

Apple Academic Press Inc.
1265 Goldenrod Circle, NE,
Palm Bay, FL 32905 USA

760 Laurentian Drive, Unit 19,
Burlington, ON L7N 0A4, CANADA

CRC Press
6000 Broken Sound Parkway NW,
Suite 300, Boca Raton, FL 33487-2742 USA

4 Park Square, Milton Park,
Abingdon, Oxon, OX14 4RN UK

Library and Archives Canada Cataloguing in Publication

Title: Quality control in fruit and vegetable processing : methods and strategies / edited by Megh R. Goyal, PhD, PE, Faizan Ahmad, MTech.
Names: Goyal, Megh R., editor. | Ahmad, Faizan, 1986- editor.
Series: Innovations in agricultural and biological engineering.
Description: First edition. | Series statement: Innovations in agricultural and biological engineering | Includes bibliographical references and index.
Identifiers: Canadiana (print) 20220399875 | Canadiana (ebook) 20220399891 | ISBN 9781774911181 (hardcover) | ISBN 9781774911198 (softcover) | ISBN 9781003304999 (ebook)
Subjects: LCSH: Fruit processing plants—Quality control. | LCSH: Vegetable processing plants—Quality control.
Classification: LCC TP440 .Q35 2023 | DDC 664/.8—dc23

Library of Congress Cataloging-in-Publication Data

CIP data on file with US Library of Congress

ISBN: 978-1-77491-118-1 (hbk)
ISBN: 978-1-77491-119-8 (pbk)
ISBN: 978-1-00330-499-9 (ebk)

ABOUT THE BOOK SERIES: INNOVATIONS IN AGRICULTURAL AND BIOLOGICAL ENGINEERING

Under this book series, Apple Academic Press Inc. is publishing book volumes over a span of 8–10 years in the specialty areas defined by the American Society of Agricultural and Biological Engineers (<asabe.org>). Apple Academic Press Inc. aims to be a principal source of books in agricultural and biological engineering. We welcome book proposals from readers in areas of their expertise.

The mission of this series is to provide knowledge and techniques for agricultural and biological engineers (ABEs). The book series offers high-quality reference and academic content on agricultural and biological engineering (ABE) that is accessible to academicians, researchers, scientists, university faculty and university-level students, and professionals around the world.

Agricultural and biological engineers ensure that the world has the necessities of life, including safe and plentiful food, clean air and water, renewable fuel and energy, safe working conditions, and a healthy environment by employing knowledge and expertise of the sciences, both pure and applied, and engineering principles. Biological engineering applies engineering practices to problems and opportunities presented by living things and the natural environment in agriculture.

ABE embraces a variety of the following specialty areas (<asabe.org>): aquaculture engineering, biological engineering, energy, farm machinery and power engineering, food, and process engineering, forest engineering, information, and electrical technologies, soil, and water conservation engineering, natural resources engineering, nursery, and greenhouse engineering, safety, and health, and structures and environment.

For this book series, we welcome chapters on the following specialty areas (but not limited to):

- Academia to industry to end-user loop in agricultural engineering;
- Agricultural mechanization;
- Aquaculture engineering;
- Biological engineering in agriculture;

- Biotechnology applications in agricultural engineering;
- Energy source engineering;
- Farm to fork technologies in agriculture;
- Food and bioprocess engineering;
- Forest engineering;
- GPS and remote sensing potential in agricultural engineering;
- Hill land agriculture;
- Human factors in engineering;
- Impact of global warming and climatic change on agriculture economy;
- Information and electrical technologies;
- Irrigation and drainage engineering;
- Nanotechnology applications in agricultural engineering;
- Natural resources engineering;
- Nursery and greenhouse engineering;
- Potential of phytochemicals from agricultural and wild plants for human health;
- Power systems and machinery design;
- Robot engineering and drones in agriculture;
- Rural electrification;
- Sanitary engineering;
- Simulation and computer modeling;
- Smart engineering applications in agriculture;
- Soil and water engineering;
- Micro-irrigation engineering;
- Structures and environment engineering;
- Waste management and recycling;
- Any other focus areas.

For more information on this series, readers may contact:

Megh R. Goyal, PhD, PE
Book Series Senior Editor-in-Chief: Innovations in Agricultural and Biological Engineering
E-mail: goyalmegh@gmail.com

OTHER BOOKS ON AGRICULTURAL AND BIOLOGICAL ENGINEERING BY APPLE ACADEMIC PRESS, INC.

Management of Drip/Trickle or Micro Irrigation
Megh R. Goyal, PhD, PE, Senior Editor-in-Chief
Evapotranspiration: Principles and Applications for Water Management
Megh R. Goyal, PhD, PE and Eric W. Harmsen, PhD Editors

Book Series: Research Advances in Sustainable Micro Irrigation
Senior Editor-in-Chief: Megh R. Goyal, PhD, PE

- **Volume 1:** Sustainable Micro Irrigation: Principles and Practices
- **Volume 2:** Sustainable Practices in Surface and Subsurface Micro Irrigation
- **Volume 3:** Sustainable Micro Irrigation Management for Trees and Vines
- **Volume 4:** Management, Performance, and Applications of Micro Irrigation Systems
- **Volume 5:** Applications of Furrow and Micro Irrigation in Arid and Semi-Arid Regions
- **Volume 6:** Best Management Practices for Drip Irrigated Crops
- **Volume 7:** Closed Circuit Micro Irrigation Design: Theory and Applications
- **Volume 8:** Wastewater Management for Irrigation: Principles and Practices
- **Volume 9:** Water and Fertigation Management in Micro Irrigation
- **Volume 10:** Innovation in Micro Irrigation Technology

Book Series: Innovations and Challenges in Micro Irrigation
Senior Editor-in-Chief: Megh R. Goyal, PhD, PE

- Engineering Interventions in Sustainable Trickle Irrigation: Water Requirements, Uniformity, Fertigation, and Crop Performance
- Management Strategies for Water Use Efficiency and Micro Irrigated Crops: Principles, Practices, and Performance

- Micro Irrigation Engineering for Horticultural Crops: Policy Options, Scheduling, and Design
- Micro Irrigation Management: Technological Advances and Their Applications
- Micro Irrigation Scheduling and Practices
- Performance Evaluation of Micro Irrigation Management: Principles and Practices
- Potential of Solar Energy and Emerging Technologies in Sustainable Micro Irrigation
- Principles and Management of Clogging in Micro Irrigation
- Sustainable Micro Irrigation Design Systems for Agricultural Crops: Methods and Practices

Book Series: Innovations in Agricultural and Biological Engineering
Senior Editor-in-Chief: Megh R. Goyal, PhD, PE

- Advanced Research Methods in Food Processing Technologies
- Advances in Food Process Engineering: Novel Processing, Preservation and Decontamination of Foods
- Advances in Green and Sustainable Nanomaterials: Applications in Energy, Biomedicine, Agriculture, and Environmental Science
- Advances in Sustainable Food Packaging Technology
- Analytical Methods for Milk and Milk Products, Volume 1: Sampling Methods, Chemical and Compositional Analysis
- Analytical Methods for Milk and Milk Products, Volume 2: Physicochemical Analysis of Concentrated, Coagulated and Fermented Products
- Biological and Chemical Hazards in Food and Food Products: Prevention, Practices, and Management
- Bioremediation and Phytoremediation Technologies in Sustainable Soil Management, Volume 1: Fundamental Aspects and Contaminated Sites
- Bioremediation and Phytoremediation Technologies in Sustainable Soil Management, Volume 2: Microbial Approaches and Recent Trends
- Bioremediation and Phytoremediation Technologies in Sustainable Soil Management, Volume 3: Inventive Techniques, Research Methods, and Case Studies
- Bioremediation and Phytoremediation Technologies in Sustainable Soil Management, Volume 4: Degradation of Pesticides and Polychlorinated Biphenyls

- Dairy Engineering: Advanced Technologies and Their Applications
- Developing Technologies in Food Science: Status, Applications, and Challenges
- Emerging Technologies in Agricultural Engineering
- Engineering Interventions in Agricultural Processing
- Engineering Interventions in Foods and Plants
- Engineering Practices for Agricultural Production and Water Conservation: An Interdisciplinary Approach
- Engineering Practices for Management of Soil Salinity: Agricultural, Physiological, and Adaptive Approaches
- Engineering Practices for Milk Products: Dairyceuticals, Novel Technologies, and Quality
- Enzyme Inactivation in Food Processing: Technologies, Materials, and Applications
- Field Practices for Wastewater Use in Agriculture: Future Trends and Use of Biological Systems
- Flood Assessment: Modeling and Parameterization
- Food Engineering: Emerging Issues, Modeling, and Applications
- Food Process Engineering: Emerging Trends in Research and Their Applications
- Food Processing and Preservation Technology: Advances, Methods, and Applications
- Food Technology: Applied Research and Production Techniques
- Functional Dairy Ingredients and Nutraceuticals: Physicochemical, Technological, and Therapeutic Aspects
- Handbook of Research on Food Processing and Preservation Technologies: Volume 1: Nonthermal and Innovative Food Processing Methods
- Handbook of Research on Food Processing and Preservation Technologies, Volume 2: Nonthermal Food Preservation and Novel Processing Strategies
- Handbook of Research on Food Processing and Preservation Technologies, Volume 3: Computer-Aided Food Processing and Quality Evaluation Techniques
- Handbook of Research on Food Processing and Preservation Technologies, Volume 4: Design and Development of Specific Foods, Packaging Systems, and Food Safety
- Handbook of Research on Food Processing and Preservation Technologies, Volume 5: Emerging Techniques for Food Processing, Quality, and Safety Assurance

- Modeling Methods and Practices in Soil and Water Engineering
- Nanotechnology and Nanomaterial Applications in Food, Health, and Biomedical Sciences
- Nanotechnology Applications in Agricultural and Bioprocess Engineering: Farm to Table
- Nanotechnology Applications in Dairy Science: Packaging, Processing, and Preservation
- Nanotechnology Horizons in Food Process Engineering, Volume 1: Food Preservation, Food Packaging, and Sustainable Agriculture
- Nanotechnology Horizons in Food Process Engineering, Volume 2: Scope, Biomaterials, and Human Health
- Nanotechnology Horizons in Food Process Engineering, Volume 3: Trends, Nanomaterials, and Food Delivery
- Novel and Alternative Methods in Food Processing: Biotechnological, Physicochemical, and Mathematical Approaches
- Novel Dairy Processing Technologies: Techniques, Management, and Energy Conservation
- Novel Processing Methods for Plant-Based Health Foods: Extraction, Encapsulation and Health Benefits of Bioactive Compounds
- Novel Strategies to Improve Shelf-Life and Quality of Foods: Quality, Safety, and Health Aspects
- Phytochemicals and Medicinal Plants in Food Design: Strategies and Technologies for Improved Healthcare
- Processing of Fruits and Vegetables: From Farm to Fork
- Processing Technologies for Milk and Milk Products: Methods, Applications, and Energy Usage
- Quality Control in Fruit and Vegetable Processing: Methods and Strategies
- Scientific and Technical Terms in Bioengineering and Biological Engineering
- Soil and Water Engineering: Principles and Applications of Modeling
- Soil Salinity Management in Agriculture: Technological Advances and Applications
- State-of-the-Art Technologies in Food Science: Human Health, Emerging Issues and Specialty Topics
- Sustainable Biological Systems for Agriculture: Emerging Issues in Nanotechnology, Biofertilizers, Wastewater, and Farm Machines

- Sustainable Nanomaterials for Biomedical Engineering: Impacts, Challenges, and Future Prospects
- Sustainable Nanomaterials for Biosystems Engineering: Trends in Renewable Energy, Environment, and Agriculture
- Technological Interventions in Dairy Science: Innovative Approaches in Processing, Preservation, and Analysis of Milk Products
- Technological Interventions in Management of Irrigated Agriculture
- Technological Interventions in the Processing of Fruits and Vegetables
- Technological Processes for Marine Foods, From Water to Fork: Bioactive Compounds, Industrial Applications, and Genomics
- The Chemistry of Milk and Milk Products: Physicochemical Properties, Therapeutic Characteristics, and Processing Methods

ABOUT SENIOR EDITOR-IN-CHIEF

Megh R. Goyal, PhD, PE

Megh R. Goyal, PhD, PE, is, currently a retired professor of agricultural and biomedical engineering from the General Engineering Department at the College of Engineering at the University of Puerto Rico–Mayaguez Campus (UPRM); and Senior Acquisitions Editor and Senior Technical Editor-in-Chief for Agricultural and Biomedical Engineering for Apple Academic Press Inc.

During his long career, he has worked as a Soil Conservation Inspector; Research Assistant at Haryana Agricultural University and Ohio State University; Research Agricultural Engineer/Professor at the Department of Agricultural Engineering of UPRM; and Professor of Agricultural and Biomedical Engineering in the General Engineering Department of UPRM. He spent a one-year sabbatical leave in 2002–2003 at the Biomedical Engineering Department of Florida International University, Miami, USA.

Dr. Goyal was the first agricultural engineer to receive the professional license in agricultural engineering from the College of Engineers and Surveyors of Puerto Rico. In 2005, he was proclaimed the "Father of Irrigation Engineering in Puerto Rico for the Twentieth Century" by the American Society of Agricultural and Biological Engineers, Puerto Rico Section, for his pioneering work on micro irrigation, evapotranspiration, agroclimatology, and soil and water engineering.

During his professional career of 52 years, he has received many awards, including Scientist of the Year, Membership Grand Prize for the American Society of Agricultural Engineers Campaign, Felix Castro Rodriguez Academic Excellence Award, Man of Drip Irrigation by the Mayor of Municipalities of Mayaguez/Caguas/Ponce and Senate/Secretary of Agriculture of ELA, Puerto Rico, and many others. He has been recognized as one of the experts "who rendered meritorious service for the development of [the] irrigation sector in India" by the Water Technology Centre of Tamil Nadu Agricultural University in Coimbatore, India, and ASABE who bestowed on him the 2018 Netafim Microirrigation Award.

Dr. Goyal has authored more than 200 journal articles and edited more than 100 books. AAP has published many of his books, including *Management of Drip/Trickle or Micro Irrigation; Evapotranspiration: Principles and Applications for Water Management;* ten-volume set on *Research Advances in Sustainable Micro Irrigation.* Dr. Goyal has also developed several book series with AAP, including Innovations in Agricultural & Biological Engineering (with over 60 titles in the series to date), Innovations and Challenges in Micro Irrigation; and Innovations in Plant Science for Better Health: From Soil to Fork.

Dr. Goyal received his BSc degree in Engineering from Punjab Agricultural University, Ludhiana, India, and his MSc and PhD degrees from the Ohio State University, Columbus, Ohio, USA. He also earned a Master of Divinity degree from the Puerto Rico Evangelical Seminary, Hato Rey, Puerto Rico, USA.

Readers may contact him at goyalmegh@gmail.com.

ABOUT THE EDITOR

Faizan Ahmad, MTech
Faizan Ahmad, MTech, Assistant Professor, Department of Post Harvest Engineering and Technology, Faculty of Agricultural Sciences at Aligarh Muslim University, Uttar Pradesh, India

Faizan Ahmad, MTech, is Assistant Professor in the Department of Post Harvest Engineering and Technology in the Faculty of Agricultural Sciences at Aligarh Muslim University, in Uttar Pradesh, India. He has more than seven years of teaching and research experience. His major area of research includes food preservation, extraction of bioactive compounds from agricultural waste, food processing, food engineering, etc. Presently he is working in food preservation using ultrasound as a pre-treatment, as well as in quality assessment of fruits and vegetables during storage and drying. Recently he designed a fully automatic vacuum dryer with a data recorder for effective and economic drying of fruits and vegetables, and the same has been published by the patent office of the Government of India.

He has published more than 10 research papers in peer-reviewed international and national journals and more than seven book chapters with various publishers, such as Elsevier, Springer, CRC Press. etc. He has presented his work at several international and national conferences and has attended more than 35 seminars, conferences, workshops, short-term courses, and faculty development program.

He has successfully guided five MTech students for their dissertation work and is presently supervising one more. He has also delivered lectures as a resource person at the Institute of Food Technology, Budelkhand University, Jhansi, India. He is a life member of the Indian Society of Technical Education (ISTE).

He received his BTech degree in Chemical Engineering from Aligarh Muslim University, Aligarh, Uttar Pradesh, India, and his MTech degree in Chemical Engineering (Process Modeling and Simulation) from Department of Chemical Engineering, Aligarh Muslim University, Aligarh, Uttar Pradesh, India.

Readers may contact him at: f4faizahmad1989@gmail.com

CONTENTS

CONTRIBUTORS

Faizan Ahmad
Department of Post Harvest Engineering and Technology, Faculty of Agricultural Sciences, A.M.U., Aligarh 202002, UP, India; E-mail: f4faizahmad1989@gmail.com

Saghir Ahmad
Department of Post Harvest Engineering and Technology, Faculty of Agricultural Sciences, A.M.U., Aligarh 202002, UP, India; E-mail: alsaghirqadri@gmail.com

Omeera Ayob
Department of Food Technology, Jamia Hamdard, New Delhi 110062, India; E-mail: umeraayob@gmail.com

Shayeeb Ahmad Bhat
Department of Post Harvest Engineering and Technology, Faculty of Agricultural Sciences, A.M.U., Aligarh 202002, UP, India; E-mail: shoaibfe91@gmail.com

Aabid Hussain Bhat
Department of Food Science, Jamia Hamdard University 110062, New Delhi, India; E-mail: aabidbhat998@gmail.com

Satyam Bansal
Department of Food Engineering and Technology, Bundelkhand University, Jhansi 248128, UP, India; E-mail: bansalsatyam0909@gmail.com

Tejaswi Boyapati
Department of Food Technology, Vignan's Foundation for Science, Technology and Research Institute, Guntur-Tenali Rd 522213, Vadlamudi, Andhra Pradesh, India; E-mail: tejaswi.tej.b@gmail.com

Ogechukwu Bose Chukwuma
Division of Environmental Technology, School of Industrial Technology, Universiti Sains Malaysia, 11800 Penang, Malaysia; E-mail: ogechukwuma@student.usm.my

Ishfaq Hamid Dar
Department of Post Harvest Engineering and Technology, Faculty of Agricultural Sciences, A.M.U., Aligarh 202002, UP, India; E-mail: ishfaqhamiddar@gmail.com

B. N. Dar
Department of Food Technology, IUST, Awantipora 192122, J&K, India; E-mail: darnabi@gmail.com

Payel Ghosh
Department of Food Technology, Vignan's Foundation for Science, Technology and Research Institute, Guntur-Tenali Rd 522213, Vadlamudi, Andhra Pradesh, India; E-mail: payelghosh89@gmail.com

Megh R. Goyal
Agric. & Biomedical Eng., University of Puerto Rico-Mayaguez, Mayaguez—Puerto Rico; E-mail: goyalmegh@gmail.com

Salam A. Ibrahim
Department of Food Microbiology and Biotechnology, North Carolina A&T State University, Greensboro 27214, NC, USA; E-mail: ibrah001@ncat.edu

Norli Ismail
Division of Environmental Technology, School of Industrial Technology,
Universiti Sains Malaysia, 11800 Penang, Malaysia; E-mail: norlii@usm.my

Bisma Jan
Department of Food Technology, Jamia Hamdard University 110062, New Delhi, India;
E-mail: Bismakaloo123@gmail.com

Pir Mohammad Junaid
Department of Post Harvest Engineering and Technology, Faculty of Agricultural Sciences,
A.M.U., Aligarh 202002, UP, India; E-mail: junaidpir24@gmail.com

Shiv Kumar Katiyar
Department of Food Engineering and Technology, Bundelkhand University, Jhansi248128, UP, India;
E-mail: skkatiyar69@gmail.com

Faisal Khan
Department of Post Harvest Engineering and Technology, Faculty of Agricultural Sciences,
A.M.U., Aligarh 202002, UP, India; E-mail: faisalkhan.khan059@gmail.com

Shams Tabrez Khan
Department of Agricultural Microbiology, Faculty of Agricultural Sciences, Aligarh Muslim University,
Aligarh 202002, UP, India; E-mail: shamsalig@yahoo.com

Arshied Manzoor
Department of Post Harvest Engineering and Technology, Aligarh Muslim University, Aligarh 202002,
UP, India; E-mail: arshidfe12@gmail.com

Shaziya Manzoor
Department of Food Science and Technology, University of Kashmir, Jammu 190006, J & K, India;
E-mail:shaziawani79@gmail.com

Zahra H. Mohammad
Conrad N. Hilton College of Hotel and Restaurant Management, UH, Houston 77001, TX - USA;
E-mail: zhmohamm@central.uh.edu

Farah Naqash
Department of Food Technology, IUST, Awantipora 192122, J&K, India;
E-mail: nakashfarah1@gmail.com

Gazia Nasir
Department of Post Harvest Engineering and Technology, Faculty of Agricultural Sciences,
A.M.U., Aligarh 202002, UP, India; E-mail: gazianasir@gmail.com

Sadaf Nazir
Department of Food Technology, BIET, Jhansi 284128, UP, India; E-mail: sadafnazirwani@yahoo.in

Junaid Ahmad Pandith
Department of Post Harvest Engineering and Technology, Faculty of Agricultural Sciences,
A.M.U., Aligarh 202002, UP, India; E-mail: junaidfe@gmail.com

Kirty Pant
Department of Food Engineering and Technology, Sant Longowal Institute of Engineering and
Technology, Longowal 148106, Punjab, India; E-mail:Kirty.pant@gmail.com

Mohd Rafatullah
Division of Environmental Technology, School of Industrial Technology, Universiti Sains Malaysia,
11800 Penang, Malaysia; E-mail: mrafatullah@.usm.my

Sandeep Singh Rana
Department of Food Technology, Vignan's Foundation for Science, Technology and Research Institute, Guntur-Tenali Rd 522213, Vadlamudi, Andhra Pradesh, India; E-mail: srana0505@gmail.com

Qurat Ul Eain Hyder Rizvi
Department of Food Technology, Eternal University, Sirmour 173025, Himachal Pradesh, India; E-mail: syeda15994@gmail.com

Shaibi Saleem
Department of Agricultural Microbiology, Faculty of Agricultural Sciences, Aligarh Muslim University, Aligarh 202002, UP, India; E-mail: zunairasaleem18@gmail.com

Asima Shafi
Department of Post Harvest Engineering and Technology, Faculty of Agricultural Sciences, A.M.U., Aligarh 202002, UP, India; E-mail: asimashafi97@gmail.com

Rafeeya Shams
Division of Food Science and Technology, Sher-e-Kashmir University of Agricultural Sciences & Technology of Jammu, Jammu 190011, J & K, India; E-mail: rafiya.shams@gmail.com

Mohit Sharma
Dr. S.S. Bhatnagar University, Institute of Chemical Engineering & Technology, Punjab University, Chandigarh 160014, India; E-mail: Mohitlomash362@gmail.com

Husnul Azan Tajarudin
Division of Bioprocess Technology, School of Industrial Technology, Universiti Sains Malaysia, 11800 Penang, Malaysia; E-mail: azan@usm.my

Mohammad Faisal Umar
Division of Environmental Technology, School of Industrial Technology, Universiti Sains Malaysia, 11800 Penang, Malaysia; E-mail: faisalumar@student.usm.my

Sadaf Zaidi
Department of Post Harvest Engineering and Technology, Faculty of Agricultural Sciences, A.M.U., Aligarh 202002, UP, India; E-mail: smsadafzaidi@gmail.com

ABBREVIATIONS & SYMBOLS

AD	anaerobic digestion
AEW	acidic electrolyzed water
AP	active packaging
AW	water activity
BOD	biological oxygen demand
C/N	carbon-nitrogen
CA	citric acid
CBP	consolidated bioprocessing
C-di-GMP	cyclic diguanylate
CFU	colony-forming unit
CH_3COOH	acetic acid
Cl	chlorine
ClO_2	chlorine dioxide
CO	chemical oxidation
Co-60	Cobalt-60
COD	chemical oxygen demand
CR	chemical reduction
Cs-137	Cesium-137
CT	computed tomography
CVMWUS	convective microwave ultrasound
CVUS	convective ultrasound
DNA	deoxyribonucleic acid
EC	edible coating
EFSA	European Food Authority
EMAP	equilibrium modified atmospheric packaging
EO	essential oil
EOs	essential oils
EPA	Environmental Protection Agency
F&V	fruits and vegetables
FAO	Food and Agricultural Organization
FAV	fruits and vegetables
FDA	Food and Drug Administration
FSSAI	Food Safety and Standards Authority of India
FVW(S)	fruits and vegetables waste(s)

GC	gas chromatography
GHG	greenhouse gas
GRAS	generally recognized as safe
H/MSI	hyperspectral/multispectral imaging
H2O2	hydrogen peroxide
HCB	hexachlorobenzene
HClO	hypochlorous acid
HEP	high electric field pulse
HHP	high hydrostatic pressure
HOCl	hypochlorite acid
HPP	high pressure processing
HPPL	high power pulsed light
HPU	high power ultrasound
HVED	high voltage electric discharge
ILP	intense light pulse
IR	infrared radiation
IUPAC	International Union of Pure and Applied Chemistry
kGy	kiloGray
KMS	potassium metabisulfite
LED	light-emitting diode
MA	malic acid
MeV	mega electron volt
MFD	microwave freeze drying
MIC	minimum inhibitory concentration
MRDL	maximum residual disinfectant level
MRI	magnetic resonance imaging
MS	mass spectroscopy
MSW	municipal solid waste
MT	microwave technology
MVD	microwave-assisted vacuum drying
MW	microwaves
NE	nanoemulsion
NGO	non-governmental organizations
NIRS	near infrared spectroscopy
NMR	nuclear magnetic resonance
NP's	nanoparticles
OD	osmotic dehydration
ORP	oxidation-reduction potential
PAA	peroxyacetic acid

PAA	peracetic acid
PEF	pulsed electric field
PL	pulse light
PPO	polyphenol oxidase
PRT	pretreatment
PSMVD	pulsed microwave vacuum drying
PV	pulsed vacuum
PVC	plasticized membrane electrodes
R&D	research and development
RNA	ribonucleic acid
ROS	reactive oxygen species
RTE	ready to eat
SDG	sustainable development goal
SHF	separate hydrolysis and fermentation
SL	shelf life
SSC	soluble solid content
SSCF	simultaneous saccharification and cofermentation
SSF	simultaneous saccharification and fermentation
STEC	Shiga toxin-producing *Escherichia coli*
TBA	thiobarbituric acid
TBARS	thiobarbituric acid reactive substances
TI	thermal imaging
TS	total solids
TSS	total soluble solids
UAOD	ultrasound-assisted osmotic dehydration
UF	ultrafiltration
UNICEF	United Nations Children's Fund
UP	ultrasound processing
US	ultrasound
USDA	United States Department of Agriculture
USFDA	United States Food and Drug Administration
USV	ultrasonic vacuum
UV	ultraviolet
VD	vacuum drying
VS	volatile solids
WHO	World Health Organization
WTE	waste to energy
XRI	x-ray imaging
ZnO	zinc oxide
γ	gamma

PREFACE

Food is one of the most essential components of life for human survival. Nowadays, people not only consume food to satiate their appetite but look for foods that are also rich in high nutritional value. Therefore, quality assessment of fruits and vegetables along with their preservation is one of the major concerns of food processing industries in recent years; since, fruits and vegetables have high moisture content and they are easily prone to degradation. In the last two decades, the preservation and quality assessment of fruits and vegetables have been done by various preservation methods including thermal methods and several destructive technologies. All these methods have their own advantages and disadvantages. Therefore, in recent years after the exhaustive research and experimentation, food researchers, scientists, and industrialists have suggested various nonthermal preservation technologies such as microwave, ultrasound, gamma irradiation, and pulse light technology, etc. in order to prolong the shelf life of food products and to reduce the textural, sensory, and nutritional losses. Other than these, scientists have suggested various novel nondestructive methods to assess the quality of fruits and vegetables. There is an urgent need for ample and scientifically correct literature to ensure food quality and food preservation, which would be helpful to students, researchers, and professional engaged in the processing and preservation of fruits and vegetables. This book is a genuine effort toward the same.

The book attempts to portray the applications of various nonthermal technologies such as microwave, ultrasound, gamma irradiation, pulsed light, and hurdle technology for improving the quality and safety of fruits and vegetables. The scope of various strategies (osmotic dehydration, ultrasound and ultrasound-assisted osmotic dehydration, nanoemulsions, and engineered nanomaterial) toward the preservation of fresh produce has also been included with a wide range of applications. Furthermore, this book also emphasizes the various nondestructive techniques that have been widely used for the quality assessment of fruits and vegetables during storage including image analysis, x-ray tomography, magnetic resonance imaging (MRI), nonmagnetic resonance imaging (NMR), color vision system, near-infrared spectroscopy (NIRS), and computerized tomography (CT). Applications of other nondestructive mechanical (E-nose, E-tongue) and dynamic methods (acoustic) for food quality and safety evaluation have also been included. At the end, potential use of fruits and vegetables waste as a viable feed stock for bioenergy and for the treatment of wastewater have also been enlightened. Thus, this book would have a significant scope in the preservation, safety, and quality evaluation of fruits and vegetables.

This book is divided into four parts. **Part I: Nonthermal Methods for Quality and Safety of Fruits and Vegetables** discusses the microwave, gamma irradiation, and pulse light technology for fruits and vegetables preservation, hurdle technology for the quality and safety of fresh produce during storage, and novel methods for decontamination of fresh produce. **Part II: Preservation Strategies for Fresh Produce** elaborates upon past, present, and future scope of preservation strategies for fruits and vegetables, application of osmotic dehydration and ultrasound technology for fruits and vegetables preservation, scope of nanoemulsions for quality and safety of fruits and vegetables, and engineered nanomaterials for food preservation and packaging. **Part III: Assessment of Quality of Fruits and Vegetables** highlights the nondestructive technologies for the quality assessment of fruits and vegetables. At the end, **Part IV: Potential Use of Fruit and Vegetable Waste** emphasizes the utilization of fruits and vegetable wastes as a viable feedstock for bioenergy and for the treatment of wastewater.

The book has taken its final shape due to the admirable/exceptional contributions by the contributing authors, who have been the soul of this book. We are highly thankful to the authors of each chapter for their sincere efforts in writing their chapters. Their names are mentioned in each chapter and also in the list of contributors.

We expect that this book will serve as knowledge capsule for all those who are working in the area of quality assessment and preservation of fruits and vegetables such as students, researchers, academicians, and food engineers.

We are also thankful to the Apple Academic Press, Inc., for giving us an opportunity to publish this book.

At last but not least we also extend our sincere thanks to our family members for their blessings, moral support, and motivation throughout this assignment, our spouses, and children for their patience, understanding, and encouragement during this project, and our institutes: Department of PHET, Aligarh Muslim University, Aligarh, UP, India, University of Puerto Rico—Mayaguez Campus, for their support during the compilation of this book.

— Editors

PART I

NONTHERMAL METHODS FOR QUALITY AND SAFETY OF FRUITS AND VEGETABLES

CHAPTER 1

QUALITY AND SAFETY OF FRESH PRODUCE DURING STORAGE: HURDLE TECHNOLOGY

ZAHRA H. MOHAMMAD, FAIZAN AHMAD, and
SALAM A. IBRAHIM

ABSTRACT

Due to the increased public awareness about the nutritional and health benefits, the quality requirements of fruits and vegetables have significantly increased. However, fruits and vegetables are highly perishable and implicated in many foods borne illnesses. Therefore, the produce industry is continually seeking effective, safe, and environmentally friendly intervention practices to decontaminate produce while preserving product quality and extending the shelf-life. However, individual chemical and nonchemical treatments show less efficiency in microbial reduction, and most of the time influence the quality and sensory attributes of produce. Hurdle technology has been shown to be more efficacious than individual treatments in diminishing the microorganisms while preserving product quality and extending shelf-life. Thus, hurdle technology is a favorable alternative technology for enhancing the quality and safety of fruits and vegetables. This chapter addresses past/current/future trends of hurdle technologies in the fruits and vegetable industry, advantages and challenges, application of ozone and electrolyzed water as promising novel disinfectants when combined with other treatments, and exploring the combination of two chemicals or organic acids in the produce sector.

Quality Control in Fruit and Vegetable Processing: Methods and Strategies. Megh R. Goyal, & Faizan Ahmad (Eds.)

1.1 INTRODUCTION

Fruits and vegetables (FAVs)[89] are highly perishable foods because they provide the ideal environment for the growth and survival of microorganisms, including spoilage, pathogenic microorganisms, yeast, and molds.[39,72,73,78] Moreover, the growth of these microorganisms affects the quality, sensory, and safety of the FAV.[39,78] As a result, these commodities require proper post-harvest handling and storage in order to prevent deterioration and to ensure product safety, thus increasing the shelf-life (SL).[57] Maintaining product quality and safety and lengthening the SL of FAV were always the major industry challenge.[56] Consequently, various preservation and intervention technologies have been utilized by the FAV processing industry in order to increase the SL and minimize the risk of spoilage and growth of pathogenic microorganisms.[21,57]

Most of the intervention methods are effective against microorganisms and have achieved the above-mentioned industry objectives.[57] However, the methods were always associated with adverse effects on the sensory quality and loss of nutrient value of the products.[57] Hurdle technology has, thus, been found to be a promising preservation method that ensures safety while maintaining the quality of products and increasing the SL.[5,21,64] Hurdle technology is the method in which two or more interventions are combined in order to improve the product quality, extend the SL, and ensure food safety.[18,53,64] Figure 1.1 represents the disinfection technologies for the FAV and the concept of hurdle technology.

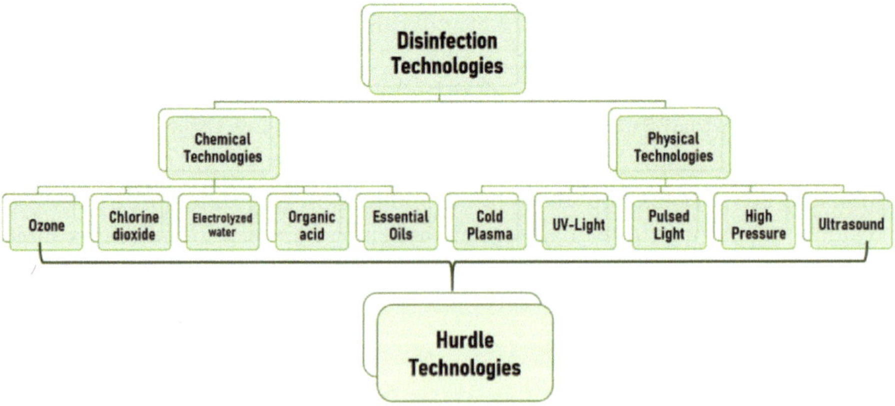

FIGURE 1.1 Disinfection technologies for fruits and vegetables.

The application of various successful hurdle technologies has, thus, been proposed as a promising alternative to individual intervention methods as a way to minimize the risk of food-borne illnesses and prolong the SL of FAV without affecting the product quality and sensory properties.[21] In this regard, Severino et al.[79] reported that the mixture of modified chitosan coating containing mandarin essential oils (EOs) treatment with ozonated water washing resulted in a higher reduction of *Listeria monocytogenes* on broccoli than the coating or ozonated water washing alone.

Ramos-Villarroel et al.[71] reported that combining plus light (PL) (180–1100 nm, 12 J/cm^2) and malic acid (MA, 2%) treatments was able to achieve up to a 5 log CFU/g reduction of *L. innocua* and *Escherichia coli* when applied to fresh-cut mushroom, watermelon, and avocado; and the results were much superior than when PL or MA acid was applied alone. Additionally, Mohammad et al.[49] applied combined treatments of ozonated water and acidic electrolyzed water (AEW) to alfalfa seeds and sprouts, which resulted in a greater reduction of *Salmonella* and Shiga toxin-producing *E. coli* (STEC) on both alfalfa seeds and sprouts than individual treatments.

Gupta et al.[31] also reported a greater reduction of microorganisms on French beans and a 1-week longer SL when they used the combined X-ray and citric acid (CA) treatments. These are some examples of findings that have demonstrated the efficacy of hurdle technologies as promising postharvest treatments for FAV.

In this chapter, the authors have addressed the scope of hurdle technologies, their potential applications in the FAV industries, and the most common technologies currently utilized along with their potential as disinfectants in this food sector.

1.2 HURDLE TECHNOLOGY: PAST, PRESENT, AND FUTURE TRENDS

Traditional food preservation methods have been utilized during the past century.[80] Hurdle technologies for food preservation have emerged more recently. However, the original concept of hurdle technology grew out of the conventional preservation methods used for food preservation and the principles behind the effect of these methods including the adjustment of water activity, pH, and temperatures.[50] As a result, successful hurdle technologies are based on temperature, relative humidity, pH, water activity, chemicals, and antimicrobials.[80] In its early stage, hurdle technology was used in food preservation by combining multiple preservation methods in order to achieve more reliable preservation goals without full knowledge of the scientific bases involved.[44,80] Later on, it was found that hurdle technology provided synergistic effects by combining two or more methods to reduce the level of various microorganisms.[71] Therefore, more attention was given to hurdle technology as a favorable method for enhancing product quality, increasing the SL, and ensuring the safety of food products using different preservation factors and by subjecting the products to milder applications.[43,50] The first hurdle technology was applied to meat in order to maintain its freshness[44] and was then extended to other food products, including processed FAV,[4] and finally, dairy products.[20]

Soliva-Fortuny et al.[82] investigated the effect of combination of methods (Hurdle Technology) to extend the SL of avocados. In their research, the authors examined different combinations, including the addition of organic acid (sorbic acid) and modified atmosphere packaging, controlling the storage temperature, reducing the pH or water activity (aw), and combining vacuum packaging and at 4°C storage temperature. These authors reported that combining sorbic acid with a modified atmosphere was sufficient to control yeasts and molds on avocados as well as the microbiological stability during storage. However, the approach of reducing water activity or pH through the addition of sugar may affect the sensory attributes of the products and negatively impact the quality. However, their attempt led to the subsequent discovery of more successful technologies.

Barwal et al.[8] also investigated hurdle technology using various combinations and different concentrations to preserve cauliflower for up to 180 days. Their hurdle combinations included CA (1.0%) and potassium metabisulfite (KMS at 0.2%) after subjecting the cauliflower to blanching. Authors found that cauliflower samples treated with a salt solution at concentrations of 10 and 15% with one-fifth percent KMS were the highest-ranked products during the complete storage period of 180 days. Moreover, the samples were ranked above the acceptable level by the sensory panel in terms of sensory attributes and quality. The hurdle combination in this study was an effective way to increase the SL of FAV and also resulted in motivating other scientists to study other options with different types of fruits or vegetables.

For example, Vibhakara et al.[88] began studying hurdle technology using chemicals, organic acids, and antimicrobials to increase the SL of grated carrots. The authors manipulated the water activity and pH in addition to antimicrobials. The experimental combinations included sodium chloride (3.0%), CA (0.9%) for water activity, and pH reduction, respectively, followed by the addition of sodium citrate (1.0%), and sodium benzoate

(0.2%) as antimicrobial agents. This was followed by partial dehydration of the carrots and storage at a temperature range of 19–33°C. The grated carrots were stored for more than 6 months at ambient temperature. Moreover, in terms of microbiological safety, the products were found to be microbiologically safe during the entire study.

Nowadays, hurdle technologies are employed in all types of food products, including fresh, minimally processed, and processed foods in order to prolong the SL and to maintain quality and microbiological safety.[21] With novel and intelligently applied hurdle technologies, food products are subjected to milder treatments that serve to retain quality while preventing spoilage and the growth of pathogenic microorganisms and extending the SL of the products.[21]

Today, food industries pay more attention to nonchemical methods such as pulse technologies, ozone, and natural antimicrobial methods.[21] However, these methods work better when combined with other methods (hurdle technology) to achieve the desired results. The food industry's current trend is to provide high-quality foods while ensuring safety.[57] Thus, in the future, there will be an increasing demand for intelligent hurdle technology as food preservation becomes more aligned with it.[57] Selected examples of past and present disinfection methods for FAVs are shown in Table 1.1.

1.3 USE OF HURDLE TECHNOLOGY IN FRUITS AND VEGETABLES: ADVANTAGES AND CHALLENGES

The hurdle technology can extend the SL of FAV, minimize spoilage and the growth of pathogenic microorganisms, and help to retain overall product quality.[21] However, to achieve these objectives, the technology must employ multitarget treatments, including postharvest handling and sanitizing procedures prior to storage. For this purpose, different disinfections and sanitizers are used, including chemical, physical, and organic acids.[21]

Some of the more common hurdle-based disinfection treatments used for FAV prior to storage utilize a combination of two or more of the following agents: chlorine (Cl), ozone, peroxyacetic acid (PAA), chlorine dioxide (ClO_2), hydrogen peroxide, natural antimicrobials, packaging, electrolyzed water, organic acids, UV light, pulse light, ultrasound (US), and high hydroelectric pressure.[44,80] In this regard, researchers are continually searching for newer, more effective hurdle combinations that will provide optimum product quality and safety.

1.3.1 COMBINING OZONE AND ELECTROLYZED WATER TO INACTIVATE PATHOGENICS ON FRESH PRODUCE

The use of a single application of ozone or AEW at standard concentration levels on fruits or vegetables has been shown to be ineffective in minimizing the growth of pathogenic microorganisms.[48] Neither the ozone application alone nor the AEW alone at low concentrations is sufficient for adequately reducing the levels of pathogenic microbes in fresh produce. Moreover, higher concentrations of ozone alone or AEW alone have resulted in a negative impact on the quality of fresh FAV. However, the combination of ozone and AEW as hurdle technology has emerged as an alternative disinfection approach to help to reduce pathogenic and spoilage microorganisms, improve product quality and extend the SL of fresh produce.[11,39] The synergistic effects that result from combining these two applications allow for milder doses of each application and more effective treatments.[21]

Considering the properties of each application and little effect that each provides alone, the combination of both provides a powerful antimicrobial treatment. Ozone is a highly reactive gas that consists of three oxygen atoms. Moreover, it is a powerful oxidative agent and has antimicrobial properties against a wide range of microorganisms including bacterial, viruses, molds, and spores.[21] Ozone has been approved by the Food and Drug Administration (FDA) for use as an antimicrobial on foods, and it is recognized as generally recognized as safe (GRAS).[10] Ozone is known for its high reactivity and destroys microorganisms via two different oxidation mechanisms.[11] Thus, it destroys microorganisms faster than common chemical disinfectants such as chlorine by targeting molecular material and damaging the microorganisms' DNA structure.[11] As a disinfection agent, ozone leaves no chemical residuals and causes irreparable damage and destruction to microorganisms. However, ozone is unstable and has a half-life of about 30 min in water. In addition, higher ozone levels and longer treatment exposure times can result in adverse effects, such as off-flavor, color change, and phytochemical degradation.[11,21,48]

Similarly, AEW possesses a robust bactericidal activity on a wide range of bacteria and has been used as an alternative disinfectant to sodium hypochlorite.[11,32] AEW is generated by producing an acidic environment using sodium chloride (NaCl) or sodium hypochlorite with a pH of 2.3–2.7 and an oxidation–reduction potential (ORP) > 1000 mV.[21,32] The bactericide effect of AEW is through the inactivation of bacterial cells by the active chlorine residuals, such as Cl_2, HOCl, and OCl that break down DNA and interrupt the metabolic process.[46] Therefore, AEW has been broadly used as a disinfection agent for the fruit and vegetable industry.[15,32,100]

In addition to its effectiveness as a strong bactericidal, AEW is safe, relatively inexpensive, easy to use, and it is eco-friendly.[25] However, some disadvantages and limitations of the application of AEW are highly adverse effects on sensory attributes and product quality as a result of low pH and long exposure time.[98] Since both ozone and AEW depend on dosing levels and exposure time, a single application is not sufficient to achieve the target goal of microbial destruction without affecting the quality of the produce. To this end, combining ozone with AEW may reduce the adverse effects caused by a single application since milder doses are used to perform the hurdle. In this regard, several hurdle applications of ozone and AEW have been conducted and found to be effective in reducing microorganisms without compromising product quality and sensory attributes.[49,90]

Wang et al.[90] investigated the application of ozonated water followed by AEW on fresh-cut cilantro leaves and the effect of the treatments on color, texture, sensory, electrolyte leakage, and microbial reduction of cilantro at 0°C during the storage of 14 days. They reported that a combination treatment using ozone and AEW was effective against total aerobic counts while maintaining low microbial counts on fresh-cut cilantro during storage. However, the overall quality of the cilantro was negatively affected.

Mohammad et al.[49] investigated the efficacy of combining ozone and AEW for the inactivation of *Salmonella* and STEC on alfalfa seeds and sprouts. The authors observed that the treatments using the mixture of ozone and AEW resulted in the reduction of both pathogens with no visual effects on quality, including SL, weight, and color of sprouts. Moreover, the mixture of ozone and AEW has been studied extensively and found to be potent against microorganisms including pathogenic *Salmonella* and *E. coli*. Consequently, this hurdle could have the potential to be used for postharvest disinfection of FAV over extended storage times.

1.3.2 COMBINING OZONE WITH ORGANIC ACIDS OR CHEMICAL TREATMENTS

Most of the FAVs are consumed raw or minimally processed without killing steps, such as cooking or processing.[63,72] As a result, these food products are subjected to contamination and growth of different types of microorganisms including pathogens during postharvest, storage, and processing, which increases the risk of food-borne illnesses.[21,37] Proper postharvest handling, processing, and storage of FAV are thus essential to ensure safety and enhance the product quality.[57] The FAV industries are continually looking for better treatments that will help to ensure product safety and satisfy consumers' demands.[37]

Nonthermal techniques have gained popularity recently among food processors and researchers.[95,96] For example, ozone applications in both gas and aqueous forms have emerged as alternative nonthermal techniques to replace chemical disinfectants for FAV.[21] Although ozone application has reduced the levels of many food-borne pathogens including *Salmonella*, *E. coli*, and *L. monocytogenes*, the level of reduction has not been sufficient.[21,48] Moreover, at higher concentrations or long exposure times, ozone application causes significant quality issues.[21] Consequently, researchers and the produce industry examined combining ozone application with organic acids along with some chemical disinfectants.[63] Combining ozone with other disinfectants might thus enhance the efficacy of ozone application against microorganisms.[37]

Yuk et al.[95,96] studied the efficacy of ozone alone and in combination with organic acids (lactic, acetic, or CAs) against *E. coli* and *LM* on lettuce and mushrooms using different concentrations and various exposure times. The results highlighted that the combined treatments were more effective in reducing pathogens than ozone alone but were not effective at controlling pathogens during storage.[95,96]

Garcia et al.[24] investigated the effect of combinations of ozone and chlorine in reducing natural aerobic microflora on lettuce in order to extend the SL. The results showed that combining ozone (7.5 mg/L) with 150 ppm choline increased the SL of lettuce salads. Whang chai et al.[91] also studied the efficacy of ozone alone or with some organic acids (citric, ascorbic, and oxalic acids) in reducing natural microflora, browning, and polyphenol oxidase (PPO) activity on longan fruit. These authors concluded that the treatment of fruits with ozone and oxalic or CA resulted in a significant reduction of natural microflora as well as less browning and less PPO activity.

The findings from these studies suggest that the combination of ozone with some organic acids or chemical disinfectants results in better efficacy than ozone alone for FAV, so combining ozone with organic acids or chemical treatment could be a potential hurdle disinfectant for FAV. Table 1.1 represents the work done by various authors in the past several years on the concept of hurdle technologies using two or multiple treatments.

TABLE 1.1 Successful Hurdle Technologies using Two or Multiple Treatments.

Combination of treatments	Commodity	Purpose	Results	Conditions	References
CA with mild doses of gamma radiation	Processed French beans	Shelf Stable	Modified atmosphere of 18% O_2 & 4% CO_2 & extend shelf-life	CA = 8.4 g/L, 0.7 kGy, at 10 °C (Irradiation dose)	[31]
Electrolyzed water and CA	Shredded Carrots	AB	3.73 Unit	1% citric & AEW at 100 part per million at 50°C for 3 min	[68]
		YAF	3.69 Unit		
		L. M	3.97 Unit		
		E. coli O157:H7	4.0 Unit *(Unit = CFU/g)*		
	Cabbage	AB	3.98 Unit	1% citric & AEW at 100 ppm at 50°C for 3 min	[69]
		YAM	3.45 Unit		
		L. M	3.99 Unit		
		E. coli O157:H7	4.19 Unit *(Unit = CFU/g)*		
Ethanol and Acetic acid	Lettuce	Antibrowning	Suppressed browning	1/4 M Acetic acid for 5 s, and 200 mL/L EtOH for 2 min	[33]
Ozone and chlorine	Lettuce	Shelf-life	Extended shelf-life for up to 25 days	Ozone 10 g/L & Chorine 0.2 mg/L	[24]
Ozone and organic acid	Lettuce	*E. coli*	2.31 CFU/g	3 ppm ozone & 1% acetic, citric	[95]
		LM	1.84 CFU/g		
	Enoki mushroom	*E. coli*	2.26 CFU/g	3 ppm ozone & 1% acetic, citric	[96]
		LM	1.32 CFU/g		

CA, citric acid; AB, aerobic bacteria; YAF, yeast and fungi; YAM, yeast and mold.

1.3.3 COMBINING ANTIMICROBIAL SUBSTANCES WITH FOOD PACKAGING TO EXTEND THE SHELF-LIFE OF FRUITS AND VEGETABLES

In recent years, combining antimicrobial agents with packaging has become popular because it can extend the SL of food products, maintain quality, and ensure protection from spoilage and pathogenic microorganisms.[40,94] The addition of antimicrobial agents into food packaging is called active antimicrobial packaging (AP) in which the packaging substances interact with the food products themselves.[94] AP has been well-studied by the scientific community.[40] The idea of adding antimicrobials to packaging grew out of the basic fact that packaging protects food. Thus, it was assumed that antimicrobials could be employed to help to prevent the deterioration caused by spoilage and pathogenic microorganisms as well as any chemical and physical hazards that might arise during storage or even during the preparation of products.[40,65] Currently, there are two types of AP-emitters and absorbers.[94]

The emitters type AP is also called an active-releasing system because it is designed to release substances into packaged food or the surrounding space. The absorbers type AP is called a scavenging system because it is designed to absorb or remove compounds such as oxygen, flavor, moisture, or any undesired compounds from the food or its surroundings.[65,94] Combining antimicrobials with food packaging as a hurdle technology prolongs the SL, inhibits microbial growth, and helps to maintain product quality.[65]

Some examples of antimicrobials used in active AP are chlorine dioxide, organic acids, plant extracts, sulfur dioxide, enzymes, bacteriocins, and EOs.[65] These antimicrobial compounds can be added to the packaging through the addition of pads or sachets containing antimicrobial volatiles, or they can be added directly to the packaging surface through either coextrusion, immobilization, or adsorption techniques.[94] For example, conventionally, active antimicrobials are added to food products, but most of the time, FAVs are contaminated with spoilage and pathogenic microbes during storage and at surfaces. Therefore, incorporating active antimicrobial agents into active packaging may results in a more effective treatment method because active packaging released antimicrobial agents directly to the surface of the food.[94]

The AP system is a promising technology that could help to prolong the SL of food products and inhibit harmful microorganisms while retaining the quality. However, AP presents some challenges, including high cost and release and absorption rates, and food composition.[40,45] For example, the release or absorption rates of active antimicrobials could be affected by fruit or vegetable composition, and in some instances, may interact with the active substance and reduce its antimicrobial efficacy. Thus, overcoming these challenges will require additional work and research. Nevertheless, interest in AP continues to grow by both the food industry and researchers.[94]

The application of these types of packing for the food industry, including FAV, is increasing as it is a promising technology for enhancing quality, ensuring safety, and extending the SL of produce. However, one of the critical remaining challenges is to develop antimicrobial active packaging systems that are pathogens specific, which would help by inhibiting specifically targeted pathogens rather than other ones.[45] In addition, AP could be enhanced with innovative features such as systems for releasing flavor, aroma, and antimicrobials.[45]

1.3.4 COMBINING CHLORINE DIOXIDE AND PEROXYACETIC ACID FOR DISINFECTION OF FRESH PRODUCE

Postharvest FAV disinfection is critical to ensure safety and product quality. Chlorine is a common commercial disinfection agent used in the FAV industry,[57] yet due to the low effectiveness of traditional chlorine and its chemical byproducts,[74] many alternative chemical and nonchemical disinfection technologies are emerging.[21] However, using these emerging methods alone at the recommended concentration levels has been shown to be ineffective in reducing microorganisms. Moreover, if used at higher concentration levels, product quality is negatively affected.[39] Therefore, combining two or multiple disinfectants at lower concentrations in order to achieve synergistic effects could be an alternative to chlorine for enhancing safety and without compromising quality.[39] This section provides detailed information on the combining of choline dioxide and PAA as a chlorine replacement for the disinfection of FAV.

Chlorine dioxide (ClO_2) is a gas that initially emerged as a water disinfection agent.[28] ClO_2 has powerful antimicrobial activity and can be used as a disinfectant for FAV.[28] It is superior to chlorine and does not leave a significant number of chemical residues. However, ClO_2 is only suitable for washing unsliced and unpeeled FAV[60] and is available in both gas and aqueous form. However, ClO_2 in its gas form is more potent as an antimicrobial than its aqueous form because the gas has greater penetrability.[28]

Several studies investigated the effectiveness of ClO_2 as a disinfectant for spinach, potatoes, mung bean sprouts, lettuce, onions, cabbage and other produce, cantaloupe, and strawberries.[47,55,66,85,94] Most of these studies found ClO_2 at high concentrations to be effective in reducing initial spoilage microflora, yeast, molds, and pathogens with minimal impact on visual quality due to high concentration.[60] The primary disadvantages of using ClO_2 include the following a short life requires on-site generation, ineffective at low concentration, creates an explosion hazard, toxicity when used at high concentration, and negatively impacts the quality of produce at higher concentrations.[21]

While a PAA is also a potent antimicrobial agent and has shown to be effective against spoilage and pathogenic microorganism,[29] combining it with ClO_2 could improve the efficiency of ClO_2 by overcoming the challenges associated with individual treatment. In conclusion, synergistic effects are expected after combining ClO_2 and aerosolized PAA. Consequently, this hurdle technology would have the potential to be used for powerful antimicrobial disinfection of FAV using the combination of two chemical disinfectants with minimal or no adverse effects on product quality.

1.3.5 COMBINATION OF NATURAL ANTIMICROBIALS FOR FRESH PRODUCE

With increasing demand and market growth, the fresh FAV industry is also facing new challenges that include maintaining the quality of fresh produce, prolonging SL, and ensuring the safety of products in order to protect consumers' health.[21] Recently, the focus of research has been on developing treatments using natural antimicrobials for fresh produce that support consumer health and meet the demands for natural products.[19] Natural antimicrobials have gained attention among the scientific community and food industries seeking natural alternatives to substitute synthetic compounds.[19]

Natural antimicrobials can be applied as natural preservatives for FAV to ensure safety, retain the quality, and prolong SL.[6,86] Natural antimicrobial agents can come from the plant, animal, and microbial sources, where they play a role in their hosts' natural defense systems.[6,86] Some antimicrobials such as EOs, phenolic compounds, flavor compounds, LAB, isothiocyanates, chitosan, and bacteriocins have been applied to fresh and processed FAV.[13]

Plant EOs and their derivative compounds have been widely investigated as natural antimicrobial, antifungals, and food preservatives[13,19] and have also been used on FAV.[39,51] EOs are derived from secondary metabolites in almost every part of plants, including the flowers, roots, leaves, seeds, buds, and stems.[13,21] The antimicrobial mechanisms of EOs are thought to be from the composition of EOs and the synergistic effects between their components.[62]

In addition to their antimicrobial properties, EOs are safe, environmentally friendly, and approved by FDA as GRAS.[13,38] However, the effectiveness of EOs applications depends on concentration levels, holding time, type of microorganisms, type of FAV, and other factors.[21] For example, to achieve the desired microbial reduction, a higher concentration of EOs is required which results in the alteration of product sensory attributes.[21]

1.4 HURDLE TECHNOLOGY USING ORGANIC ACIDS

Historically, organic acids have been used as food preservatives or food additives to prolong the SL of food products[21] and to disinfect FAV.[21] Organic acids are effective in reducing microorganisms from FAVs by environmental and microbial cellular pH.[74] However, the efficacy of organic acids is dependent on the type of organic acid, concentration, treatment time, targeted microorganism, and type of treated produce.[21] Treatment with low concentration or short exposure time has shown little effect in reducing microorganisms.[77]

Additionally, using organic acids in higher concentrations resulted in adverse effects on the quality of FAV, including flavor and aroma.[39] Therefore, researchers have examined the hurdle technology by combining organic acids with chemical or physical disinfectants that could increase the efficacy of organic acids and thus enhance the quality of the treated produce. This section addresses some of the successful applications of hurdle technologies using organic acids.

1.4.1 ACETIC ACID

Acetic acid is a monocarboxylic acid known for its antimicrobial activity against various microorganisms including bacteria and yeast, and some types of molds. This acid is recognized as GRAS[35] and can thus be used as a sanitizer on FAV.[35] In the past, researchers have investigated the efficacy of combining acetic acid with other compounds in reducing different microorganisms on FAV.[98]

For example, Zhang and Farber[98] evaluated the efficacy of acetic acid or lactic acid combined with chlorine against *L. monocytogenes* on lettuce. The results demonstrated that both combinations were effective in reducing these pathogenic bacteria on lettuce. However, lactic acid was more effective than acetic acid in this regard.[98] Recently, acetic acid was evaluated for antibrowning to inhibit the enzymatic browning of lettuce by being combined with ethanol.[33] The results showed that the combination of acetic acid and ethanol was effective at reducing enzymatic browning on lettuce. Therefore, the combining of acetic acid with other chemicals can be used for antimicrobial and antibrowning purposes. To this end, combining acetic acid with other compounds has great potential for application in the food industry and food processing.[32]

1.4.2 LACTIC ACID

Lactic acid is a widely used organic acid antimicrobial that reduces microbial pathogens in foods.[74] The inhibitory capacity of lactic acid is based on its pH. The pH is usually lower than bacterial pH, the pH at which the bacteria cannot grow.[83] Lactic acid is also GRAS and is thus a suitable alternative to chemical or synthetic techniques for FAV disinfection as it is a natural compound.[83] In addition, it has antimicrobial activity and has been used as a preservative and antioxidant.[83] Therefore, the combination of lactic acid and other chemicals has the potential to be used for FAV as a sanitizer to ensure food safety.[83]

In this regard, researchers have evaluated the combination of lactic acid with other natural antimicrobials.[20,87] Venkitanarayanan et al.[87] investigated the antimicrobial efficacy of combining lactic acid with hydrogen peroxide against *E. coli* O157: H7, *Salmonella Enteritidis*, and *L. monocytogenes* on oranges, apples, and tomatoes. The results showed that lactic acid effectively reduced the number of *Salmonella*, *E. coli*, and LM on FAV. Another study evaluated the efficacy of combining lactic acid with chlorine, benzalkonium chloride, and lactic acid against *E. coli* O157: H7 and *Yersinia enterocolitica* on fresh vegetables.[20] The authors found the treatment to be useful in reducing the number of pathogens. Therefore, lactic acid could be used for FAV as a sanitizer to ensure protection from pathogenic microorganisms.[20]

1.4.3 CITRIC ACID (CA)

CA is another organic acid that has been used as a hurdle technology to disinfect FAV.[59,61,69] Park et al.[59] evaluated the effect of the combination of electrolyzed water and CA against *L. monocytogenes* on lettuce. The authors reported that the mixture of alkaline electrolyzed water and 1% CA was effective in reducing L. *monocytogenes* on lettuce compared to individual treatment.

Park et al.[60,61] also evaluated the effect of electrolyzed water and CA against *Bacillus cereus* cells and spores on cereal grains. These authors found that combining 1% CA and electrolyzed water was effective against *B. cereus* on cereal grains. Additionally, they reported that the combination resulted in synergistic effects of CA and electrolyzed water that were more potent than what was seen with individual treatments.

Rahman et al.[69] investigated the effect of electrolyzed water and CAs with mild heat to reduce the growth of microorganisms on cabbage. Their results exhibited a significant decrease in total bacterial counts, yeast and mold, *L. monocytogenes*, and *E. coli.* Those same authors[68] also investigated the efficacy of combining alkaline electrolyzed water and CA with mild heat on shredded carrots to extend the SL and retain the quality and ensure product safety. The results exhibited that the combination caused a significant decrease in total bacterial counts, yeast, and molds. Gupta et al.[31] found that combining CA with mild doses of gamma radiation and modified atmosphere packaging as hurdles was effective in increasing the SL of French beans and maintaining product quality during storage. Thus, the combining of CA with other existing disinfection methods could be another potentially effective hurdle technology.

1.5 HURDLE TECHNOLOGY (NON-THERMAL METHODS) FOR THE PRESERVATION OF FRUIT JUICES

In today's world, consumers are looking not only for food products that satisfy their appetite but that are also natural and GRAS status. Thus, functional foods including fruit juices are of great interest because they contain various types of bioactive compounds, such as polyphenols. Generally, polyphenols are antioxidants that have anti-inflammatory activities. It has been approximately calculated that the consumption of functional foods in adequate amounts not only provides a sufficient amount of nutrition but also has a significant impact on human health and can even alter the risk of appearing various severe diseases such as cancer.[67] However, in terms of cost, newly developed functional foods are much more costly than traditional foods, and it had been calculated that the global market for functional foods was about $304.5 billion in 2020. Moreover, in subsequent years, this market is projected to increase by an average annual growth of nearly 8%.[54]

Therefore, in this regard, most consumers have only one affordable option that would serve to satisfy their daily nutrient requirements and improve their well-being quickly and easily, they can purchase fruit or vegetable juices or beverages. One of the agencies in the UK has suggested that 150 mL of pure fruit or vegetable juices is equivalent to five servings of actual FAV (*NHS-UK. 5 A Day Portion Sizes; NHS: Les, UK, 2015)*. As compared to the consumption of whole FAV, the consumption of the juice transferred the bioactive compounds more effectively into the human body where the compounds are easily absorbed and digested. Hence, juices are an attractive alternative transport medium for necessary bioactive ingredients. For these reasons plus today's busy lifestyles, people are preferring fruit juices instead of consuming fresh fruits. With the rapidly increasing demand for fruit juices, food industries are now focusing on designing and producing new and different fruit juice products as well as the production of functional fruit juices.

In order to fulfill the nutritional requirements and increasing demand for juices with increasing levels of bioactive compounds, industries have begun adding several nutritious ingredients such as noni, acai, areola, goji berries, as well as probiotics, prebiotics, and omega-3 fatty acids.[67] For example, in one study, researchers developed a functional fruit juice by adding pear juice with cranberry bush juice in different concentrations. Their results showed that most consumers preferred the combination juice that included 97.5% pear juice and 2.5% cranberry juice.[41]

In the past several years, food industries have processed fruit juices by using the traditional heat treatment method (pasteurization) or by either the squeezing or cold pressing method with storage at a temperature of 2–5°C. As a result, the produced juices have very short SL (1–3) days. In addition, such traditional methods disrupt the stability of the thermolabile bioactive compounds in the juice. Today, consumers prefer fruit juices that contain a number of nutrients, and photochemical compounds, minimally processed, and have an increased SL. Therefore, the use of hurdle technology is an alternative and promising approach in order to fulfill the customer benchmark of quality, nutritional requirement, and the safety of the product.[39]

Generally, hurdles themselves are the food preservation factors such as temperature, preservatives, pH, water activity, and microorganisms. In several cases, these factors are frequently integrate with different hurdles such as ultraviolet or pulsed light, and antimicrobial additives in order to attain a certain level of food quality and stability. To date, more than 60 hurdles are known, but not all of them are used to preserve a single food product.[42] As per the requirement, two or more hurdle technologies can be combined together to accomplish food quality and microbial stability. In this respect, these days most hurdle technology studies are conducted on pulsed electric field (PEF), high-pressure processing (HPP), high-power US, cold plasma, and high voltage electric discharge.[9]

These methods require shorter processing times and are performed at a lower temperature. In addition, these methods assure inactivation of enzymes, enhanced microbial protection, and higher stability of the bioactive compounds in the food. It has been noted that, compared to the traditional pasteurization methods, hurdle technology provide a better-extended SL of juices along with retention of nutritional and sensory values.[39] Therefore, nonthermal hurdle technologies, such as HPP, UP, and PEF, are discussed in this section.

1.5.1 HIGH-PRESSURE PROCESSING (HPP)

The research studies on high-pressure processing of foods have demonstrated that HPP technology is one of the best technologies among nonthermal technologies for packed foods. Packed food products such as solids and semisolids having low moisture content (vacuum packed), vacuum-packed solids, and packed liquid foods (bottled) can be easily processed by HPP technology. However, prior to processing food items using this technology, the items are initially packed in suitable packing to decrease the risk/possibility of post contamination. And then, the packed items are placed into a basket and lowered into a jar. After that, the jar is filled completely with water, and the food items are exposed to high hydrostatic pressure (400–800 MPa).[3]

The products are considered to be adequately processed when the pressure is uniformly spread throughout the jar. The application of HPP to food products results in microbial inactivation and the rupturing of cell membranes, as well as rupturing of noncovalent bonds present in enzymes and ribosomes.[76] Sometimes this technology can integrate with high temperature in order to deactivate bacterial spores more efficiently as bacterial spores are slightly resistant to HPP. HPP technology has gained international recognition and has received approval from the governmental authorities of Canada, USA, and Europe for use in killing pathogens in juices and for increasing SL. In recent years, food processing industries have started to embrace HPP technology over the traditional pasteurization methods, particularly with juices. Moreover, various researchers have studied the implementation of HPP for the pasteurization of fruit juices, such as mango juice, apple juice, kiwi juice, pineapple, etc.[1,99].

For example, one of the study's authors compared the pasteurization (90°C, 5 min) of mango juice with HPP (190 MPa, 60°C) and found that HPP reduced the number of bacteria and yeast and increased the SL of the juice up to 60 days when it stored at 4°C. In addition, compared to the control sample, HPP treated samples have a higher concentration of total phenolic compounds, carotenoids, and ascorbic acid.[76]

Similarly, other studies have reported the implementation of HPP for the preservation of different juices, some of which are represented in Table 1.2. Some of the process variables such as pH, temperature, time, and pressure affect the efficiency of HPP technology. Like other technologies, HPP technology also has some limitations, such as high initial investment costs for the equipment and batch operations.

1.5.2 ULTRASOUND (US) PROCESSING

Recently, the application of US processing technology has gained significant potential for the processing of fruit juices. This technology is also considered to be the best alternative option for the pasteurization of fruit juices. During the application of US, the released acoustic energy is rapidly transferred into the entire volume of juice, which generates compressions and rarefaction in the medium particles such as juice, which results in the increased mass transfer, which automatically reduces the processing/treatment time. Moreover, due to low energy consumption and probably due to the involvement of low costs, this technology has also been recommended for industrial usage.[84]

Generally, for the processing of fruit juices, several studies have recommended combinations of high power of US with ultrasonic waves of high intensity at low frequencies. The implementation of high-power US in juice processing causes several physicochemical changes, expedites biochemical reactions, and also results in microbial and enzymatic inactivation. When ultrasonic waves are passed through juice or any liquid, it causes the generation of tiny bubbles and negative pressure which can thus be defined as a cavitation phenomenon.[36] This phenomenon of collapsing the cavitation bubbles inside and outside of the microbial cells increases the pressure around the cells, which results in mechanical damage to the cells. Several studies reported that the use of high-power US technology provided microbiologically stable fruit juices with better retention of bioactive compounds.

In addition, it has been reported that in some cases, the application of US in juice provided a maximum of 5-log reduction in food-borne pathogens.[16] The results of several studies further revealed that the high-power US processing of juice enhanced physical properties such as rheological properties, cloudiness, water-soluble pectin, and color. This may be achieved due to the physical forces of US causing excessive injury to the cellular structures. For example, one of studies conducted on the implementation of high-power US on the juice of strawberries revealed that US increased the acidity sweetness, and antioxidant capacity of the sample by 75%, and the values of these parameters continuously increased during the storage period. Moreover, the application of US did not alter the color of the juice.

It has been reported that a higher value of phenolic compounds is present in treated juice as compared to untreated juice.[27] Some of the parameters that affect the efficacy of high-power US technology in the processing of fruit juices are US power, frequency, amplitude, and strength, processing time and temperature, and the diameter of the ultrasonic probe. Table 1.2 indicates the results of some of the studies that have been carried out over the last few years on the implementation of US technology for the preservation of fruit juices.

1.5.3 PULSED ELECTRIC FIELD PROCESSING (PEF)

Several studies have determined that PEF processing technology is one of the best alternative processes for the pasteurization of fruit juices. PEF processing has shown outstanding outcomes without altering the quality of juice (nutritional or sensory).[23] In PEF processing, the sample of the food is set down between two electrodes, and a voltage of about 50 kV/cm is applied for a shorter period of time (μs to ms). Variables such as time, temperature, intensity of electric field, and energy input play a key role during this type of processing and need to be optimized for the selected food.[81]

During PEF processing, microbial inactivation occurs due to the potential differences between the outside and inside of the cells. When the potential difference reaches a critical value, electroporation occurs, including irreversible structural changes and it results the death of the microorganisms. The process of PEF depends on several parameters such as applied temperature, processing time, the intensity of the electric field, pulse shape, width, and frequency.

The efficacy of the PEF process is also depends on several conditions, such as composition of foods and the electrical conductivity of the food. It has been suggested that food is more difficult to process with this method if the food conductivity is greater than 6 μs.[7]

Moreover, several studies have indicated that under mild processing conditions, the PEF is not effectively killing the spores which mean that it can be combined with other methods of preservation to attain the desirable outcome. However, various other studies have indicated that the PEF method is more suitable for the processing of liquid foods without bubbles because it contains charge molecules.[23] Thus, it can be concluded that apart from juices, the PEF can also be applied to the dairy products processing. Various studies have shown that some of the physical and chemical parameters of juices such as total soluble solids, water activity, and pH affect the PEF process with regard to microbial inactivation. Moreover, these studies also confirmed that as compared to thermally treated juices, PEF-treated juices have a large amounts of minerals, polyphenols, content of free amino acids, and better retention of the original color.[75]

For example, one of the studies on the extraction of carotenoids from the tomato matrix confirmed that the application of PEF technology releases carotenoids from the tomato matrix more readily compared to other technologies. In addition, the results of that study also revealed that the preprocessing application of PEF to raw tomatoes increased the concentration of individual carotenoids, such as β-carotene, γ-carotene, δ-carotene, and lycopene.[30] The results of the studies on the implementation of PEF technology for the preservation of fruit juices are shown in Table 1.2.

1.6 SUMMARY

FAVs are an essential source of nutrition and energy. However, they are highly perishable food and they have also been associated with a large number of food-borne outbreaks. Moreover, fresh food can become contaminated with pathogenic microorganisms at various points during handling, storage, and processing and is also subject to spoilage. The contamination of FAV causes deterioration and negatively affects the product quality Thus, proper handling, storage, and processing are critical in order to ensure safety and to retain quality. Postharvest cleaning and sanitizing are key to preventing such contamination and ensuring safety, thus improving quality. Various thermal and nonthermal disinfection technologies have been evaluated and applied to FAV. However, single disinfection technology has been shown to be insufficient in reducing microorganisms. Additionally, at higher concentrations, most of the applied technologies had an adverse impact on product quality or resulted in harmful chemical byproducts.

TABLE 1.2 Application of US, PEF, HPP Methods for the Preservation of Fruit Juices.

Method	Treatment conditions	Type of juice	Observed quality parameters	Findings	References
US	600 W, 28 kHz, power radiation 70%, 20°C, 30, 60, and 90 min	Grapefruit Juice	• Individual phenolic compounds • Total lycopene • Total carotenoids • Total anthocyanins • Viscosity • Sugar content • Microbiological factors	• The values of total carotenoids, phenolic compounds, total lycopene, and sugar content were found to be significantly increased • Viscosity and microbial activity were found to be decreased • Given treatment of 90 min provides better results	[2]
	25kHz, 70% amplitude (Sonication) 5 and 10 min (ultra-sonication); 100°C, 4 min (Blanching)	Blend of apple and grape juice (1:1)	• TPC • TFC • Total flavonoids • TAA • Viscosity, turbidity, pH, and acidity	• As compared to other treatment ultrasonication (5 and 10 min) showed the increment in AA and AC • Blanching and ultrasonication treatments reflect a significant effect on color parameters, turbidity, and viscosity	[1]
	500W, 20kHz, 19 mm diameter probe, Less than 35°C, Amplitude: 50, 75, and 100%, Time: 0–30 min, 5 s on and 5 s off	Pomegranate juice	• TPC • TMC • SSC • pH	• TPC did not change significantly • At the time of greater than 18 min the content of anthocyanins was found to be decreased • US treatment enhances the quality of pomegranate juice	[58]
	100 W, 30 kHz, 10 and 20 mm diameter probe, 3, 6, 9 min, 40 and 80% of amplitude, 7 days at 4°C (storage)	Apple juice	• TPC • Total Flavan-3-ols • AA	• US treatment reduces the TPC and total flavan-3-ols content by 32.94%, and 21.66% in comparison to untreated samples • Similarly values of AC decline by 23.76%	[12]

TABLE 1.2 *(Continued)*

Method	Treatment conditions	Type of juice	Observed quality parameters	Findings	References
PEF	Square wave pulses (bipolar) of 2.5 μs, 20–40 kV/cm, 0–2.5% (w/v) Stevie Leaf infusion	Mango and papaya juices with added Stevie infusion	• Total carotenoids and anthocyanins • Ascorbic acid content • TSS • Steviol glycosides	• Higher value of electric field showed stevia beverages with higher value of total carotenoid and anthocyanins • Higher value of electric field showed a greater reduction in ascorbic acid content	[14]
	PEF Treatment: pulse applied for a 3 μs, with a delay of 20 μs, frequency 500 pps, flow rate: 60 mL/min, Temperature: 5–35°C	Pomegranate juice	• TPC • TAC • TAA	• Increase in temperature decreases the AA and phenolic content	[22]
HPP	Pressure: 200, 300, 400 MPa Temperature: 5, 15, 25°C Time: 5 min	Pomegranate juice	• Ascorbic acid-mannitol • TPC • TAA	• As compared to control lower value of mannitol was detected in HPP treated juice • No significant decrease in TPC and AA was observed	[17]
	High Hydrostatic Pressure: 200–600 MPa for 15 min	Keitt mango juice	• Volatile compounds • Sensory evaluation	• In Keitt mango juice around 35 volatile compounds were detected • As compared to fresh juice HHP treatment alter the aroma of treated juice	[99]
	Pressure: 200, 300, 600 MPa, Temperature: 5, 25, 45°C, Time: 1, 5, 15 min, storage at 4°C for 2–12 weeks	Cloudy apple juice	• TP • TSS • pH • sugar content and Vitamin C	• HPP decreased the TPC	[93]

TABLE 1.2 *(Continued)*

Method	Treatment conditions	Type of juice	Observed quality parameters	Findings	References
	High-pressure thermal processing: 400–600 MPa, 30–60°C, 0–15 min	Mixed juice of aloe vera and litchi	• Physicochemical Properties • Content of ascorbic acid	• High-pressure thermal processing minimally affected physicochemical properties • High pressure treatment reduces the ascorbic acid content by up to 40%	[34]

TPC, total phenolic content; TMA, total monomeric anthocyanins; SSC, soluble solid content; TSS, total soluble solid; TP, total polyphenols; TAC, total anthocyanins content; AA, antioxidant activity; TP, total polyphenolic; AC, anthocyanin content; TAA, total antioxidant activity; TFC, total flavonoics content.

Hurdle technology was developed in order to overcome the above-mentioned challenges, improve overall product quality, and meet the increasing customer demand of healthy and safer food items. Hurdle technologies have already been implemented to several food items and have demonstrated efficient disinfection and preservative approaches for the food industry. In addition, hurdle technologies have been shown to provide greater efficiency than individual treatments. For example, hurdle technologies minimize the intensities associated with each individual technology such as power, concentration, doses, or exposure times. As a result, the application of hurdle technology is beneficial to the food industry in that it can provide cost-effective technology, extend SL, and ensure safety while maintaining product quality. A combination of specific hurdles technologies is thus promising for the future of the fruit and vegetable industry. In conclusion, more attention should be given to this innovative technology, and additional research to examine thermal and nonthermal techniques in hurdle technology is highly warranted.

KEYWORDS

- **combining antimicrobial**
- **disinfection**
- **fruits and vegetables**
- **hurdle technology**
- **novel disinfection**

REFERENCES

1. Aadil, R. M.; Khalil, A. A.; Rehman, A.; Khalid, A.; Inam-ur-Raheem, M.; Karim, A.; Gill, A. A.; Abid, M.; Afraz, M. T. Assessing the Impact of Ultra-Sonication and Thermo-Ultrasound on Antioxidant Indices and Polyphenolic Profile of Apple-Grape Juice Blend. *J. Food Process. Preserv.* **2020**, *44* (5), E-article 14406.
2. Aadil, R. M.; Zeng, X. A.; Wang, M. S.; Liu, Z. W.; Han, Z.; Zhang, Z. H.; Hong, J.; Jabbar, S. A Potential of Ultrasound on Minerals, Micro-Organisms, Phenolic Compounds and Colouring Pigments of Grapefruit Juice. *Int. J. Food Sci. Technol.* **2015**, *50* (5), 1144–1150.
3. Abera, G. Review on High-Pressure Processing of Foods. *Cogent Food Agric.* **2019**, *5* (1), 1568725.

4. Alzamora, S. M.; Guerrero, S. N.; Raffellini, S.; Ferrario, M.; Schenk, M. Hurdle Technology in Fruit Processing. In *Fresh-Cut Fruits and Vegetables*; CRC Press; 2016; pp 131–168.
5. Alzamora, S. M.; Tapia, M. S.; Argaiz, A.; Welli, J. Application of Combined Methods Technology in Minimally Processed Fruits. *Food Res. Int.* **1993**, *26* (2), 125–130.
6. Ayala-Zavala, J. F.; González-Aguilar, G. A.; Del-Toro-Sánchez, L. Enhancing Safety and Aroma Appealing of Fresh-Cut Fruits and Vegetables Using the Antimicrobial and Aromatic Power of Essential Oils. *J. Food Sci.* **2009**, *74* (7), R84–R91.
7. Barba, F. J.; Parniakov, O.; Pereira, S. A.; Wiktor, A.; Grimi, N.; Boussetta, N.; Saraiva, J. A.; Raso, J.; Martin-Belloso, O.; Witrowa-Rajchert, D.; Lebovka, N. Current Applications and New Opportunities for the Use of Pulsed Electric Fields in Food Science and Industry. *Food Res. Int.* **2015**, *77*, 773–798.
8. Barwal, V. S.; Sharma, R.; Singh, R. Preservation of Cauliflower by Hurdle Technology. *J. Food Sci. Technol. Mysore* **2005**, *42* (1), 26–31.
9. Bevilacqua, A.; Petruzzi, L.; Perricone, M.; Speranza, B.; Campaniello, D.; Sinigaglia, M.; Corbo, M. R. Nonthermal Technologies for Fruit and Vegetable Juices and Beverages: Overview and Advances. *Comprehen. Rev. Food Sci. Food Safet.* **2018**, *17* (1), 2–62.
10. Botondi, R.; De Sanctis, F.; Moscatelli, N.; Vettraino, A. M.; Catelli, C.; Mencarelli, F. Ozone Fumigation for Safety and Quality of Wine Grapes in Postharvest Dehydration. *Food Chem.* **2015**, *188*, 641–647.
11. Brodowska, A. J.; Nowak, A.; Śmigielski, K. Ozone in the food industry: Principles of Ozone Treatment, Mechanisms of Action, and Applications: An Overview. *Crit. Rev. Food Sci. Nutr.* **2015**, *58* (13), 2176–2201.
12. BursaćKovačević, D.; Bilobrk, J.; Buntić, B.; Bosiljkov, T.; Karlović, S.; Rocchetti, G.; Lucini, L.; Barba, F. J.; Lorenzo, J. M., Putnik, P. High-Power Ultrasound Altered the Polyphenolic Content and Antioxidant Capacity in Cloudy Apple Juice During Storage. *J. Food Process. Preserv.* **2019**, *43* (8), E-article 14023.
13. Burt, S. Essential Oils: Their Antibacterial Properties and Potential Applications in Foods—A Review. *Food Microbiol.Int. J. Food Microbiol.* **2004**, *94* (3), 223–253.
14. Carbonell-Capella, J. M.; Buniowska, M.; Cortes, C.; Zulueta, A.; Frigola, A.; Esteve, M. J. Influence of Pulsed Electric Field Processing on the Quality of Fruit Juice Beverages Sweetened with Stevia Rebaudiana. *Food Bioprod. Process.* **2017**, *101*, 214–222.
15. Chen, X.; Xue, S. J.; Shi, J.; Kostrzynska, M.; Tang, J.; Guévremont, E.; Mondor, M. Red Cabbage Washing with Acidic Electrolysed Water: Effects on Microbial Quality and Physicochemical Properties. *Food Qual. Safet.* **2018**, *2* (4), 229–237.
16. Condón-Abanto, S.; Arroyo, C.; Álvarez, I.; Condón, S.; Lyng, J. G. Application of Ultrasound in Combination with Heat and Pressure for the Inactivation of Spore Forming Bacteria Isolated from Edible Crab (*Cancer pagurus*). *Food Microbiol.Int. J. Food Microbiol.* **2016**, *223*, 9–16.
17. da Silveira, T. F. F.; Cristianini, M.; Kuhnle, G. G.; Ribeiro, A. B.; Teixeira Filho, J.; Godoy, H. T. Anthocyanins, Non-Anthocyanin Phenolics, Tocopherols and Antioxidant Capacity of açaí Juice (*Euterpe oleracea*) as Affected by High Pressure Processing and Thermal Pasteurization. *Innov. Food Sci. Emerg. Technol.* **2019**, *55*, 88–96.
18. Dabas, K.; Khan, K. A. *Use of Hurdle Technology in Processing of Fruits and Vegetables*; Apple Academic Press: New Jersey, 2019; pp 113–138.
19. Davidson, P. M.; Critzer, F. J.; Taylor, T. M. Naturally Occurring Antimicrobials for Minimally Processed Foods. *Annu. Rev. Food Sci. Technol.* **2013**, *4*, 163–190.

20. del Carmen Velázquez, L.; Barbini, N. B.; Escudero, M. E.; Estrada, C. L.; de Guzmán, A. M. S. Evaluation of Chlorine, Benzalkonium Chloride and Lactic Acid as Sanitizers for Reducing *Escherichia coli* O157: H7 and Yersinia Enterocolitica on Fresh Vegetables. *Food Control* 2009, *20* (3), 262–268.
21. Deng, L. Z.; Mujumdar, A. S.; Pan, Z.; Vidyarthi, S. K.; Xu, J.; Zielinska, M.; Xiao, H. W. Emerging Chemical and Physical Disinfection Technologies of Fruits and Vegetables: A Comprehensive Review. *Crit. Rev. Food Sci. Nutr.* **2019**, *60* (15), 2481–2508.
22. Evrendilek, G. A. Impacts of Pulsed Electric Field and Heat Treatment on Quality and Sensory Properties and Microbial Inactivation of Pomegranate Juice. *Food Sci. Technol. Int.* **2017**, *23* (8), 668–680.
23. Gabrić, D.; Barba, F.; Roohinejad, S.; Gharibzahedi, S. M. T.; Radojčin, M.; Putnik, P.; BursaćKovačević, D. Pulsed Electric Fields as an Alternative to Thermal Processing for Preservation of Nutritive and Physicochemical Properties of Beverages: A Review. *J. Food Process Eng.* **2018**, *41* (1), E-article 12638.
24. Garcia, A.; Mount, J. R.; Davidson, P. M. Ozone and Chlorine Treatment of Minimally Processed Lettuce. *J. Food Sci.* 2003, *68* (9), 2747–2751.
25. Gómez-López, V. M. (Ed.). *Decontamination of Fresh and Minimally Processed Produce*; John Wiley & Sons: New York, 2012; p 323.
26. Gómez-López, V. M., Gil, M. I.; Allende, A. A Novel Electrochemical Device as a Disinfection System to Maintain Water Quality During Washing of Ready to Eat Fresh Produce. *Food Control* **2017**, *71*, 242–247.
27. Gómez-López, V. M.; Buitrago, M. E.; Tapia, M. S.; Martínez-Yépez, A. M. A. U. R. Y. Effect of Ultrasonication on Microbial Quality, Colour, and Ascorbic Acid Content of Passion-Fruit Juice During Storage. *Acta Alimentaria* **2017**, *46* (4), 470–480.
28. Gómez-López, V. M.; Rajkovic, A.; Ragaert, P.; Smigic, N.; Devlieghere, F. Chlorine Dioxide for Minimally Processed Produce Preservation: A Review. *Trend Food Sci. Technol.*, **2009**, *20* (1), 17–26.
29. González-Aguilar, G.; Ayala-Zavala, J. F.; Ruiz-Cruz, S.; Acedo-Félix, E. Effect of Temperature and Modified Atmosphere Packaging on Overall Quality of Fresh-Cut Bell Peppers. *LWT-Food Sci. Technol.* **2004**, *37* (8), 817–826.
30. González-Casado, S.; Martín-Belloso, O.; Elez-Martínez, P.; Soliva-Fortuny, R. Application of Pulsed Electric Fields to Tomato Fruit for Enhancing the Bioaccessibility of Carotenoids in Derived Products. *Food Funct.* **2018**, *9* (4), 2282–2289.
31. Gupta, S.; Chatterjee, S.; Vaishnav, J.; Kumar, V.; Variyar, P. S.; Sharma, A. Hurdle Technology for Shelf Stable Minimally Processed French Beans (Phaseolus vulgaris): A Response Surface Methodology Approach. *LWT-Food Sci. Technol.* 2012, *48* (2), 182–189.
32. Hao, J.; Wang, Q. Application of Electrolyzed Water in Fruits and Vegetables Industry. In *Electrolyzed Water in Food: Fundamentals and Applications*; Springer: Singapore, 2019; pp 67–111.
33. Huang, S. J.; Lin, S. Y.; Wang, T. T.; Hsu, F. C. Combining Acetic Acid and Ethanol as an Anti-Browning Treatment for Lettuce Butt Discoloration Through Repression of the Activity and Expression of Phenylalanine Ammonia Lyase. *Postharv. Biol. Technol.* **2020**, *164*, E-article 111151.
34. Hulle, N. R. S., Rao, P. S. Effect of High Pressure and Thermal Processing on Quality Changes of Aloe Vera-Litchi Mixed Beverage (ALMB) During Storage. *J. Food Sci. Technol.* **2016**, *53* (1), 359–369.

35. In, Y. W.; Kim, J. J.; Kim, H. J.; Oh, S. W. Antimicrobial Activities of Acetic Acid, Citric Acid and Lactic Acid Against *Shigella* Species. *J. Food Safet.* **2013**, *33* (1), 79–85.
36. J Mason, T.; Chemat, F.; Vinatoru, M. The Extraction of Natural Products Using Ultrasound or Microwaves. *Curr. Org. Chem.* **2011**, *15* (2), 237–247.
37. Joshi, K.; Mahendran, R.; Alagusundaram, K.; Norton, T.; Tiwari, B. K. Novel Disinfectants for Fresh Produce. *Trend Food Sci. Technol.*, **2013**, *34* (1), 54–61.
38. Kang, J. H.; Song, K. B. Inhibitory Effect of Plant Essential Oil Nanoemulsions Against *Listeria monocytogenes*, *Escherichia coli* O157: H7, and *Salmonella* typhimurium on Red Mustard Leaves. *Innov. Food Sci. Emerg. Technol.* **2018**, *45*, 447–454.
39. Khan, I.; Tango, C. N.; Miskeen, S.; Lee, B. H.; Oh, D. H. Hurdle Technology: A Novel Approach for Enhanced Food Quality and Safety–A Review. *Food Control*, **2017**, *73*, 1426–1444.
40. Khaneghah, A. M.; Hashemi, S. M. B.; Limbo, S. Antimicrobial Agents and Packaging Systems in Antimicrobial Active Food Packaging: An Overview of Approaches and Interactions. *Food Bioprod. Process*. 2018, *111*, 1–19.
41. Lachowicz, S.; Oszmiański, J. The Influence of Addition of Cranberrybush Juice to Pear Juice on Chemical Composition and Antioxidant Properties. *J. Food Sci. Technol.* **2018**, *55* (9), 3399–3407.
42. Leistner, L. Basic Aspects of Food Preservation by Hurdle Technology. *Food Microbiol. Int. J. Food Microbiol.* **2000**, *55* (1–3), 181–186.
43. Leistner, L. Introduction to Hurdle Technology. *Food Preservation by Combined Processes. FLAIR Report by European Commission. DX X*; **1994**, *11*, 89.
44. Leistner, L. Principles and Applications of Hurdle Technology. In *New Methods of Food Preservation*; Springer: Boston, MA, 1995; pp 1–21.
45. Limbo, S.; Khaneghah, A. M. Active Packaging of Foods and Its Combination with Electron Beam Processing. In *Electron Beam Pasteurization and Complementary Food Processing Technologies*; Woodhead Publishing: London, 2015; pp 195–217.
46. Liu, Q.; Wu, J. E.; Lim, Z. Y.; Lai, S.; Lee, N.; Yang, H. Metabolite Profiling of *Listeriainnocua* for Unraveling the Inactivation Mechanism of Electrolyzed Water by Nuclear Magnetic Resonance Spectroscopy. *Food Microbiol.Int. J. Food Microbiol.* **2018**, *271*, 24–32.
47. Mahmoud, B. S. M.; Vaidya, N. A.; Corvalan, C. M.; Linton, R. H. Inactivation Kinetics of Inoculated *Escherichiacoli* O157: H7, Listeria Monocytogenes and *Salmonella* Poona on Whole Cantaloupe by Chlorine Dioxide Gas. *Food Microbiol.* **2008**, 25 (7), 857–865.
48. Mohammad, Z.; Kalbasi-Ashtari, A.; Riskowski, G.; Castillo, A. Reduction of *Salmonella* and Shiga Toxin-Producing *Escherichiacoli* on Alfalfa Seeds and Sprouts Using an Ozone Generating System. *Food Microbiol.Int. J. Food Microbiol.* 2019, *289*, 57–63.
49. Mohammad, Z.; Kalbasi-Ashtari, A.; Riskowski, G.; Juneja, V.; Castillo, A. Inactivation of *Salmonella* and Shiga Toxin-Producing *Escherichia coli* (STEC) from the Surface of Alfalfa Seeds and Sprouts by Combined Antimicrobial Treatments Using Ozone and Electrolyzed Water. *Food Res. Int.* **2020**, *136*, E-article 109488.
50. Mohapatra, D.; Mishra, S.; Giri, S.; Kar, A. Application of Hurdles for Extending the Shelf-Life of Fresh Fruits. *Trend Postharv. Technol.* **2013**, *1*, 37–54.
51. Moore-Neibel, K.; Gerber, C.; Patel, J.; Friedman, M.; Jaroni, D.; Ravishankar, S. Antimicrobial Activity of Oregano Oil Against Antibiotic-Resistant *Salmonella* Enterica on Organic Leafy Greens at Varying Exposure Times and Storage Temperatures. *Food Microbiol.* **2013**, *34* (1), 123–129.

52. Mtaoua, H.; Sánchez-Vega, R.; Ferchichi, A.; Martín-Belloso, O. Impact of High-Intensity Pulsed Electric Fields or Thermal Treatment on the Quality Attributes of Date Juice Through Storage. *J. Food Process. Preserv.* **2017**, *41* (4), E-article 13052.
53. Mukhopadhyay, S.; Tomasula, P. M.; Luchansky, J. B.; Porto-Fett, A. C. S. Removal of *Salmonella Enteritidis* from Commercial Unpasteurized Liquid Egg White Using Pilot Scale Cross Flow Tangential Microfiltration. *Food Microbiol. Int. J. Food Microbiol.* **2010**, *142* (668), 309–317.
54. Musina, O.; Putnik, P.; Koubaa, M.; Barba, F. J.; Greiner, R.; Granato, D.; Roohinejad, S. Application of Modern Computer Algebra Systems in Food Formulations and Development: A Case Study. *Trend Food Sci. Technol.* **2017**, *64*, 48–59.
55. Neal, J. A.; Marquez-Gonzalez, M.; Cabrera-Diaz, E.; Lucia, L. M.; O'Bryan, C. A.; Crandall, P. G. et al. Comparison of Multiple Chemical Sanitizers for Reducing *Salmonella* and *Escherichia coli* O157: H7 on Spinach (Spinacia oleracea) Leaves. *Food Res. Int.* **2012**, *45* (2), 1123–1128.
56. Ngnitcho, P. F. K.; Khan, I.; Tango, C. N.; Hussain, M. S.; Oh, D. H. Inactivation of Bacterial Pathogens on Lettuce, Sprouts, and Spinach Using Hurdle Technology. *Innov. Food Sci. Emerg. Technol.* **2017**, *43*, 68–76.
57. Pal, M.; Shimelis, A.; Mamo, W.; Barot, A. M.; Pinto, S. V.; Prajapati, J. P. Hurdle Technology: A Novel Approach for Food Preservation. *Bev. Food World* **2017**, *44* (1), 20–26.
58. Pala, Ç. U.; Zorba, N. N. D.; Özcan, G. Microbial Inactivation and Physicochemical Properties of Ultrasound Processed Pomegranate Juice. *J. Food Protection* **2015**, *78* (3), 531–539.
59. Park, B. K.; Oh, M. H.; Oh, D. H. Effect of Electrolyzed Water and Organic Acids on the Growth Inhibition of Listeria Monocytogenes on Lettuce. *Korean J. Food Preserv.* **2004**, *11* (4), 530–537.
60. Park, S. H.; Kang, D. H. Combination Treatment of Chlorine Dioxide Gas and Aerosolized Sanitizer for Inactivating Foodborne Pathogens on Spinach Leaves and Tomatoes. *Food Microbiol.Int. J. Food Microbiol.* **2015**, *207*, 103–108.
61. Park, Y. B.; Guo, J. Y.; Rahman, S. M. E.; Ahn, J.; Oh, D. H. Synergistic Effect of Electrolyzed Water and Citric Acid Against *Bacillus cereus* Cells and Spores on Cereal Grains. *J. Food Sci.* **2009**, *74* (4), M185–M189.
62. Patrignani, F.; Siroli, L.; Serrazanetti, D. I.; Gardini, F.; Lanciotti, R. Innovative Strategies Based on the Use of Essential Oils and Their Components to Improve Safety, Shelf-Life and Quality of Minimally Processed Fruits and Vegetables. *Trend Food Sci. Technol.* **2015**, *46* (2), 311–319.
63. Petri, E.; Rodríguez, M.; García, S. Evaluation of Combined Disinfection Methods for Reducing *Escherichia coli* O157: H7 Population on Fresh-Cut Vegetables. *Int. J. Environ. Res. Public Health* **2015**, *12* (8), 8678–8690.
64. Piližota, V. *Food Safety Management.* Chapter 9; In: *Fruits and Vegetables (including Herbs)*. Netherlands: Elsevier Inc., 2013.
65. Prasad, P.; Kochhar, A. Active Packaging in Food Industry: A Review. *J. Environ. Sci. Toxicol. Food Technol.*, **2014**, *8* (5), 1–7.
66. Prodduk, V.; Annous, B. A.; Liu, L.; Yam, K. L. Evaluation of Chlorine Dioxide Gas Treatment to Inactivate Salmonella Enterica on Mung Bean Sprouts. *J. Food Protection* **2014**, *77* (11), 1876–1881.

67. Putnik, P.; Pavlić, B.; Šojić, B.; Zavadlav, S.; Žuntar, I.; Kao, L.; Kitonić, D.; Kovačević, D. B. Innovative Hurdle Technologies for the Preservation of Functional Fruit juices. *Foods* **2020**, *9* (6), 699.
68. Rahman, S. M. E.; Jin, Y. G.; Oh, D. H. Combination Treatment of Alkaline Electrolyzed Water and Citric Acid with Mild Heat to Ensure Microbial Safety, Shelf-Life and Sensory Quality of Shredded Carrots. *Food Microbiol.* **2011,** *28* (3), 484–491.
69. Rahman, S. M. E.; Jin, Y. G.; Oh, D. H. Combined Effects of Alkaline Electrolyzed Water and Citric Acid with Mild Heat to Control Microorganisms on Cabbage. *J. Food Sci.* **2010**, *75* (2), M111–M115.
70. Rahman, S. M. E.; Khan, I.; Oh, D. H. Electrolyzed Water as a Novel Sanitizer in the Food Industry: Current Trends and Future Perspectives. *Comprehen. Rev. Food Sci. Food Safet.* **2016,** *15* (3), 471–490.
71. Ramos-Villarroel, A. Y.; Martín-Belloso, O.; Soliva-Fortuny, R. Combined Effects of Malic Acid Dip and Pulsed Light Treatments on the Inactivation of *Listeriainnocua* and *Escherichiacoli* on Fresh-Cut Produce. *Food Control* 2015, *52*, 112–118.
72. Ramos, B.; Miller, F. A.; Brandão, T. R.; Teixeira, P.; Silva, C. L. Fresh Fruits and Vegetables: An Overview on Applied Methodologies to Improve Its Quality and Safety. *Innov. Food Sci. Emerg. Technol.* **2013**, *20*, 1–15.
73. Rawat, S. Food Spoilage: Microorganisms and Their Prevention. *Asian J. Plant Sci. Res.* 2015, *5* (4), 47–56.
74. Rico, D.; Martin-Diana, A. B.; Barat, J. M.; Barry-Ryan, C. Extending and Measuring the Quality of Fresh-Cut Fruit and Vegetables: A Review. *Trend Food Sci. Technol.*, **2007**, *18* (7), 373–386.
75. Rodríguez-Roque, M. J.; de Ancos, B.; Sánchez-Moreno, C.; Cano, M. P.; Elez-Martínez, P.; Martín-Belloso, O. Impact of Food Matrix and Processing on the In Vitro Bioaccessibility of Vitamin C, Phenolic Compounds, and Hydrophilic Antioxidant Activity from Fruit Juice-Based Beverages. *J. Funct. Foods* **2015**, *14*, 33–43.
76. Roobab, U. Aadil, R. M., Madni, G. M., Bekhit, A. E. D. The Impact of Nonthermal Technologies on the Microbiological Quality of Juices: A Review. *Comprehen. Rev. Food Sci. Food Safet.* **2018**, *17* (2), 437–457.
77. Sagong, H. G.; Lee, S. Y.; Chang, P. S.; Heu, S.; Ryu, S.; Choi, Y. J.; Kang, D. H. Combined Effect of Ultrasound and Organic Acids to Reduce *Escherichia coli* O157:H7, *Salmonella Typhimurium*, and *Listeria monocytogenes* on Organic Fresh Lettuce. *Food Microbiol.Int. J. Food Microbiol.* **2011,** *145* (1), 287–292.
78. Sahu, M.; Bala, S. Food Processing, Food Spoilage and Their Prevention: An Overview. *Int. J. Life. Sci. Scienti. Res.* **2017,** *3* (1), 753–759.
79. Severino, R.; Vu, K. D.; Donsì, F.; Salmieri, S.; Ferrari, G.; Lacroix, M. Antimicrobial Effects of Different Combined Non-Thermal Treatments Against *Listeriamonocytogenes* in Broccoli Florets. *J. Food Eng.* **2014,** *124*, 1–10.
80. Singh, S.; Shalini, R. Effect of hurdle technology in food preservation: a review. *Crit. Rev. Food Sci. Nutr.* **2016**, *56* (4), 641–649.
81. Snyder, A. B.; Worobo, R. W. The Incidence and Impact of Microbial Spoilage in the Production of Fruit and Vegetable Juices as Reported by Juice Manufacturers. *Food Control* **2018**, *85*, 144–150.
82. Soliva-Fortuny, R. C.; Elez-Martínez, P.; Sebastianán-Calderó, M.; Martín-Belloso, O. Effect of Combined Methods of Preservation on the Naturally Occurring Microflora of Avocado Purée. *Food Control* **2004**, *15* (1), 11–17.

83. Stanojević-Nikolić, S.; Dimić, G.; Mojović, L.; Pejin, J.; Djukić-Vuković, A.; Kocić-Tanackov, S. Antimicrobial Activity of Lactic Acid Against Pathogen and Spoilage Microorganisms. *J. Food Process. Preserv.* **2016**, *40* (5), 990–998.
84. Tiwari, B. K. Ultrasound: A clean, green extraction technology. *TrAC Trends Analyt. Chem.* **2015**, *71*, 100–109.
85. Trinetta, V.; Linton, R. H.; Morgan, M. T. The Application of High-Concentration Short-Time Chlorine Dioxide Treatment for Selected Specialty Crops Including Roma Tomatoes (Lycopersiconesculentum), Cantaloupes (Cucumismelo ssp. melo) and Strawberries (Fragaria× ananassa). *Food Microbiol.* **2013**, *34* (2), 296–302.
86. Tzortzakis, N. G. Impact of cinnamon oil-enrichment on microbial spoilage of fresh produce. *Innov. Food Sci. Emerg. Technol.* **2009**, *10* (1), 97–102.
87. Venkitanarayanan, K. S.; Lin, C. M.; Bailey, H.; Doyle, M. P. Inactivation of *Escherichiacoli* O157: H7, *Salmonella* Enteritidis, and *Listeriamonocytogenes* on Apples, Oranges, and Tomatoes by Lactic Acid with Hydrogen Peroxide. *J. Food Protection* **2002,** *65* (1), 100–105.
88. Vibhakara, H. S. J.; Gupta, D. K. D.; Jayaraman, K. S.; Mohan, M. S. Development of a High-Moisture Shelf-Stable Grated Carrot Product Using Hurdle Technology. *J. Food Process. Preserv.* **2006**, *30* (2), 134–144.
89. Vincente, A. R.; Manganaris, G. A.; Ortiz, C. M.; Sozzi, G. O.; Crisosto, C. H. Nutritional Quality of Fruits and Vegetables. In *Postharvest Handling*; Academic Press: New York, 2014; pp 69–122.
90. Wang, H.; Feng, H.; Luo, Y. Microbial Reduction and Storage Quality of Fresh-Cut Cilantro Washed with Acidic Electrolyzed Water and Aqueous Ozone. *Food Res. Int.* **2004**, *37* (10), 949–956.
91. Whangchai, K.; Saengnil, K.; Uthaibutra, J. Effect of Ozone in Combination with Some Organic Acids on the Control of Postharvest Decay and Pericarp Browning of Longan Fruit. *Crop Protection* **2006,** *25* (8), 821–825.
92. Wibowo, S.; Essel, E. A.; De Man, S.; Bernaert, N.; Van Droogenbroeck, B.; Grauwet, T.; Van Loey, A.; Hendrickx, M. Comparing the Impact of High Pressure, Pulsed Electric Field and Thermal Pasteurization on Quality Attributes of Cloudy Apple Juice Using Targeted and Untargeted Analyses. *Innov. Food Sci. Emerg. Technol.* **2019**, *54*, 64–77.
93. Wu, V. C.; Rioux, A. A Simple Instrument-Free Gaseous Chlorine Dioxide Method for Microbial Decontamination of Potatoes During Storage. *Food Microbiol.* **2010**, *27* (1), 179–184.
94. Yildirim, S.; Röcker, B.; Pettersen, M. K.; Nilsen-Nygaard, J.; Ayhan, Z.; Rutkaite, R.; et al. Active Packaging Applications for Food. *Comprehen. Rev. Food Sci. Food Safet.* **2018**, *17* (1), 165–199.
95. Yuk, H. G.; Yoo, M. Y.; Yoon, J. W.; Marshall, D. L.; Oh, D. H. Effect of Combined Ozone and Organic Acid Treatment for Control of *Escherichia coli* O157: H7 and *Listeria monocytogenes* on enoki Mushroom. *Food Control* **2007**, *18* (5), 548–553.
96. Yuk, H. G.; Yoo, M. Y.; Yoon, J. W.; Moon, K. D.; Marshall, D. L.; Oh, D. H. Effect of Combined Ozone and Organic Acid Treatment for Control of *Escherichia coli* O157: H7 and *Listeria monocytogenes* on Lettuce. *J. Food Sci.* **2006**, *71* (3), M83–M87.
97. Zhang, J.; Yang, H.; Chan, J. Z. Y. Development of Portable Flow-Through Electrochemical Sanitizing Unit to Generate Near Neutral Electrolyzed Water. *J. Food Sci.* **2018**, *83* (3), 780–790.

98. Zhang, S.; Farber, J. M. The Effects of Various Disinfectants Against *Listeria monocytogenes* on Fresh-Cut Vegetables. *Food Microbiol.* **1996**, *13* (4), 311–321.
99. Zhang, W.; Dong, P.; Lao, F.; Liu, J.; Liao, X.; Wu, J. Characterization of the Major Aroma-Active Compounds in Keitt Mango Juice: Comparison Among Fresh, Pasteurization and High Hydrostatic Pressure Processing Juices. *Food Chem.* **2019**, *289*, 215–222.
100. Zhao, L.; Zhao, M. Y.; Phey, C. P.; Yang, H. Efficacy of Low Concentration Acidic Electrolyzed Water and Levulinic Acid Combination on Fresh Organic Lettuce (*Lactuca sativa* Var. Crispa L.) and Its Antimicrobial Mechanism. *Food Control* **2019**, *101*, 241–250.

CHAPTER 2

MICROWAVE TECHNOLOGY: PRESERVATION OF FRUITS AND VEGETABLES

BISMA JAN, ARSHIED MANZOOR, RAFEEYA SHAMS, and
QURAT UL EAIN HYDER RIZVI

ABSTRACT

Fruits and vegetables (FAVs) hold high water and moisture content and are vital sources of numerous nutrients including vitamins, minerals, fiber, and significant bioactive compounds. Combination methods for the microwave drying of FAV need to be considered to guarantee a good appearance and to preserve the nutritional significance of dried products. The exploitation of microwaves in the drying of FAVs has been growing at an alarming rate in recent years. The most important application of preservation lies in precise processing time. When compared to other drying methods, due to its substantial decrease in energy consumption and cooking time and energy consumption, microwave heating technology is most useful for product quality. Compared to traditional method, cooking or heating with microwave technology will typically maintain elevated levels of photochemical constituent's antioxidant activity and significant color in FAVs. Microwave sterilization effectively inactivates the enzyme activity and, compared to traditional pasteurization, has a negligible impact on the texture, color, and radical scavenging activity of FAVs.

Quality Control in Fruit and Vegetable Processing: Methods and Strategies. Megh R. Goyal, & Faizan Ahmad (Eds.)

2.1 INTRODUCTION

The better living standard and increasing consumer income have contributed to the need for the modern application of food processing. The present advancement of science and technology has brought tremendous changes in human production and lifestyle. Microwave technology (MT) is one of the typical representatives. It refers to an electromagnetic wave within a frequency between 300 and 300 GHz and an electromagnetic wave with a wavelength between 1 mm and 1 m, and between the infrared radiation of the electromagnetic spectrum and the radio wave.[9] Furthermore, it is a process of volumetric heating, wherein heat is produced uniformly over whole food material. This attribute makes it quite different from conventional heating. This is due to the complete interaction of food ions in microwaves, polar water molecules, and charged ions.[32]

Microwave heating technology can provide rapid heat in continuous mode.[53] The key factor promoting the recognition of microwave heating is its rapidity which suits the present rapid standard of living of individuals very well.

This novel technology has come up among the foremost hopeful methods for food processing and production skill. Owing to its substantial importance over traditional heating methods, it has gained immense popularity. It has now been used in various industries for food production, such as blanching, frying, thawing, baking, sterilization, pasteurization, and drying of food products.[23,32] It is also famous for its equipped nutrient and protection ability, with the least loss of heat-sensitive nutrients, such as antioxidants, phenols, flavonoids, Vitamin B complex, vitamins C, and some carotenoids.[70]

The invention of the cavity magnetron by duo British physicists, namely, John T. Randall and Henry A. Boot,[7] contributed to discovering MT during World War II. During World War II, this technology was primarily employed for communication through the production of radar. MT emerged in 1946 when a pioneer scientist, namely, Dr. Percy Le Baron Spencer Inadvertently, found that a candy bar in his pocket melted when subjected to microwave radiation while conducting laboratory tests for an innovative vacuum tube which he named as magnetron tube (apparatus that produces an electromagnetic field). He improved the idea further and proved that it was possible to use microwaves as a heating process. This pioneer scientist is credited to have many patents on his name for many microwave heating applications in vast areas. As a result of this microwave energy has been incorporated into the food processing and preservation industry.[16]

Microwave oven acceptance has grown significantly and today 70% of the household of American use microwave oven.[25] In addition to household use, some uses for the special heating properties of microwaves have been discovered by the industry, such as tempering, cooking bacon, and drying pasta. In countries such as Canada, United States, and several European countries, these microwave ovens have now turned out to be an important domestic device for households in the 21st century. The growing urbanization and accessibility to electricity in developing countries have contributed to its raised use worldwide. Microwave ovens are presently utilized mostly for used for defrosting and cooking besides heating and reheating of food.[74]

Moreover, high-quality, minimally processed goods are demanded by modern food consumers[26] and have contributed to the advancement of new technologies for microwave processing. Because of its broad variety of benefits, customers have generally embraced MT, which has allowed them to resolve the distrust and taboos that existed in the beginning.

The present chapter focuses on the mechanism and the principle of working of MT and its recent application as a preservation technique in the fruit and vegetable sector.

2.2 PRINCIPLE AND MECHANISM OF MICROWAVE TECHNOLOGY

Various food components function differently when kept in a microwave oven. The primary ingredient that allows microwaves to heat food is water. Higher water content results in a higher heating rate in food products. Water in food behaves like a magnet due to the presence of two oppositely charged ends of oxygen molecules and hydrogen atoms. Similarly, owing to two separate water poles, the water molecules rotate as microwaves oscillate. This is due to the water's negatively charged end drawn toward the positively charged microwave while the water's positive end is drawn toward the negatively charged end of microwaves. At an exceptionally higher speed of 2450 times/s, the microwaves rotate. This incredibly elevated rotation results in the collision of water molecules at a rapid rate that induces friction among water molecules and hence heat is created. Through conduction, convection, or radiation, the heat flows through the food. Food warms up, thus another microwave heating theory is that dipole rotation occurs under microwave radiation in dielectric materials comprising polar molecules having an electrical dipole moment.

Datta and Davidson, 2002, postulate that microwave heating can also happen owing to the oscillatory migration of food ions that produce heat in the occurrence of higher frequency electric field oscillation. The energy equation for microwave heating is given as follows:

$$\partial T / \partial t = \alpha[\nabla^2 T] + Q/[\rho Cp] \quad (2.1)$$

where T = temperature, t = time, α = thermal diffusivity, ρ = density, Cp = specific heat of the material, and Q = the heat generated.

2.3 FACTORS AFFECTING MICROWAVE HEATING

Several studies have been conducted over the 30 years to examine variables that affect bacterial destruction and temperatures attained in food products heated through applied microwave energy.[24] The key elements influencing microwave heating are penetration depth and dielectric properties. Dielectric properties demonstrate a material's ability to transform microwave heat and control the effectiveness and consistency of the heating process.[26] The real component of dielectrics known as a dielectric constant is its capability to store electricity while the theoretical component of dielectrics, known as dielectric loss, employs its capability to transfer electric energy into heat.[10,17] These characteristics can be split into the dielectric loss factor and dielectric constant meanwhile these features are responsible for storing and dissipating electrical energy.[46]

When a polar dielectric material is displayed to the oscillating electric field, dipoles inside the substance effort to align themselves with the applied field. The depth of penetration is illustrated as the distance away from surface of the substance from where power drops to 1/e (about 37%)[70] or depth of penetration, which depends on food composition, is a kind of capacity that passes across the interior of foods through the electromagnetic wave.

$$Z = [\lambda_0 / 2\pi](2\varepsilon'')^{0.5}\{[1 + (\varepsilon'/\varepsilon'')^2]^{0.5} - 1\}^{-0.5} \quad (2.2)$$

where Z = penetration depth, λ_0 = wave length in free space, ε' = dielectric constant, and ε'' = dielectric loss factor.

It is apparent from the above equation that the depth of penetration is increased with decreasing frequency. At frequencies below 100 MHz, penetration depth usually poses few issues, except for the high loss factors.[44]

2.3.1 LOCATION AND GEOMETRY OF FOODS

The shape is a critical factor of the food material that mainly influences the microwave penetration depth, heating rate, and uniformity. The difference in product thickness and uneven-shaped products undergo nonuniform heating. The closer the size and wavelength ratio, the higher will be the center temperature. For example, cylindrical or spherical shape foods are heated more uniformly than a cubical food. The heating rate is improved by a greater surface-to-volume ratio.[6]

2.3.2 TEMPERATURE

Depending upon the material, dielectric properties may differ with temperature. Both moisture content and temperature can change during heating and therefore show joint result on dielectric loss factor, and dielectric constant, consequently on the heating characteristics.[35]

2.3.3 MOISTURE CONTENT

This property determines the dielectric characteristics of the food material and microwave penetration depth. Foods with high moisture content show low microwave penetration depth, and thus results in, uneven heating rate unlike, minimal moisture content foods that have a more even heating rate due to microwave penetration that is deeper.[46]

2.3.4 MASS

There is a direct correlation between the amounts of microwave power and mass-consumed that might be added to attain preferred heating power. In case of a less significant mass generally, the batch oven is used, however, for large capacity equipment conveyors are used. These types of equipment provide uniform heating by moving the product throughout the microwave field.

2.4 ENZYME DEACTIVATION AND MICROBIAL DESTRUCTION BY MICROWAVE HEATING IN FRUITS AND VEGETABLES

Taking into consideration the rising market need for high-quality food that is microbiologically secure with well retained sensory and characteristics, efforts are taken during processing to reduce heating of food products. This could be achieved through, for example, reducing the heating period by using advanced processing techniques to kill microorganisms and tissue enzymes or using methods other than traditional hot water or steam.[38] The application of microwave energy is one of these new approaches. Microwave heating is currently utilized in various food technology processes, namely, freeze-drying, vacuum drying, tempering, dehydration, pasteurization, and sterilization.[50]

In food, enzymes are inherently there and can result in the degradation of the food product. For that purpose, most food processing steps attempted to decrease the activity of enzymes. The enzymes like polyphenol oxidase (PPO) and peroxidase (POD) are mainly accountable for the color deterioration and nutritional significance of the majority of fruits and vegetables (FAVs).[56,21] An investigation was attempted to observe the enzymatic inactivation impact of POD, pectin methylesterase (PME), and PPO in kiwi fruit. Microwave heating resulted in 90% inactivation in POD whilst maintaining the overall antioxidant activity of kiwi fruit. In case of cloudy apple juice, microwave heating also results in the destruction of enzymes like POD, PME, and PPO.[62] Likewise, the growth of spoilage-causing microorganisms, namely, *Saccharomyces cerevisiae* and *Lactobacillus plantarum* was destructed by the continuous flow microwave heating (CFMH) process as studied in apple juice.[67]

Kinetics of degradation of *L. plantarum* and *S. Cerevisiae* in CFMH followed standard first-order reaction with a linear destruction rate versus times on logarithmic survivor plot. Collectively, the analysis indicated that the conditions of continuous microwave heating killed the microorganisms faster than traditional batch heating conditions by an order of magnitude. Microwave heating may result in 5-log 10 reduction in *Escherichia coli* and *Listeria monocytogenes* with power levels of 400, 600, 800, and 1000 W in apple juice.[61]

Coconut water is safe and sterile; however, a few reactions may happen, mostly oxidation that is assisted by enzymes like POD and PPO that are physically there in coconut when exposed to air. So it is important to deactivate these enzymes. A study was conducted to inactivate the enzymes in coconut water to render it sterile. It was observed by applying microwave heating the thermal tolerance of the enzymes was greater than that of marketable ones applied to sterile coconut water. In this study, the first-order kinetic model satisfactorily represented the experimental results.[42]

Marszałek et al.[38] found applying microwave radiation at 120°C, PPO and POD activity on strawberry puree reduced by 98 and 100%, correspondingly. The beverages like lemon, guava, pineapple, and nopal, juice when treated with microwaves of 490 W, 2450 MHz resulted in PME at 90°C at 12 s.[22]

Hence it may be concluded that microwave heating can destroy microorganisms and efficiently inactivate the enzymatic activity.

2.5 ADVANTAGES OF MICROWAVE HEATING

- Owing to its source inside the food substance, rapid temperature rise, controllable heat deposition, and simple cleanup, microwave heating is gaining popularity for food heating.
- Minimum fouling depositions are observed due to the removal of the hot heat transfer surfaces.
- High heating efficiency in microwaves (80% or greater effectiveness can be attained).
- For clean-in-place systems, they have ideal geometry.
- Microwaves are ideal for heat-liable, high-viscous, and multiphase fluids as well as low maintenance costs for systems.
- The influence of minimal heat conductivities and coefficient of heat transfer does not exhibit a vital role. As a result, in a little time and with a more uniform distribution of temperature, larger parts can be heated. Such advantages also lead to no-exhaust gas emissions, increased demand, and minimal harmful unwanted waste to the surroundings.
- Sample heating is also carried out inside the food, since the microwave penetrates.
- The quality attributes of FAVs mainly color, taste, texture, and vitamin content are preserved better when heated quickly by microwave.
- Inactivation of enzyme complexes.

2.6 ROLE OF MICROWAVE AND COMBINED TECHNOLOGIES IN THE PRESERVATION OF FRUITS AND VEGETABLES

The microwave is an essential mechanism for enzyme inactivation and safe food storage and consumption. The effects of the inactivation of enzymes by oven heating and microwave on quality parameters of tea being processed and harvested in the autumn and spring seasons were determined in an investigation. The importance of microwave heating technology is given in Table 2.1.

A study come up with the results that no major difference was noted between oven heating and microwave in autumn and spring seasons in polyphenol content of tea. The study concluded that characteristics of green tea

TABLE 2.1 Role of Microwave Heating in Selected Fruits and Vegetables.

Fruit name	Major findings	References
Apple	Microwave heat treatment resulted in improved extraction of flavonoids and phenolics from apple juice and mash.	[20]
Apple puree	Heating by microwave technology could effectively reduce the growth of pathogenic microorganisms (*E. coli* O157:H7) and shelf-life prolonged by 14 days at refrigerated conditions.	[53]
Apple juice	Continuous flow microwave heating resulted in the enhanced destruction of *L. plantarum* and *S. cerevisiae* by following a first-order reaction.	[67]
Apple juice	Pasteurization of apple juice by microwave processing resulted in the destruction of *E. coli.*	[8]
Pear	Intermittent microwave-vacuum drying in pears resulted in raised mineral and protein amounts with higher drying rates.	[68]
	Microwave heating caused in increased concentration of phenolics like 3,4-dihydroxybenzoic acid, gallic acid, apigenin 7 glucoside, rutin, caffeic acid, resveratrol, etc.	[30]
	Microwave convective drying of pear resulted in increased retention of calcium, phosphorus, and potassium than convective drying and preserved sample characteristics.	[18]
	Microwave heating at 2450 MHz for 2–3 min resulted in complete inhibition of *Penicillium expansum,* thereby controlling the postharvest blue mold rot on pear fruits.	[77]
Banana	The power output of 650 w in microwave vacuum drying resulted in rapid drying at low temperatures hence can be regarded as an efficient method of drying and preservation method.	[45]
	This study resulted in a shortening of drying time and increased the maturity of banana slices. Parameters like sugar content, color, hardness, fracturability displayed better performance.	[28]
	In contrast to hot air and microwave drying techniques, hot air followed by microwave finish drying resulted in a higher drying rate, decreased drying time, and had a limited impact on the rehydration potential and color of end products.	[40]
	Increased microwave power caused an increased drying rate and reduction in the processing period. Furthermore combined effect of infrared and microwaves resulted in the increased rate of moisture removal from banana samples with a time saving of about 98%.	[51]
	Microwave-assisted vacuum frying could offer a considerably reduced frying time over vacuum frying, from 75 min to 50 min, and has the ability to maintain the quality characteristics of banana chips comparable to that of vacuum frying technology.	[31]

TABLE 2.1 *(Continued)*

Fruit name	Major findings	References
Tomatoes	Microwave heating of tomatoes resulted in higher retention of polyphenols like lycopene, gallic acid, and some flavonoids in tomato slices.	[37]
Potatoes	According to the findings, as the increased power level resulted in reduced mass load and thickness increases, and hence rate of dehydration increases, resulting in deceased consumption of drying energy.	[72]
Cauliflower	When 550 g of broccoli by-products were used, microwave hydro-diffusion permitted a dried sample with 12% moisture to be obtained in 43 min while preserving proteins and polysaccharides.	[15]

harvested in autumn and spring were significantly improved using microwave heating.[27] Microwave heating was applied to different fruits like kiwi, strawberries, and papaya to deactivate enzymes including oxydo reductases peroxidase and polyphenol oxidase. There was no significant change noted in the original color of fruits; however, small modifications of the qualitative and quantitative composition were observed.[11] Microwave power was used to control the postharvest pathogens in peaches. Heating with MT for 2 min resulted in a significant reduction percentage of infected wounds caused by *Botrytis cinerea* and *Penicillium expansum*.[34] Similarly, the growth of *Rizophus stolonifer* was entirely inhibited by 2450 MHz microwave heating for a time duration of 2 min or more.[75] Recently a study observed that heating via MT at 1–2 min in peaches was way better over remaining existing treatments and also improved the bioactive functional ingredients of beverage counting total phenolic compounds to 33% and antioxidants to 24% (ferric reducing antioxidant power) and 16% (2,2-diphenyl-1-picrylhydrazyl), respectively.[57]

2.6.1 MICROWAVE BLANCHING

In general, blanching is done by dipping food substances either in boiling or hot water solutions or steam-containing acids or salts for color preservation and enzyme inactivation. The benefits of microwave blanching comprise energy savings, reliable process controls, speed of operation, and rapid start-up as well as shut-down durations compared to traditional heating techniques.[6] Proctor and Goldblith[54] were the first to report microwave blanching for several green vegetables using 3000 MHz in 1948, and it was discovered that vegetables could retain maximum vitamin C amounts by microwave heating. This study, however, reported that treatment by microwave blanching resulted in preserving water-soluble vitamins in vegetables comparatively better to traditional blanching techniques.

Muftugil[47] found that microwave-blanched green beans had better retention of vitamin C than hot water-blanched beans. For effective heat transfer in food, microwave blanching requires a minute or no water thus could comparatively decrease the quantity of nutrients and vitamins loss due to leaching than hot water immersion.[36] Another study reported that microwave blanching of peanuts was effective as comparatively to conventional blanching techniques in terms of time and energy. Moreover, present study also reported high-temperature microwave blanching of peanuts resulted in incidence of ashy and floral flavors.[58]

Patricia et al.[52] observed impact of microwave blanching on the nutritional attributes of broccoli and reported that the amounts of iron, phosphorus, ashes, and vitamin C in microwave blanched broccoli was higher than those processed by traditional hot water blanching; moreover, these attributes were much nearer to fresh broccoli. Another study reported little difference in sensory attributes, volatile aroma compounds, between dried and fresh dill which was hot water blanched at 90°C for a duration of 30 s followed by pretreatment with microwave for 30 s.[63]

2.6.2 MICROWAVE-ASSISTED PASTEURIZATION AND STERILIZATION

The main purpose of heat treatments like pasteurization and sterilization are to destroy or inactivate microbes to guarantee prolonged storage life and the protection of food products. Pasteurization is a technique wherein heat treatment destroys disease-causing microbes or pathogens in vegetative forms, such as bacteria. In general, fruit juices are not counted among foodborne disease carriers, but recent studies indicate that three disease-causing bacteria, namely, *E. coli*, *Salmonella enteric*, and *Crytosporidium parvum* are linked by foodborne diseases in acidic fruit juices.

Traditionally, pasteurization of fruit juices is performed using a plate heat exchanger via the high-temperature short-time heating process, accompanied by precise holding time and cooling. Another study was conducted to examine the *E. coli* exposure to microwave heating in apple juice. Results obtained demonstrated a significant decrease in the microbial growth in apple juice that was pasteurized at 720–900 W for 60–90 s than conventional pasteurization (83°C for 30 s).[8]

Similarly, another study investigated the kinetics destruction under continuous-flow microwave heating conditions of *S. cerevisiae* and *L. plantarum* in apple juice and measured to that of traditional batch heating in a circulatory water bath. The Z values for *L. Plantarum* and *S. cerevisiae* under microwave heating were 4.5 and 7°C, correspondingly, whereas traditional heating values were 15.9 and 13.4°C, respectively. Under microwave heating, microbial degradation was therefore highly susceptible to temperature than under thermal heating.[19] The inactivation kinetics of *S. typhimurium* and *E. coli* O157:H7 have recently been investigated for microwave pasteurization at a temperature of 80 and 90°C which means same conditions used for conventional pasteurization. Results indicated that inactivation increased to a microbial decrease of up to seven cycles of Log10 with temperature, power level, and treatment time. For the food industry, the study is important because it provides a quantitative method for predicting pathogens' survival characteristics under other conditions of nonisothermal processing.[43]

Different blends of acidic and nonacidic juices were made by continuous-flow microwave system (CFMS) (2 kW) at 250 mL/min. Biochemical and chemical attributes like protein and vitamins and nutritional characteristics of microwave blend juices were similar to fresh juice upon 1 year of storage.[41]

An undesired enzyme, PME causes loss of cloudiness and spoilage during the storage period in orange juice. Furthermore, this enzyme is highly resistant to heat than the spoilage microbes and therefore regarded as an adequacy index for pasteurization. In a study, orange CFMS was used to pasteurize orange juice by PME deactivation and microbial count as an index. As compared to traditional heating, deactivation of *L. plantarum* and PME was found to be prominent by employing microwaves.[49]

2.7 RECENT DEVELOPMENTS IN MICROWAVE DRYING TECHNOLOGY FOR THE PRESERVATION OF FRUITS AND VEGETABLES

2.7.1 MICROWAVE FREEZE-DRYING

The increased perishability of FAV leads to different deterioration and undesirable rapid reactions. Ohmic heating, radiofrequency, high-pressure processing, infrared, microwave heating ultrasound are several novel technologies gaining attention in food processing and technology.

Microwave freeze drying (MFD) is a famous process that prevents chemical decomposition and promotes rapid rehydration by removing moisture from heat-sensitive food products, such as raspberries or tomatoes, and many more. Ice converts water during MFD by local overheating. It is a well-known fact that the capability of water to take up microwave power is quite moreover ice. Hence, a microwave will be concentrated at the melting point, contributing to inconsistent drying. This process is known as a thermal runaway in MFD.[14]

Compared to traditional freeze-dried products, the primary advantage of MFD is its quick energy dissipation into the material, substantial brief processing time, effective drying in the dropping rate duration, volatile constituent retention, and energy-saving.[29] In a vacuum environment, microwaves will heat the material volumetrically, thereby increasing the rate of freeze-drying. There are limited studies conducted on MFD among which most have focused on mass and heat transfer. Earlier in 1999 MFD attributes of unsaturated raw beef were investigated experimentally for different levels of vacuum pressure, sample thickness, electric field strength, and initial saturation. The results indicated the drying time is directly correlated to primary saturation and inversely correlative to strength of the electric field and meanwhile, the consistency of freeze-dried raw beef freeze-dried was superior to traditional freeze-dried beef.[71]

Duan et al. investigated the MFD process and sterilization properties.[12] In their study, the thickness of material to be dried, input microwave power, and the impacts of pressure, on indices like several microorganisms and drying time were examined. In sea cucumber (*Stichopus japonicus*), MFD decreased the drying time by approximately half the traditional FD system and offered a comparable high-quality product.[13]

Wang et al.[73] researched the impact of vacuum and MFD on the quality attributes microstructure of potato slices and observed that during MFD, both blanching and calcium ion treatments were effective methods to prevent variations in shape.

2.7.2 MICROWAVE DRYING

In era of food preservation and processing technology, microwave heating has a wide approach like cooking, drying, pasteurization, and maintenance of nutrients in their original form in food materials. Microwave drying techniques have recently been studied as possible methods to produce dehydrated food products of high quality. In comparison to traditional hot air drying, microwave drying is energy-efficient evenly distributed, and quick, as the microwaves penetrate into the food within, resulting water to heat up inside the food. This leads to substantially raised variation in vapor pressure amongst the surface and center of the food material, allowing moisture to be extracted rapidly from the food.

Techniques like microwave vacuum drying and microwave convective have successfully been involved in drying some FAVs. One of the drawbacks of microwave drying is that extremely high temperatures, particularly during the final phases of drying, lead to the burning and production of undesirable off-flavors along the edges or corners of the food sample. This is because of the complexity of monitoring the temperature of the finished material in microwave drying, although the product temperature does not reach the hot air temperature in hot air drying.[33] It is therefore essential to merge microwave drying with conventional drying to increase the rate of drying and preserving the excellence of the food substance. Combined with other drying technologies, microwave energy increases the drying performance besides the consistency of food products, which is much greater than that which can only be accomplished by microwave drying or other traditional procedures.[76]

2.7.3 MICROWAVE-ASSISTED AIR DRYING

It is a technique wherein hot air drying is merged with microwave heating so as to increase the drying rate.[77] It is difficult to dry the banana as it falls under the dropping rate era. In comparison to convective air drying, in kiwifruits, combined impact of microwave finish and hot air drying shortened the drying rate by almost 64%.[40]

Kiwifruits when dried by combined hot air MT also demonstrated greater dehydration potential than those dried by single hot air or microwave drying.[39] In the study, garlic cloves were dehydrated using both microwave and hot air drying methods. Parameters like drying time, color, and flavor strength were involved to elevate the effectiveness of both techniques. Microwave hot air drying was found to minimize the time of drying by 80–90% compared to traditional drying, and the consistency of the end product was relatively better.[59]

Another research compared the cranberries' microwave-convective and microwave-vacuum drying. The drying performance depicted that microwave-vacuum drying was highly energy-saving over microwave convective drying in case of the quantity of water vaporized per unit of energy applied. Nonetheless, the sensory study, on the other hand, revealed that microwave-convective dried cranberries (color, texture, taste, overall appearance) were somewhat superior to the microwave-vacuum dried cranberries.[66]

Another study was conducted to investigate dry pumpkin slices using different drying methods. The suitable drying technique for pumpkin slices was combined microwave–air-drying technology in case of color and consumption of energy during drying.[3]

2.7.4 MICROWAVE-ASSISTED VACUUM DRYING

Microwave-assisted vacuum drying (MVD) is a novel technique with the benefits of microwave heating, namely, better controllability, performance, and hygiene and quick heating rates. This novel technology also decreases the boiling point of water induced by the vacuum environment, therefore increasing energy efficacy and decreasing the creation of scorched stains on the surface of a finished product.[76] Initially in the late nineteenth-century vacuum drying of banana slices was studied experimentally. Outcome of the current research depicted that microwave vacuum drying might be utilized to preserve perishable food foodstuffs like banana.

Another study demonstrated that high-quality dried banana slices can be produced by microwave that can be achieved by varying duty cycle time and power density.[45,48] The quality attributes like darkness, color, crystallized sugar, stickiness, and uniformity in microwave dried raisins were superior to hot dried raisins.[69] In one of the studies, durian chips were developed and were compared with chips developed by conventional fried methods. The structure and hardness of the durian chips that were microwave dried were close to that of the durian chips that were conventionally fried.

Furthermore, microwave vacuum drying decreased the durian chips' fat content by almost 90%, as compared to conventionally deep-fried durian chips.[5] The impact of microwave power on drying kinetics and consistency of tomato pieces in microwave-vacuum dryer at different applied vacuum pressures were explored in another study. The results showed improvement in conditions for microwave-vacuum drying rates and drying time from 84 to 14 min.[1]

2.7.5 MICROWAVE-ASSISTED VACUUM FRYING

Vacuum frying is amongst important food industry's frying technique that fries food products under reduced pressures probably below atmospheric pressure thus decreasing the boiling point of frying oil and water content of food product. The main advantages of vacuum frying include lower acrylamide and oil content, reduced unfavorable consequences on oil consistency, and better preservation of the product's natural color and taste because of minimal oxygen and temperature content during the process. The implementation of microwave vacuum frying (MVF) following prefrying is highly recommended in order to additionally improve the characteristics of fried food stuffs, as it raises the heating rate significantly and minimizes oil uptake and frying time.[55]

In one of the study potato chips were subjected to two distinct frying methods, namely, vacuum frying and MVF. Based on the conclusion of the investigation, MVF can appreciably decrease oil uptake in potato chips from an initial content of 39.14–29.35 g oil/100 g d. Meanwhile quality parameters like crunchiness, crispiness was enhanced and the color was retained to a significant extent, mainly owing to increased moisture loss, than those samples developed using vacuum frying technology.[64] However, it is essential to note that temperature is one important attribute to affect the consistency of microwave-vacuum fried potato chips, so manufacturer's requisite to discover optimum temperature environments to produce better food products with excellent quality attributes.[65]

Furthermore, the combination of vacuum prefrying and pulsed microwave vacuum drying (PSMVD) implies a new method to incorporate microwave irradiation into the production process to achieve premium-quality French fries. A comparative study produced French fries with an oil content of 25% in combined technique PSMVD and vacuum-fried French fries with an oil content of 40%. However textural qualities were comparable in both techniques.[55] Recently ultrasound combined MVF technology was employed to fry apple slices to improve overall frying time, crispiness, and color. Microwave power levels utilized were 800–1000 W with fixed ultrasound power of 600 W and frequency range of 28 kHz. The application of ultrasound combined MVF technology appreciably improved the moisture removal rate, enhanced crispiness and crunchiness, reduced the frying time, and yielded more pleasing yellow-colored apple slices compared to MVF technology.[2]

2.7.6 MICROWAVE-ASSISTED INFRARED DRYING

In a large range of agricultural products, the popularity of infrared drying (IRD) has improved immensely as another alternate technique for conventional drying. However, its low penetrating power is the core limitation so the combined effect of microwave energy and infrared radiation (IR) radiation can overcome this difficulty. The connection between microwave–IR heating and dielectric properties on the drying properties of kiwifruit and bananas has recently been investigated.

The combined drying technique was found to have faster drying rates (approximately 98% time savings) with reduced absolute moisture content (0.011–0.15 kg water/kg dry matter) as compared to traditional drying.[51] The quality and drying kinetics of raspberries under microwave-vacuum IR drying at various vacuum pressures and microwave powers were recorded in another study. Bioactive chemical constituents like anthocyanins and DPPH radical-scavenging activity were 17.5 and 21.21% higher than IRD at optimum conditions, respectively. Besides, this study suggested that IR–microwave-vacuum drying could be recommended for usage in the drying industry because of its positive advantages in nutrient maintenance time saving.[60]

Another study utilized eggplants to examine the influence of microwave IR at diverse microwave powers (30, 40, and 50%) and diverse IR powers (10, 20, and 30%). Microwave–IR combination offered considerably lesser drying duration over hot-air drying. Furthermore, microwave–IR-dried eggplants had low shrinkage, more porous structure, and high rehydration ratio as compared to conventional hot-air dried samples.[4]

2.8 SUMMARY

Currently, people are becoming more health conscious, preferring to consume food that contains all the necessary nutrients and fewer preservatives. Microwave ovens have an ordinary place in households and are utilized these days as a device of day-to-day use. The microwave heating method has been effectively used for the maintenance important bioactive constituents of numerous FAVs in food industries. In order to design microwave oven, knowledge of dielectric properties and penetration depth is very useful. Uneven distribution of temperature throughout microwave heating has been observed to be influenced by the size, shape, and location of a food substance.

The microwave pasteurization and sterilization of various FAVs have claimed to successfully kill pathogenic microbes and appreciably decrease processing time without severe harm to the excellence of a final foodstuff than conventional techniques. Likewise, this premium heating technology used in the fruit and vegetable sector such as blanching has the significance of preserving and retaining original quality in terms of nutritional importance, color, taste, with better preservation than the traditional method.

Compared to traditional techniques, microwave heating has greatly decreased energy consumption during the drying of foods. The combined microwave heating system including infrared or freeze and vacuum drying supplemented with traditional methods has shown the greatest benefits. In the case of microwave-assisted freeze-drying, the advantages include lesser drying time, uniform heating of particulate foods, improved thermal efficacy, greater rehydration ability, and preservation of heat-sensitive food material components. While microwaves are commonly used in different food processes, there is a need for substantial research aimed at improving those areas. Specifically, it is important to explore methods for obtaining final FAVs with better nutritional and sensor qualities characteristics.

KEYWORDS

- **bioactive constituent**
- **microwave technology**
- **perishable**
- **preservation**
- **sensory attributes**

REFERENCES

1. Abano, E. E.; Ma, H.; Qu, W. Influence of Combined Microwave-Vacuum Drying on Drying Kinetics and Quality of Dried Tomato Slices. *J. Food Qual.* **2012,** *35* (3), 159–168.
2. Al Faruq, A.; Zhang, M.; Adhikari, B. A. Novel Vacuum Frying Technology of Apple Slices Combined with Ultrasound and Microwave. *Ultrasonics Sonochem.* **2019,** *52*, 522–529.
3. Alibas, I. Microwave, Air and Combined Microwave–Air-Drying Parameters of Pumpkin Slices. *LWT-Food Sci. Technol.* **2007,** *40* (8), 1445–1451.
4. Aydogdu, A.; Sumnu, G.; Sahin, S. Effects of Microwave-Infrared Combination Drying on Quality of Eggplants. *Food Bioprocess Technol.* **2015,** *8* (6), 1198–1210.
5. Bai-Ngew S.; Therdthai, N.; Dhamvithee, P. Characterization of Microwave Vacuum-Dried Durian Chips. *J. Food Eng.* **2011,** *104* (1), 114–122.
6. Bhatt, K.; Vaidya, D.; Kaushal, M.; Gupta, A.; Soni, P.; Arya, P.; Sharma, C. Microwaves and Radio Waves: In *Food Processing and Preservation. Int. J. Curr. Microbiol. Appl. Sci.* **2020,** *9* (9), 118–131.

7. Brittain, J. E. The Magnetron and the Beginnings of the Microwave Age. *Phys. Today* **1985,** *38* (7), 60–67.
8. Canumir, J. A.; Celis, J. E.; de Bruijn, J.; Vidal, L. V. Pasteurization of Apple Juice by Using Microwaves. *LWT-Food Sci. Technol.* **2002,** *35* (5), 389–392.
9. Chandrasekaran, S.; Ramanathan, S.; Basak, T. Microwave Food Processing—A Review. *Food Res. Int.* **2013,** *52* (1), 243–261.
10. Curet, S.; Rouaud, O.; Boillereaux, L. Estimation of Dielectric Properties of Food Materials During Microwave Tempering and Heating. *Food Bioprocess Technol.* **2014,** *7* (2), 371–384.
11. de Ancos, B.; Cano, M. P.; Hernandez, A.; Monreal, M. Effects of Microwave Heating on Pigment Composition and Color of Fruit Purees. *J. Sci. Food Agric.* **1999,** *79* (5), 663–670.
12. Duan, X.; Zhang, M.; Mujumdar, A. S. Studies on the Microwave Freeze-Drying Technique and Sterilization Characteristic of Cabbage. *Dry. Technol.* **2007,** *25* (10), 1725–1731.
13. Duan, X.; Zhang, M.; Mujumdar, A. S.; Wang, S. J. Microwave Freeze Drying of Sea Cucumber (*Stichopus japonicus*). *J. Food Eng.* **2010,** *96* (4), 491–497.
14. Duan, X.; Zhang, M.; Mujumdar, A. S.; Wang, R. Trends in Microwave-Assisted Freeze Drying of Foods. *Drying Technol.* **2010,** *28* (4), 444–453.
15. Ferreira, S. S.; Passos, C. P.; Cardoso, S. M.; Wessel, D. F.; Coimbra, M. A. Microwave Assisted Dehydration of Broccoli By-Products and Simultaneous Extraction of Bioactive Compounds. *Food Chem.* **2018,** *246*, 386–393.
16. Fito, P.; Chiralt, A.; Eugenia Martin, M. E. Current State of Microwave Applications to Food Processing. In *Novel Food Processing Technologies*; Barbosa-canovas, G. V., Tapia, M. S., Cano, M. P., Eds.; CRC Press: Boca Raton, FL, 2005; pp 525–537.
17. Franco, A. P.; Yamamoto, L. Y.; Tadini, C. C.; Gut, J. A. W. Dielectric Properties of Green Coconut Water Relevant to Microwave Processing: Effect of Temperature and Field Frequency. *J. Food Eng.* **2015,** *155*, 69–78.
18. Fumagalli, F.; Silveira, A. M. Quality Evaluation of Microwave-Dried Packham's Triumph Pear. *Dry. Technol.* **2005,** *23* (9–11), 2215–2226.
19. Gentry, T. S.; Roberts, J. S. Design and Evaluation of a Continuous Flow Microwave Pasteurization System for Apple Cider. *LWT-Food Sci. Technol.* **2005,** *38* (3), 227–238.
20. Gerard, K. A. and Roberts, J. S. Microwave Heating of Apple Mash to Improve Juice Yield and Quality. *LWT-Food Sci. Technol.* **2004,** *37* (5), 551–557.
21. Gonçalves, E.; Pinheiro, M.; Abreu, J.; Brandao, M.; T. R. S.; Silva, C. L. M. Carrot (Daucuscarota L.) Peroxidase Inactivation, Phenolic Content and Physical Changes Kinetics Due to Blanching. *J. Food Eng.* **2010,** *97*, 574–581.
22. González-Monroy, A. D.; Rodríguez-Hernández, G.; Ozuna, C. Sosa-Morales, M. E. Microwave-Assisted Pasteurization of Beverages (Tamarind and Green) and Their Quality During Refrigerated Storage. *Innov. Food Sci. Emerg. Technol.* **2018,** *49*, 51–57.
23. Gupta, M.; Wong, W. L. E. *Microwaves and Metals*; John Wiley & Sons (Asia) Pvt. Ltd.: Singapore, 2007; E-book; pp 256.
24. Heddleson, R. A.; Doores, S. Factors Affecting Microwave Heating of Foods and Microwave Induced Destruction of Food Borne Pathogens–A Review. *J. Food Protection* **1994,** *57* (11), 1025–1037.
25. Hegenbart, S. Microwaveable Foods: Making Faster Better. *Prepared Foods* **1989,** *159* (12), 87–90.

26. Hong, X.; Wang, J. Use of Electronic Nose and Tongue to Track Freshness of Cherry Tomatoes Squeezed for Juice Consumption: Comparison of Different Sensor Fusion Approaches. *Food Bioprocess Technol.* **2015,** *8* (1), 158–170.
27. Huang, Y.; Sheng, J.; Yang, F. Hu, Q. Effect of Enzyme Inactivation by Microwave and Oven Heating on Preservation Quality of Green Tea. *J. Food Eng.* **2007,** *78* (2), 687–692.
28. Jiang, H.; Zhang, M.; Liu, Y.; Mujumdar, A. S.; Liu, H. The Energy Consumption and Color Analysis of Freeze/Microwave Freeze Banana Chips. *Food Byproducts Process.* **2013,** *91* (4), 464–472.
29. Jiang, N.; Liu, C., Li, D.; Zhang, Z.; Liu, C.; Wang, D.; Niu, L. Zhang, M. Evaluation of Freeze Drying Combined with Microwave Vacuum Drying for Functional Okra Snacks: Antioxidant Properties, Sensory Quality, and Energy Consumption. *LWT-Food Sci. Technol.* **2017,** *82*, 216–226.
30. Juhaimi, F.; Özcan, M. M.; Uslu, N.; Ghafoor, K. Babiker, E. E. Effect of Microwave Heating on Phenolic Compounds of Prickly Pear (*Opuntia ficus-indica* L.) Seeds. *J. Food Process. Preserv.* **2018,** *42* (2), 13437.
31. Jumras, B.; Inprasit, C.; Suwannapum, N. Effect of Microwave-Assisted Vacuum Frying on the Quality of Banana Chips. *Songklanakarin J. Sci. Technol.* **2020,** *42*(1).
32. Kalla, A. M.; Devaraju, R. Microwave Energy and Its Application in Food Industry: A Review. *Asian J. Dairy Food Res.* **2017,** *36* (1), 37–44.
33. Kandasamy, P.; Varadharaju, N. Assessment of Biochemical Characteristics of Foam-Mat Dried Papaya Powder. *Int. J. Agric. Food Sci.* **2014,** *4* (1), 54–58.
34. Karabulut, O. A.; Baykal, N. Evaluation of the Use of Microwave Power for the Control of Postharvest Diseases of Peaches. *Postharvest Biol. Technol.* **2002,** *26* (2), 237–240.
35. Khattarpaul, N. Food Processing and Preservation. In *Microwave Application to Food*; Daya Publishing House: New Delhi, 2005; pp 110–120.
36. Lin. S.; Brewer, M. S. Effect of Blanching Method on the Quality Characteristics of Frozen Peas. *J. Food Qual.* **2005,** 28, 350–360.
37. Mahieddine, B.; Amina, B.; Faouzi, S. M.; Sana, B. and Wided, D. Effects of Microwave Heating on the Antioxidant Activities of Tomato (*Solanum lycopersicum*). *Anna. Agric. Sci.* **2018,** *63* (2), 135–139.
38. Marszałek, K.; Mitek, M.; Skąpska, S. Effect of Continuous Flow Microwave and Conventional Heating on the Bioactive Compounds, Color, Enzymes Activity, Microbial and Sensory Quality of Strawberry Purée. *Food Bioprocess Technol.* **2015,** *8* (9), 1864–1876.
39. Maskan, M. Drying, Shrinkage and Rehydration Characteristics of Kiwifruits During Hot Air and Microwave Drying. *J. Food Eng.* **2001,** *48*, 177–182.
40. Maskan, M. Microwave/Air and Microwave Finish Drying of Banana. *J. Food Eng.* **2000,** *44*, 71–78.
41. Math, R.; Nagender, A.; Nayani, S.; Satyanarayana, A. Continuous Microwave Processing and Preservation of Acidic and Non Acidic Juice Blends. *Int. J. Agric. Food Sci.* **2014,** *2*, 81–90.
42. Matsui, K. N.; Gut, J. A. W.; De Oliveira, P. V.; Tadini, C. C. Inactivation Kinetics of Polyphenol Oxidase and Peroxidase in Green Coconut Water by Microwave Processing. *J. Food Eng.* **2008,** *88* (2), 169–176.
43. Mendes-Oliveira, G.; Deering, A. J.; San Martin-Gonzalez, M. F.; Campanella, O. H. Microwave Pasteurization of Apple Juice: Modeling the Inactivation of *Escherichia coli* O157: H7 and *Salmonella Typhimurium* at 80–90 C. *Food Microbiol.* **2020,** *87*, 10338.

44. Metaxas, A. C.; Meredith, R. J. Industrial Microwave Heating, 2nd ed.; IEEE Press Power Systems Engineering Series, Volume 4; London: Peter Peregrinus Ltd. (on Behalf of the Institution of Electrical Engineers); 1993; online.
45. Mousa, N.; Farid, M. Microwave Vacuum Drying of Banana Slices. *Dry. Technol.* **2002,** *20* (10), 2055–2066.
46. Mudgett, R. E. Electrical Properties of Foods in Microwave Processing. *Food Technol.*, **1982,** *36*, 109–115.
47. Muftugil, N. Effect of Different Types of Blanching on the Color and the Ascorbic Acid and Chlorophyll Contents of Green Beans. *J. Food Process. Preserv.* **1986,** *10*, 69–76.
48. Nemes, S. M.; Sivakumar, S. S.; Gariepy, Y.; Raghavan, G. S. V. Microwave Vacuum Drying of Banana Slices Under Different Power Levels. Unpublished Paper at 3rd Inter-American Drying Conference; Montreal, Canada, 2005.
49. Nikdel, S.; Chen, C. S.; Parish, M. E.; MacKellar, D. G.; Friedrich, L. M. Pasteurization of Citrus Juice with Microwave Energy in a Continuous-Flow Unit. *J. Agric. Food Chem.* **1993,** *41* (11), 2116–2119.
50. Orsat, V.; Raghavan, V.; Meda, V. Microwave Technology for Food Processing: An Overview. In *The Microwave Processing of Foods*; Schubert, H., Regier, M., Eds.; Taylor and Francis/CRC Press, Boca Raton, 2005; pp 105–118.
51. Öztürk, S.; Şakıyan, Ö.; Özlem-Alifakı, Y. Dielectric Properties and Microwave and Infrared-Microwave Combination Drying Characteristics of Banana and Kiwi Fruit. *J. Food Process Eng.* **2017,** *40* (3), E-article 12502.
52. Patricia, C. M.; Bibiana, D. Y.; Jose, P. M. Evaluation of Microwave Technology in Blanching of Broccoli (*Brassica oleraceae* L.) as a Substitute for Conventional Blanching. *Procedia Food Sci.* **2011,** *1*, 426–432.
53. Picouet, P. A.; Landl, A.; Abadias, M.; Castellari, M.; Viñas, I. Minimal Processing of a Granny Smith Apple Purée by Microwave Heating. *Innov. Food Sci. Emerg. Technol.* **2009,** *10* (4), 545–550.
54. Proctor, B. E.; Goldblith, S. A. Radar Energy for Rapid Food Cooking and Blanching and Its Effect on Vitamin Content. *Food Technol.* **1948,** *2*, 95–104.
55. Quan, X.; Zhang, M.; Fang, Z.; Liu, H.; Shen, Q.; Gao Z. Low oil French Fries Produced by Combined Pre-Frying and Pulsed-Spouted Microwave Vacuum Drying Method. *Food Bioproducts Process.* **2016,** *99*, 109–115.
56. Queiroz C.; Mendes, M. L.; Fialho, E.; Valente-Mesquita, V. L. Polyphenol Oxidase: Characteristics and Mechanisms of Browning Control. *Food Rev. Int.* **2008,** *24*, 361–375.
57. Sattar, S.; Ahmad, T.; Nisa, M. U.; Imran, M.; Holmes, M.; Maycock, J.; Nadeem, M.; Khan, M. K. Microwave Processing Impact on Physicochemical and Bioactive Attributes of Optimized Peach Functional Beverage. *J. Food Process. Preserv.* **2019,** *43* (7), E-article 13952.
58. Schirack, A. V.; Sanders, T. H.; Sandeep, K. P. Effect of Processing Parameters on the Temperature and Moisture Content of Microwave-Blanched Peanuts. *J. Food Process Eng.* **2007,** 30 (2), 225–240.
59. Sharma, G. P.; Prasad, S. Drying of Garlic (*Allium sativum*) Cloves by Microwave-Hot Air Combination. *J. Food Eng.* **2001,** *50*, 99–105.
60. Si, X.; Chen, Q.; Bi, J.; Yi, J.; Zhou, L.; Wu, X. Infrared Radiation and Microwave Vacuum Combined Drying Kinetics and Quality of Raspberry. *J. Food Process Eng.* **2016,** *39* (4), 377–390.

61. Siguemoto, É. S.; Gut, J. A. W.; Martinez, A.; Rodrigo, D. Inactivation Kinetics of *Escherichia coli* O157: H7 and *Listeria monocytogenes* in Apple Juice by Microwave and Conventional Thermal Processing. *Innov. Food Sci. Emerg. Technol.* **2018,** *45*, 84–91.
62. Siguemoto, É. S.; Pereira, L. J.; Gut, J. A. W. Inactivation Kinetics of Pectin Methylesterase, Polyphenol Oxidase, and Peroxidase in Cloudy Apple Juice Under Microwave and Conventional Heating to Evaluate Non-Thermal Microwave Effects. *Food Bioprocess Technol.* **2018,** *11* (7), 1359–1369.
63. Straumite E.; Kruma, Z.; Galoburda, R.; Saulite, K. Effect of Blanching on the Quality of Microwave Vacuum Dried Dill (*Anethumgraveolens* L.). *World Acad. Sci. Eng. Technol.* **2012,** *64*, 756–762.
64. Su, Y.; Zhang, M.; Zhang, W.; Adhikari, B.; Yang, Z. Application of Novel Microwave-Assisted Vacuum Frying to Reduce the Oil Uptake and Improve the Quality of Potato Chips. *LWT*, **2016,** *73*, 490–497.
65. Su, Y.; Zhang, M.; Zhang, W. Effect of Low Temperature on the Microwave-Assisted Vacuum Frying of Potato Chips. *Dry. Technol.* **2016,** *34* (2), 227–234.
66. Sunjka, P. S.; Rennie, T. J.; Beaudry, C.; Raghavan, G. S. V. Microwave-Convective and Microwave Vacuum Drying of Cranberries: A Comparative Study. *Dry. Technol. Int. J.* **2004,** *22*, 1217–1231.
67. Tajchakavit, S.; Ramaswamy, H. S.; Fustier, P. Enhanced Destruction of Spoilage Microorganisms in Apple Juice During Continuous Flow Microwave Heating. *Food Res. Int.* **1998,** *31* (10), 713–722.
68. Taskin, O.; Polat, A.; Izli, N.; Asik, B. B. Intermittent Microwave-Vacuum Drying Effects on Pears. *Polish J. Food Nutr. Sci.* **2019,** *69* (1), 101–108.
69. Tulasidas, T. N.; Raghavan, G. S. V.; Mujumdar, A. S. Microwave Drying of Grapes in a Single Mode Cavity at 2450 MHz –II: Quality and Energy Aspects. *Dry. Technol.* **1995,** *13* (8/9), 1973–1992.
70. Vadivambal, R.; Jayas, D. S. Non-Uniform Temperature Distribution During Microwave Heating of Food Materials—A Review. *Food Bioprocess Technol.* **2010,** *3*, 161–171.
71. Wang, Z.H, Shi, M. H. Microwave Freeze Drying Characteristics of Beef. *Dry. Technol.* **1999,** *17* (3), 434–447.
72. Wang, J.; Xi, Y. S.; Yu, Y. Microwave drying characteristics of potato and the effect of different microwave powers on the dried quality of potato. *Eur. Food Res. Technol.* **2004,** *219* (5), 500–506.
73. Wang, R.; Zhang, M.; Mujumdar, A. S. Effects of vacuum and microwave freeze drying on microstructure and quality of potato slices. *J. Food Eng.* **2010,** *101*, 131–139.
74. WHO. World Health Day. Food Safety: What You should Know. http://www.searo. who. int/entity/world_health_day/2015/whd-what-you-should-know/en/ (accessed on July 10, 2021).
75. Zhang, H.; Fu, C.; Zheng, X., Xi, Y.; Jiang, W. Wang, Y. Control of Postharvest Rhizopus Rot of Peach by Microwave Treatment and Yeast Antagonist. *Eur. Food Res. Technol.* **2004,** *218* (6), 568–572.
76. Zhang, M.; Jiang, H.; Lim, R. X. Recent Developments in Microwave-Assisted Drying of Vegetables, Fruits, and Aquatic Products: Drying Kinetics and Quality Considerations. *Dry. Technol.* **2010,** *28* (11), 1307–1316.
77. Zhang, M.; Tang, J.; Mujumdar, A. S.; Wang, S. Trends in Microwave-Related Drying of Fruits and Vegetables. *Trends Food Sci. Technol.* **2006,** *17*, 524–534.

CHAPTER 3

ROLE OF GAMMA IRRADIATION IN THE PRESERVATION OF FRUITS AND VEGETABLES

RAFEEYA SHAMS, QURAT UL EAIN HYDER RIZVI, BISMA JAN, and ARSHIED MANZOOR

ABSTRACT

Gamma irradiation from cobalt-60 or cesium-137can be used as a food protection and preservation method to enhance the food shelf-life, safety, and security devoid of changing the nutritional, toxicological, or biological food quality. Various food products use gamma irradiation to reduce microbial growth and disinfestations. This method is economically and technically feasible and physically a safe technique having a strong antimicrobial effect. These radiations are also used to treat large volumes of food products or sealed containers. Different varieties of irradiation have the potential to enhance the phenolic content and the free radical scavenging activity of food and food by-products. Far-infrared has been utilized to release antioxidant-active low-molecular-weight phenolics. Recommended doses of gamma radiation for inhibiting sprouts doses used are <0.5 kGy for hindering ripening; lower or equal than 1 kGy for pest disinfestations; 3–5 kGy to increase shelf-life and for pathogenic microbes; 5 kGy or higher if the food material supports it without deteriorating its properties.

Quality Control in Fruit and Vegetable Processing: Methods and Strategies. Megh R. Goyal, & Faizan Ahmad (Eds.)

3.1 INTRODUCTION

A physical treatment wherein food material is exposed to a specific radiation dosage is known as gamma irradiation. Irradiation method is used in more than 80 countries according to Codex Alimentarius Commission (CAC).[41] Some of the natural processes are delayed by the food irradiation depending upon the dosage level, like sprouting, ripening besides control of insect infestation, spoilage and pathogenic microbes are decreased (*Aspergillus, Escherichia coli, Salmonella*) and also for the commercial foods sterility. There are only three kinds of radiations approved according to CAC, which are used at commercial scale for irradiation of food such as accelerated gamma X-ray sand electrons rays.[32]

Irradiation is classified into three main groups depending upon the dosage level, such as: radicidation (2.5–10 kGy), radappertization (30–40 kGy), and radurization (0.75–2.5 kGy). It is effective for the removal and inhibition of antinutritional factors and food allergens.[8] Some of the basic benefits of irradiation techniques are: fast acting, minimal sample preparation, nonthermal process, no catalyst used, and high power of penetration (in the final package, products can be treated), no residues, samples that are treated can be instantly distributed with no rise in temperature (radiation dose of 1 kGy raises the product temperature only by 0.36°C). The label on the irradiated foods must withhold the international icon for irradiation, such as: "Treated with radiation" or "Treated by irradiation".[46]

This chapter discusses effects of various irradiation techniques on the shelf-life, physicochemical parameters, and other chemical and biochemical properties of the food produce. Further the chapter also focuses on the effect of gamma radiation on the quality attributes of various fruits and vegetables.

3.2 PRINCIPLES OF FOOD IRRADIATION

The principle of irradiation of foods identifies the process by which irradiation particles interacts with the food product. This may provide insights in the energetic electron arrangement at arbitrary throughout the material, which in turn may cause the energetic molecular particles arrangement. Such particles can be used to capture electron and segregation, and quick change via ion-molecule responses, or can segregate with time depending on the particle complexity.

Radiations impact on material depends on the type of radiation, its level of energy and the physical state, composition; climatic conditions and the temperature of the retaining substance. The irradiated foods can also be defined as the foods that are exposed to various forms of irradiations and these radiations depend on the cause for exposure. This process is a physical treatment process where food products are exposed to electrons, electromagnetic beams so as to prolong the shelf-life of these products. During the process of irradiation, the source of irradiation and treated food substance does not come in direct contact with each other and it is inconceivable to attempt radioactivity within the food substance by using gamma beams or electron beams at a limit of 10 MeV.[14]

The ionization emission is a part of the electromagnetic continuum of electric waves on two sides, with the heavy energy radiations with gamma rays on the other. There are visible light rays from all sides of ultraviolet and infrared waves in the middle. The microwaves, which are used in standard kitchen appliances or domestic, are unlike lower frequency radio waves. These both waves are members of the electromagnetic spectrum, and possess some definite issues. Both the waves have the normal frequency, wavelength, and certain related strength. When the wavelength is more the lesser is the related energy. A limited energy is associated with radio waves having long wavelengths of 30 cm to 3 km.

3.3 MECHANISM OF IRRADIATION TO INHIBIT MICROBIAL PATHOGENS

Ionizing radiations are used to irradiate foods and provide indirect and direct impact on the substance which probably becomes the target. The target includes the irradiated microbes and insect pathogen that destroys the food texture and structure. Irradiation causes the incidence of dissociation otherwise ionization or electron-release of hydrogen as well as excitation.[3] Irradiation also shows the immediate effects of DNA bonds in insect cells and microbes so as the capability of cells to multiply and survive will be affected. While the indirect effects of irradiation on the water, which is the major constituent of the cell, are affected by radiolysis where decomposition of water molecules and the formation of free radicals occurs.[45]

Overall irradiation causes the physical, chemical changes in the cells of microbes. Hydrogen bonding termination and sugar phosphate chain termination from each polinukleotida of DNA termination of adjacent chains on both polinukleotida of DNA loss of nitrogen bases until cross-ties intramolecular (base damage) is formed.[50] Every microorganism has differing resistance and sensitivity against gamma radiation; thus the success rate of irradiation depends on the resistance of microbes. Some microorganisms are hard to kill or inhibit, but most of the microorganisms are sensitive and simple to destroy or inhibit.

D_{10} reveals the microbial cell damage level related to the resistance against irradiation. D_{10} is a radiation dose that is required to destroy microorganisms by 10- to 90-folds of the whole disease-causing microorganisms. Resistance of microbes against irradiation depends on various factors, such as composition and the microbial DNA's size, DNA-related components, O_2, temperature, H_2O, and post-radiation and media.[19]

3.4 SAFETY OF IRRADIATION TECHNIQUE ON FOODS

The irradiation technique is a safer method to extend the power save food as stated by World Health Organization (WHO), and it also reported that the recommended irradiation doses to a food product do not damage the toxic hazards and nutritive value.[12] If the recommended radiation dosage is extended, it can cause some undesirable effects to food product. The recommended dosage of gamma irradiation is a maximum of 5 MeV. If the gamma irradiation dosage exceeds 5 MeV, then there will be the negative impact on the food material. The hazardous radioactive material is obtained if the radiation doses exceeded the recommended threshold.

The substance comprises high radioactive components that are able to change the genetic order. There will be no radioactive residue left after irradiation because of the radionuclide from which the metal-coated capsules store radiation perfectly. Irradiation technology plays a great role in food preservation, where the certain dose of radiations are used in the form of rays and these rays penetrate through the food product so as to kill the pathogenic microorganisms, insects, pests, etc. These radiations are safe that do not reduce the nutrition quality and leaves no residue against food so the irradiation technique is completely safe, clean, and environmentally friendly.[40]

Nutritional value and chemical changes are estimated by the quantity of the radiation doses. The food substances that are irradiated with low dose of energy up to 1 kGy do not experience any change in its nutritional or chemical properties. The food substances that are irradiated with energy doses of 1–10 kGy) experience the loss of vitamin. Substances that are irradiated by higher doses (10–50 kGy) still experience vitamin depletion and the vitamins, such as K, A, B1, B, and E are easily lost and certain vitamins, such as riboflavin, calciferol, and niacin are retained.[20] Furthermore, if the dose (high) of radiations is not appropriate to cause protein denaturation, and to affect the fatty acid/ carbohydrate composition; and if radiation doses are given at low temperature, then chemical changes in the food product may be reduced.[18]

3.5 STATUS AND REGULATORY ASPECTS OF FOOD IRRADIATION

Around 60 diverse irradiated foods are available in the marketplace currently. Potato, meat, onion, spices, etc., are major products. The irradiated ground beef in the United States was sold by almost 7000 supermarkets and several other retail outlet stores. The latest status on irradiated foods across the globe is given in Table 3.1. Irradiation is used at a large scale and Asia is number one followed by USA and Africa. The commercial use of food irradiation is lagging in Europe because of its strict legislation.[28]

TABLE 3.1 Global Status of Food Irradiation During 2015–2016.

Country	Food items
Brazil	Grains, spices, herbs, fruits, vegetables
China	Meat, spices, grains, garlic, onion, etc.
France	Gums, vegetables, poultry, spices, etc.
Germany	Herbs, spices
India	Fruits, potato, onion, spices, etc.
Indonesia	Spices, sea foods, cocoa, etc.
Malaysia	Vegetables, spices, etc.
Netherland	Spices, frog legs, egg white, poultry, sea foods
Pakistan	Spices, fruits, pulses, etc.
Spain	Herbs, vegetables, etc.
USA	Sea food, spices, grains, meat, fruits and vegetables
Vietnam	Fruits, vegetables, sea foods

During 1992, the expert missions were sent to Vietnam and Pakistan by International Atomic Energy Agency (IAEA) to update the guidelines on food irradiation based on recommendation of a "1992 (20 January–24 January) Seminar held on Harmonization of Food Irradiation Regulations at Kuala Lumpur – Malaysia." The food irradiation regulations approved by the government are abetted by IAEA professionals in Philippines, Sri Lanka, and Malaysia.[21]

Several international and national regulatory authorities approved food irradiation and these regulatory authorities are the Institute of Food

Technologists, WHO, American Medical Association, *CODEX*, Food Standards Australia–New Zealand, Food Safety and Standards Authority of India (FSSAI), and Food and Agriculture Organization (FAO), etc..[28]

In Canada, the irradiated foods include: potatoes, wheat flour, spices, and onions. In 2002, an organizational offer was submitted for mangoes; however, it was suspended because of consumer's refusal. Irradiation was permitted for onion and potato in 1965 and 1960, respectively, at 0.15 kGy dose. Irradiation of spices and wheat flour was permitted at a dose of 10 kGy and 0.75 kGy, respectively, in 1984 and 1969.[28]

For ground beef, irradiation was permitted in 2013 to inactivate *Salmonella* and *E. coli*. Spices that are annually irradiated are approximately 2000 tons in Canada.[33] In USA, Brazil, and Canada, the total quantity of irradiated foods in 2014 was around 7120 tons of fruits, 9000 tons of meat, and 101,800 tons of spices.[33] The sterilization of bacon by irradiation was approved by US FDA in 1963. Since then, the irradiation technology impact on various foods has been extensively studied.[13]

The motivation for irradiation acceptance in the United States ensued in 1993 after 4 deaths and various victims with permanent kidney damage occurred due to outbreak *E. coli* O157:H7. It was later identified that the source was undercooked hamburgers and because of this outbreak, the food safety and quality regulations were examined for the first time in a century and then irradiation was approved for frozen meat and chilled meat, where frozen meat was irradiated up to a dose of 7 kGy and chilled meat was irradiated to a highest dosage of 4.5 kGy.[28]

Currently, 50 commercial irradiation plants have been established in USA. In 2005, the US FDA approved the irradiation in shell fishes, mussels, oysters, and clams. Irradiation was also allowed in shrimp, lobsters, crabs, and RTE foods (like packed salads, deli meats, baby food, and frozen meals).[28] During 2011–2012, radiated foods sols in USA were about 1200 tons of other fruits, 550 tons of guava, 8000 tons of meat, and 3300 tons sweet potato.[9] In the USA during 2016–2017, the irradiated foods for animal and human consumption were more than 150,000 tons, which included pet treats (20,000 tons), spices (80,000 tons), vegetables and fruits (30,000 tons), ground beef (910 tons), and sweet potato (3800 tons).

In 2015, more than 23 million tons of irradiated food products were imported from seven countries. The irradiated mango and longan were exported from Thailand to USA in 2007 and other fruits were included in 2010. The total export in 2010 was 595 tons of longan, 330 tons of mangoes, 8 tons of rambutan, and 18 tons of litchis. Pakistan exported about 152 tons of mangoes to USA in 2015, whereas South Africa exported 200 tons of persimmons and litchi. Nearly 257 tons of irradiated guava were exported from Mexico to USA in 2008 as compared to 3521 tons in 2009. Moreover, 10,318 tons of fruits were recorded (mango, guava, sweet lime, and pepper) in 2010 as compared to 12,000 tons in 2016. Mexico has the benefit for export to USA because of lower transport distance and cost compared to other countries.[13]

Food irradiation dates back to 1967 in India, after establishing Processing Laboratory at Trombay; and food irradiation for treating potatoes and wheat with the main purpose of sprout inhibition and disinfestation. The numerous other studies explored the feasibility of irradiation technique in processed foods, cereals, vegetables, fruits, and pulses. Finally, the irradiation treatment was permitted in 1994 to numerous other food products, such as onions, spices, and potatoes, etc. The irradiation plants in Navi Mumbai during 2000 and in Maharashtra during 2003 were established having capacity of 30 tons per day and 10 tons per day, respectively. These were established for onions potatoes, some dry vegetables, and spices. About 2500 tons of onions, spices, and dried vegetables were exposed to irradiation, and approximately 730 tons of mangoes were exported to USA.[21]

In USDA-approved plants, irradiation of mangoes is only done there. There were 15 functional commercial irradiation plants in India during 2016. The Hindustan Agro Cooperation Ltd., Indian Agricultural Association, and United Innovation Corporation signed a bilateral agreement in 2016 for the expansion of 25 irradiation centers in India.[21]

In 1987, the irradiation facility was started in China. There were only seven centers for irradiations till 2003 compared to 78 radiation centers in 2006, 78 centers for irradiation, and 115 centers today. In the beginning, only hogs and potatoes were permitted for irradiation; however, in 2010, irradiation was extended to fruits, pork, nuts, poultry grains, spices, beef, vegetables, and cooked meat.[21] The three largest gamma irradiators using Cobalt-6 were operated by Chinese Company JPY each with 0.7, 2.6, and 2 million tons capacity of curies.[13]

3.6 IMPACT OF GAMMA IRRADIATION ON PHYSICOCHEMICAL AND SENSORY PROPERTIES OF FRUITS AND VEGETABLES

Gamma irradiation of fruit and vegetables has been successful for preserving these without altering their properties and has been studied for several years.

The research studies on irradiation of fruits and vegetables (FAV) have been compiled by Arvanitoyannis et al.,[7] who concluded that this technology can extend the shelf-life of FAV at low cost, without changing organoleptic and physicochemical characteristics compared to other traditional preservation techniques. This method has been commercially approved in South Korea with the aim of rooting inhibition or sprouting at 0.25 kGy[10] and the dose for disinfestations is 0.50 kGy.[29] However, it is significant to estimate the impact of gamma radiation on antioxidant parameters, natural, nutritional, and physicochemical parameters.

Najafabadi et al.[38] evaluated the gamma irradiation effect at diverse doses (0.0, 0.5, 1.0, 2.5, and 5.0 kGy) on some physicochemical characteristics of jujube fruits. At these doses, the total soluble solids values did not change, but a slight enhancement of total acidity level at doses ≥ 2.5 kGy ($p \leq 0.05$) was observed. The irradiation dose caused significant rise in color values of a^* and b^* values ($p \leq 0.05$) with significant decline in L^* value ($p \leq 0.05$); also, color changes were slight upto the dose of 5 kGy. There was no major impact on the absorption of succinic citric and malic acids due to gamma irradiation upto 2.5 kGy, while a considerable decrease in the ascorbic acid level was observed at different doses of irradiation (0–5 kGy).

The main anthocyanin (about 68% in jujube) is Cyanidin-3,5-diglucoside that was decreased significantly at an irradiation of 5 kGy.[38] The water-soluble vitamins in jujube fruits were vitamins B2, B1, and C. Vitamins B1 and C content were declined at all applied doses (0–5 kGy), whereas B2 values at a dose ≤2.5 kGy was not affected. Another investigation mentioned that irradiation dose at ≤2.5 kGy has been used considerably to enhance the quality parameters of jujube fruits.

Kim and Yook[27] observed that three pathogenic microbes (*Diaportheactinidiae, Botrytis cinerea,* and *Botryos phaeriadothidea*) were inactivated by ionizing radiation in kiwi fruits. It was also observed that irradiated kiwi fruits were softer as compared to nontreated kiwi fruits. The irradiation treatment minimally affected the organic acid content and color of kiwi fruits. With increasing irradiation dose, the total soluble solids content was reduced in irradiated kiwi fruits. The kiwi fruits irradiated upto 3 kGy dose had harmful impact on antioxidant activity and vitamin C content, but had beneficial impact on hygienic and sensory quality by destroying the increased growth of harmful microorganisms.

Jiang et al.[23] concluded that Shiitake (*Lentinula edodes*) mushrooms were kept in polypropylene bags and then were allowed to expose at three irradiation doses of 1.0, 1.5, and 2.0 kGy and finally were stored at 4°C for 20 days.

The firmness was maintained at 1.0 kGy among these three doses. Products treated with 1.0 kGy showed a less decline in soluble protein, and lower malondialdehyde accumulation levels. However at 1.0 kGy, the accumulation of phenolic components was promoted, and high radical scavenging activity was exhibited during storage. At high irradiation dose of 2.0 kGy, both positive and negative impacts were observed: the higher the dose, the greater is the microbial reduction; and negative impact was observed on chemical, textural, and functional properties. All the gamma irradiations were efficient in retaining the sensory characteristics of mushrooms. Therefore, gamma irradiation with MAP can prolong the shelf-life of shiitake mushrooms upto 20 days.

Arjeh et al.[6] found that storage time and irradiation treatment led to a significant decline in a^* values and increase in L^* and b^* values. The concentrations of oxalic and malic acids were increased significantly after irradiation at 0–6 kGy; but citric, ascorbic, succinic, and fumaric acids were decreased significantly. Gamma irradiations were able to enhance the shelf-life of sour cherry juice at 3 kGy followed by a general decrease in microbial growth. Owing to the outcome of microbial analysis and physicochemical characteristics, irradiation of sour cherry juice at dose >3.0 kGy is not permitted.

Maraei and Elsawy[35] observed the impact of different irradiation doses (0, 0.3, 0.6, and 0.9 kGy) on phytochemical content and quality parameters of strawberry fruits during storage period at 10°C. The irradiation decreased the decaying and weight loss significantly compared to the control (nonirradiated sample) treatments; thus showing the highest contents of these quality attributes. Both storage period and irradiation did not show any considerable impact on pH and acidity of fruits. Each treatment decreased the vitamin C content during storage. In another study,[49] the effect of radiation on fresh vegetable juice was observed. Due to irradiation, amino acids were stable at 5 kGy but there was decrease in the vitamin C content. At 3 days of storage, the irradiated juice's sensory quality characteristics were not deteriorated, while the quality was deteriorated in the nonirradiated juice.[49]

3.7 EFFECTS OF GAMMA IRRADIATION ON MICROBIOLOGICAL ATTRIBUTES OF MINIMALLY PROCESSED FRUITS AND VEGETABLES

Minimally processed fruits and vegetables are nutrient-rich produce. Gamma irradiation provides a safe alternative method for decontamination of food stuffs. Landgraf et al.[30] concluded that irradiation on the minimally processed food products (such as iceberg lettuce, watercress, mango cubes, pineapple, and watermelon) was able to reduce the presence of food borne pathogens, such as *Listeria monocytogenes* and *Salmonella* spp.

It was found that *Listeria monocyte genisis* was not recovered at dose equivalent to 5 and 8 D_{10}. Also, the doses of 1, 3, and 4kGy for *Salmonella* spp. did not impair sensorial characteristics of the watercress. When cubes of pineapple and watermelon were exposed to irradiation dose of one kGy and 2.5 kGy, there was no change in sensory qualities (e.g., water melon's sweetness or pineapple's sourness). Irradiation is a low-cost method to enhance the hygienic conditions and it protects the minimally processed vegetables and fruits from microbial infestation. When blanched cherry tomatoes and pre-cut carrots were irradiated at doses <20 kGy, their quality parameters were not reduced. Also, in the case of both blanched vegetables and fresh pre-cut vegetables, quality parameter was maintained. There was no significant effect on sensory quality of blanched vegetables. In both cases, irradiation dose of 2 kGy ensured 5 log reductions in the population of *Salmonella enteritidis*.[47]

Yun et al.[54] studied the soybean (*Glycine max* L. Merrill) that was exposed to different doses of 0.0, 1.0, 3.0, 5.0, and 10.0 kGy of gamma irradiation. The microorganism growth on soybeans, sensory properties and color, tocopherol contents, raffinose family oligosaccharides were estimated as a function of irradiation dose. The results showed that these radiations decreased the fungal and aerobic bacterial growth. The isoflavone contents of soybeans were not changed at these irradiation doses ($p < 0.05$), but tocopherol content was reduced. The key flatulence components in treated soybeans (at 10 kGy) were declined by 82.1% compared to the control treatment.

Another study exhibited[49] that microbiological quality of fresh vegetable juice was enhanced due to the radiation pasteurization process. Kale and carrot juices were exposed to irradiations to enhance their sensory, nutritional, and microbiological characteristics. Before the irradiation, the contaminating microbes present in the juices varied from 106 to 107 CFU/ml; but after exposure to irradiation (3 kGy), all the coliform and aerobic bacteria in the carrot juice were eradicated, whereas 102 CFU/mL bacteria endured in the kale juice that was exposed to irradiation at 5 kGy. Moreover in the kale juice, the survived bacteria from irradiation did not expand further, and in nontreated samples, it extended up to 109 CFU/mL after three days of storage at 10°C.[49]

3.8 EFFECTS OF GAMMA IRRADIATION ON ANTIOXIDANT PROPERTIES

The antioxidant activity of phenolic components in plants is mostly due to the redox actions, decomposing peroxides, quenching singlet and triplet oxygen, neutralizing free radicals, etc. Phenolic components consist of a group of bioactive molecules, being valued for their health benefits due to antiatherogenic, vasodilatory effects, antiallergic, anti-inflammatory, antimicrobial, cardio-protective, and antioxidant activities.[36]

As an antioxidant, the mechanism of action is an essential by decreasing the oxidation processes by providing bio-macromolecules and cell protection; and hence, intervening against inflammatory diseases, heart disease, cancer, neurological degeneration, etc.[1] Therefore, the most important intervention of antioxidant potential of phenolic components is their capability to chelate metal ions and to scavenge reactive oxygen species.[43]

Phenolic acids help to maintain the quality of various vegetables and fruits and these are widespread in nature. These acids are responsible for the difference in the relative astringencies of peach and also contribute to the bitterness, sourness, and astringency of wheat and soy flours. The protective effect of flavonoids is because of several mechanisms, such as enzymes inhibition, free-radicals trapping, and chelation of metallic ions. These characteristics are based on the flavonoid structure and the degree of saturation and substitution. Vegetables and fruits contain sufficient amounts of flavonoids.[22] The electron-donating capacity of an antioxidant is calculated by reducing power assay. The ferric ions (Fe^{3+}) reduction to ferrous ions (Fe^{2+}) is calculated by the intensity of blue-green solution, which is absorbed at 700 nm and high reducing power is indicated by an enhanced absorbance.[37]

Several biological reactions produce superoxide radical scavenging activity being very toxic species, although they directly cannot cause lipid per oxidation, the precursor of damaging oxygen species are superoxide anion radicals. The superoxide anion is produced by dissolved oxygen by reduction in NBT due to PMS/NADH-NBT coupling reaction.[4] In the reaction mixture, the superoxide anions are consumed when the reduction in absorbance at 560 nm with antioxidants are shown.[17] Aldehydes and malondialdehyde (MDA) are recognized as lipid peroxidation products, which react with TBA to produce a species with pink color and gets absorbed at 532 nm.

The MDA and TBARS percentage are inversely proportional to the antioxidant activities, that is, the antioxidant activity of fruit will be reduced with rise in TBARS content.[11] The impact of irradiation (0.25–1.0 kGy) on antioxidant property of ber fruits was investigated by Kavitha et al..[26] These fruits are rich in phenolics, vitamin C content, and flavonoids. Due to irradiation, superoxide anion radical activity, flavonoid content, and DPPH radical activity were enhanced with slight decrease in reducing power activity and total phenolics. In the irradiated fruits, the total antioxidant activity (TBARS) was diminished from 64 to 167% compared to raw fruits at 0.25–1.0 kGy. Therefore, the optimum doses to maintain the natural antioxidant activity in ber fruits are 0.25–0.5 kGy.

Pereira et al.[42] evaluated the irradiation impact on the phenolic components of *Ginkgo biloba* L. The 25 components were identified consisting of 18 flavonoids, 5 terpene lactones, one phenolic acid, and one unknown biocomponent. Flavonoids were mainly present among the quantified phenolic components, being two kaempferol derivatives and the main components, such as: kaempferol-3-*O*-rutinoside and kaempferol-3-*O*-dirhamnosylglucoside. The maximum dose (10 kGy) of irradiation can disinfect the product and microbial decontamination, and also can increase the extractability of phenolic components, both in water or methanol and infusion preparations.

In jujube fruit, distinct information is lacked on the effect of irradiation on profile of bioactive components (such as water-soluble vitamins, organic acids, and anthocyanins) and their changes. Najafabadi et al.[38] investigated the impact of irradiation at various doses of 0.0, 0.5, 1.0, 2.5, and 5.0 kGy on the bioactive components of jujube fruit. Irradiation up to 2.5 kGy significantly increased the total phenolic content and monomeric anthocyanin (6% and 12%, respectively); but after irradiation at 5 kGy, a rapid decline was observed in both parameters. Sour cherry juice was irradiated at dosage >3 kGy that caused antioxidant activity loss due to the ascorbic acid content and anthocyanins degradation. Thus for sour cherry juice, a dosage level of <3 kGy is recommended.[6]

Maraei and Elsawy[35] found that strawberry fruits irradiated at 0.6 kGy showed maximum antioxidant activity and total phenolics followed by 0.3 kGy. Irradiation also triggered synthesis of certain phenolic components, such as gallic, pyrogallol, ellagic, chlorogenic acid, catechol, etc.

3.9 ROLE OF GAMMA IRRADIATION ON SHELF-LIFE AND PRESERVATION OF FRUITS AND VEGETABLES

One of the prime focuses during the process of irradiation in fresh-cut fruits and vegetables is the shelf-life extension by changing the chemical and microbiological properties. Majeed et al.[34] concluded that an extension of 3–8 days in shelf-life of strawberries was observed due to 0.5–1.0 kGy of irradiation dose. Similarly in citrus fruits and Shantag mandarin, an exposure to radiation dose from 0.2 to 0.6 kGy led to remarkable extension in their shelf-life.[56] Moreover, a significant enhancement in fruit juice quality without altering the chemical characteristics was recorded.[24] Preclimacteric mangoes were exposed to low irradiation dose of 0.2 kGy for 8–10 days and its shelf-life was significantly improved.

Zaman et al.[55] concluded the extension of shelf-life in banana using gamma irradiation, in which bananas were stored at dry place at ambient temperature after exposing them to 0.5, 0.4, and 0.3 kGy of gamma irradiation doses. The authors also recorded the extension of shelf-life by 20 days compared to the control samples coupled with a delay in ripening. In another study, the shelf-life was seen to be extended significantly in pears due to gamma irradiation,[52] where the irradiation doses were maintained at 0.8–2.0 kGy during the storage at ambient temperature. The exposure to the irradiation doses at 1.5–1.7 kGy resulted in the decay inhibition of pears up to 16 days. It was found by Singh and Pal[48] that gamma rays at a dose of 0.25 kGy enhanced the postharvest life of guava fruits by 3–4 days.

The growth of spoilage microorganisms is reduced due to which the shelf-life of fresh produce is increased. A reduction of one log cycle of pathogenic microbial count can be obtained by exposure of fruits to a radiation dose of 0.2–0.8 kGy. Comparatively, fungi and pathogenic viruses displayed better resistance against the radiation at 1–3 kGy for obtaining one log reduction.[39] However, various investigations suggested that irradiation at low doses of 1 kGy can help in the shelf-life enhancement of fresh vegetables by inhibiting ripening technique and by deactivating the spoilage microbes.[53]

3.10 APPLICATIONS OF GAMMA IRRADIATION IN FRUIT AND VEGETABLE INDUSTRIES

Various kinds of irradiation methods can be categorized based on the processing intention and the amount of dose used (Table 3.2). Although, it is technically conceivable that the meats and other food products can be disinfected using gamma irradiations[31] that may render foods unacceptable based on organoleptic characteristics and spore formation in spices and herbs by heat-resistant microorganisms. A dose of 7 ± 10 kGy resulted in the elimination of the microbes from certain foods thereby preserving their quality attributes and volatile oils. Due to security issues to deal with the gases and the buildups in the food items, scientists in EU recommended to ban the use of chemical sterilization by ethylene oxide in 1991, which has paved the way for its substitution by irradiation, particularly in spices.[2]

TABLE 3.2 Food Irradiation and Doses: Applications in Foods.

Application	Irradiation dose (kGy)	Food items
<1 kGy (Low dose)		
Delay in ripening	0.2–1.0	Vegetables and fresh fruits
Disinfestation	0.15–0.5	Cocoa beans, fruits, flours, grains, dry foods (like meats, fruits, fish)
Inhibiting sprout formation	0.05–0.15	Garlic, potatoes, root ginger, onions, yam
1–10 kGy (Medium dose)		
Food sterilization	7–10	Herbs, meat, spices, sea foods, poultry
Mold control	2.5	Fresh fruit storage gets extended
Shelf-life extension	1–3	Fresh meat and fish at 0–4°C, mushrooms, strawberries.
>10 kGy (High dose)		
Food additives; Decontamination	10–50	Natural gums, enzyme preparations
Hospital diets Sterilization	30–50	Ready-to-eat meals
Packaging materials Sterilization	10–25	Wine corks

3.10.1 REDUCTION OF PATHOGENS

A dosage of the order of 2.5±10 kGy resulted in the removal of food poisoning bacteria.[16] Rajkowski and Fan[44] concluded that the rate of food poisoning from numerous fresh vegetables and fruits and fruit juices can be controlled using irradiation, and pathogenic microbes can also be controlled by utilizing irradiation. According to them, irradiation can increase the security and safety of fresh fruits, which is a far better method compared to the chemical treatments.

3.10.2 EXTENSION OF SHELF-LIFE

Irradiation results in the elimination of foodborne diseases coupled with the shelf-life extension of foods to improve the health of consumers. For example, maturation can be delayed in fruits and vegetables using an irradiation dose of 1 kGy; however, yeasts, parasites, and nonspore-forming microbes can be killed using low doses (Table 3.2). Also, irradiation method reduces the number of vegetative cells resulting in their shelf-life extension. Microbes are easily killed by heat and the combination of heat treatment with irradiation compared to either treatment alone.[15]

Venugopal et al.[51] showed that radurization is utilized to enhance the shelf-life of shellfish and fresh fish. Few vegetables and fruits (such as tomatoes and strawberries) could be irradiated to expand the shelf-life by two or three times when left at 10°C. Measurements of 2 ± 3 kGy could enhance twofolds shelf-life of mushrooms and can restraint of cap opening. A synergistic impact was obtained by combination of irradiation and MAP; and also low radiations gave similar results.

3.10.3 CONTROL OF RIPENING

Fruits and vegetables under irradiation must be checked first for the stage of maturity, as ripe fruits and vegetables are mostly recommended. This is due to the fact that it is the ability of irradiation process to restrain the generation of hormones coupled with the retardation of development and cell division that finally restrains maturing. However, irradiation does not completely result in the enzymatic deterioration of foods, which need to be followed by heat treatment for drawn-out capacity.[2]

3.10.4 DISINFESTATION OF FOOD PRODUCTS

There was ban on selected fumigants (such as: ethylene dichloride, ethylene oxide, and ethylene dibromide) for disinfestation of grains and tropical fruits and vegetables in various countries. For example in 2005, methyl bromide ban was followed in various industrialized countries that was expected to be followed in several other developing countries in 2015.[5]

Irradiation provides an alternative for the pesticides and fumigants that could otherwise lead to the buildups on foodstuffs. For example, a dose of 0.25±1.0 kGy in helminths was helpful in fresh fish, parasitic protozoa and to avoid the advancement of insects in dried fish,[51] such as: the inactivation of *Trichinella spiralis* and *Toxoplasma gondii* at 0.3 and 0.25 kGy. The irradiations at low doses <1 kGy for disinfestations are also successful to improve the shelf-life by delaying ageing and by enhancing growth. The irradiations are also used for the treatment of dried nuts and fruits.[25] Irradiations have been used to reduce the sprouting in potatoes.[2]

3.11 SUMMARY

Marketing surveys have concluded that the purchase of irradiated foods by consumers has increased because of its positive effects on our health. The USDA and WHO have also promoted this technique, because no chemical residues, no toxic substances and no nutrient losses were found. Food irradiation is considered to be a new step in food production and food technology and should be made as smooth and straightforward as possible.

ACKNOWLEDGMENTS

Arshied Manzoor acknowledges the award of ICMR-Senior Research Fellowship from the Indian Council of Medical Research, New Delhi, India.

KEYWORDS

- **fruits and vegetables**
- **irradiation detection**
- **physicochemical**
- **parameters**
- **shelf-life**
- **gamma radiation**

REFERENCES

1. Acosta-Estrada, B. A.; Gutierrez-Uribe, J. A.; Serna-Saldivar, S. O. Bound Phenolics in Foods: A Review. *Food Chem.* **2014,** *152*, 46–55.
2. Agbaka, J. I.; Ibrahim, A. N. Irradiation: Utilization, Advances, Safety, Acceptance, Future Trends, and a Means to Enhance Food Security. *Adv. Appl. Sci. Res.* **2020,** *11* (3), 1–9.
3. Ahn, D. U.; Lee, E. J. Mechanisms and Prevention of Quality Changes in Meat by Irradiation. Chapter 8. In *Food Irradiation Research and Technology*; Sommers, C. H., Fan, X., Eds., 2nd ed.; Blackwell Publishing: Ames–IA, 2013; online. <https://doi.org/10.1002/9780470277638.ch8> (accessed on July 31, 2021).
4. Anand, P. K.; Policegoudra, R. S.; Aradhya, S. M. Chemical Composition and Antioxidant Activity of Sapota (*Achras sapota Linn.)* Fruit. *J. Food Biochem.* **2007,** *31* (3), 399–414.

5. Anonymous. Alternatives to Methyl bromide, 2004. https://www.yumpu.com/en/document/view/8154567/alternatives-to-methyl-bromide-unep-dtie (accessed on July 31, 2021).
6. Arjeh, E.; Barzegar, M.; Ali Sahari, M. Effects of Gamma Irradiation on Physicochemical Properties, Antioxidant and Microbial Activities of Sour Cherry Juice. *Radiat. Phys. Chem.* **2015,** *114*, 18–24.
7. Arvanitoyannis, I. S.; Stratakos, A. C.; Tsarouhas, P. Irradiation Applications in Vegetables and Fruits: A Review. *Crit. Rev. Food Sci. Nutr.* **2009,** *49*, 427–462.
8. Bashir, K.; Aggarwal, M. Effects of Gamma Irradiation on Cereals and Pulses-a Review. *Int. J. Recent Sci. Res.* **2016,** *7* (12), 14680–14686.
9. Bustos-Griffin, E.; Hallman, G. J.; Griffin, R. L. Current and Potential Trade in Horticultural Products Irradiated for Phytosanitary Purposes. *Radiat. Phys. Chem.* **2012,** *81*, 1203–1207.
10. Chung, H. W.; Delincée, H.; Han, S. B.; Hong, J. H.; Kim, H. Y.; Kim, M. C.; Kwon, J. H. Trials to Identify Irradiated Chestnut (*Castanea bungena*) with Different Analytical Techniques. *Radiat. Phys. Chem.* **2004,** *71,* 179–182.
11. Devasagayam, T. P. A.; Boloor, K. K.; Ramasarma, T. Methods for Estimating Lipid Peroxidation: An Analysis of Merits and Demerits. *Indian J. Biochem. Biophys.* **2003,** *40*, 300–308.
12. Dwiloka, B. *Course on Food Irradiation: Class Notes*; Semarang University: Semarang, **2002**.
13. Eustice, R. F. Global Status and Commercial Applications of Food Irradiation. *Food Irradiat. Technol.* **2017,** 2017, 397–424.
14. Farkas, J. Charged Particle and Photon Interactions with Matter. *J. Am. Chem. Soc.* **2004,** *10*, 785–812.
15. Gould, G. W. Food Irradiation: Microbiological Aspects. *J. Food Sci. Technol.* **1986,** *19*, 175–180.
16. Guise, B. Irradiation Waits in the Wings. *Food Eur.* **1986,** *6*, 7–9.
17. Gulcin, I.; Uguz, M. T.; Oktay, M.; Beydemirs, K. O. I. Evaluation of Antioxidant Antimicrobial Activities of Clary Sage (*salvia sclarea* l.). *Turkish J. Agric.* **2004,** *28*, 25–33.
18. Harris, R. S.; Karmas E. Evaluation of Nutrition in Food Processing. Unpublished Report; Bandung Institute of Technology: Bandung, 1989; p 34.
19. Harsojo Andini, L. S. Decontamination of Some Pathogenic Bacteria on Meat and Gamma Irradiation with Buffalo Entrails. *Proc. National Workshop Buffalo BATAN, Jakarta*, 2010; pp 116–120.
20. Hermana. Food Irradiation. Unpublished report; Bandung Institute of Technology: Bandung, 1991; p 21.
21. Ihsanullah, I.; Rashid, A. Current Activities in Food Irradiation as a Sanitary and Phytosanitary Treatment in the Asia and the Pacific Region and a Comparison with Advanced Countries. *Food Control* **2017,** *72*, 345–359.
22. Irina, I.; Mohamed, G. Biological Activities and Effects of Food Processing on Flavonoids as Phenolic Antioxidants. *Adv. Appl. Biotechnol.* **2012,** 2012, 101–124.
23. Jiang, T.; Luo, S.; Chen, Q.; Shen, L.; Ying, T. Effect of Integrated Application of Gamma Irradiation and Modified Atmosphere Packaging on Physicochemical and Microbiological Properties of Shiitake Mushroom (*Lentinusedodes*). *Food Chem.* **2010,** *122* (3), 761–767.
24. Jo, C.; Ahn, D. U.; Lee, K. H. Effect of Gamma Irradiation on Microbiological, Chemical, and Sensory Properties of Fresh Kale Juices. *Radiat. Phys. Chem.* **2012,** *81* (8), 1076–1078.

25. Johnson, J.; Marcotte, M. Irradiation Control of Insect Pests of Dried Fruits and Walnuts. *Food Technol.* **1999,** *53*, 46–53.
26. Kavitha, C.; Kuna, A.; Supraja, T.; Sagar, S. B.; Padmavathi, T. V. N.; Prabhakar, N. Effect of Gamma Irradiation on Antioxidant Properties of ber (*Zizyphus mauritiana*) fruit. *J. Food Sci. Technol.* **2014,** *52* (5), *3123–3128.*
27. Kim, K. H.; Yook, H. S. Effect of Gamma Irradiation on Quality of Kiwifruit (*Actinidia deliciosa*). *Radiat. Phys. Chem.* **2009,** *78 (6),* 414–421.
28. Kume, T.; Todoriki, S. Food Irradiation in Asia, the European Union and the United States: A Status Update. *Radioisotopes* **2013,** *62*, 291–299.
29. Kwon, J. H.; Kwon, Y. J.; Byun, M. W.; Kim, K. S. Competitiveness of Gamma Irradiation with Fumigation for Chestnuts Associated with Quarantine and Quality Security. *Radiat. Phys. Chem.* **2004,** *71*, 41–44.
30. Landgraf, M.; Goularte, L.; Martins, C.; Cestari, A. J. R.; Nunes, T.; Aragonalegro, L.; Hutzler, B. Use of Irradiation to Improve the Microbiological Safety of Minimally Processed Fruits and Vegetables. In *Use of Irradiation to Ensure the Hygienic Quality of Fresh, Pre-Cut Fruits and Vegetables and Other Minimally Processed Food of Plant Origin*; International Atomic Energy Agency (IAEA): Vienna, Austria, 2006; pp 41–60.
31. Lewis, M. J. *Physical Properties of Foods and Food Processing Systems*; Elsevier: New York, 1990; p 480.
32. Lusk, J. L.; Roosen, J.; Bieberstein, A. Consumer Acceptance of New Food Technologies: Causes and Roots of Controversies. *Annu. Rev. Resour. Econ.* **2014,** *6*, 381–405.
33. Maherani, B.; Hossain, F.; Criado, P.; Ben-Fadhel, Y.; Salmieri, S.; Lacroix, M. World Market Development and Consumer Acceptance of Irradiation Technology: A Review. *Foods* **2016,** *5* (79), 1–21.
34. Majeed, A., Muhammad, Z., Majid, A., Shah, A. H., & Hussain, M. Impact of Low Doses of Gamma Irradiation on Shelf-Life and Chemical Quality of Strawberry (*Fragaria ananassa*). *J. Anim. Plant Sci.* **2014,** *24*(5), 1531–1536.
35. Maraei, R. W.; Elsawy, K. M. Chemical quality and nutrient composition of strawberry fruits treated by γ-irradiation. *J. Radiat. Res. Appl. Sci.* **2017,** *10* (1), 80–87.
36. Martins, S.; Mussatto, S. I.; Martínez-Avila, G.; Montanez-Saenz, J.; Aguilar, C. N.; Teixeira, J. A. Bioactive Phenolic Compounds: Production and Extraction by Solid-State Fermentation. A Review. *Biotechnol. Adv.* **2011,** *29*, 365–373.
37. Nagendran, B.; Tan, Y. A.; Ravigadevi, S.; Kalyana, S.; Samir, S. Antioxidant properties of palm fruit extracts. *Asia Pacific J. Clin. Nutr.* **2005,** *4* (4), 319–324.
38. Najafabadi, N. S., Sahari, M. A., Barzegar, M., Esfahani, Z. H. Effect of Gamma Irradiation on Some Physicochemical Properties and Bioactive Compounds of Jujube (*Ziziphusjujubavar vulgaris*) Fruit. *Radiat. Phys. Chem.* **2017,** *130*, 62–68.
39. Niemira, B. A.; Fan, X. Low-Dose Irradiation of Fresh and Fresh-Cut Produce: Safety, Sensory and Shelf-Life. *Food Irradiat. Res. Technol.* **2006,** *32* (5), 45–52.
40. Nuclear Spain. Nuclear Technology Applied to Food Preservation and Sterilization. *Nucl. Espana* **1996,** *316*, 43–48.
41. Parlato, A.; Giacomarra, M.; Galati, A.; Crescimanno, M. ISO-14470: 2011 and EU Legislative Background on Food Irradiation Technology: The Italian Attitude. *Trends Food Sci. Technol.* **2014,** *38* (1), 60–74.
42. Pereira, E.; Barros, L.; Dueñas, M.; Antonio, A. L.; Santos-Buelga, C.; Ferreira, I. C. Gamma Irradiation Improves the Extractability of Phenolic Compounds in *Ginkgo biloba*. *Ind. Crops Prod.* **2015,** *74*, 144–149.

43. Port's, P. S.; Chisté, R. C.; Godoy, H. T.; Prado, M. A. The Phenolic Compounds and the Antioxidant Potential of Infusion of Herbs from the Brazilian Amazonian Region. *Food Res. Int.* **2013,** *53*, 875–881.
44. Rajkowski, K. T.; Fan, X. Ionizing Radiation of Seeds and Sprouts: A Review. *Am. Chem. Soc.* (ACS), **2004,** *875*, 107–116.
45. Ray, B. *Fundamental Food Microbiology*, 3rd ed.; CRC Press: Boca Raton, 2005; p 625.
46. Sádecká, J. Irradiation of Spices. *Czech J. Food Sci.* **2007,** *25* (5), 231–242.
47. Shurong, L.; Meixu, G.; Chuanyao, W. Use of Irradiation to Ensure Hygienic Quality of Fresh Pre-Cut and Blanched Vegetables and Tofu. In *Use of Irradiation to Ensure the Hygienic Quality of Fresh, Pre-Cut Fruits and Vegetables and Other Minimally Processed Food of Plant Origin*; Proceedings of a final research coordination meeting organized by the Joint FAO/IAEA Programme of Nuclear Techniques in Food and Agriculture and held in Islamabad, Pakistan, 22–30 July; Vienna: International Atomic Energy Agency, 2005; pp 87–105.
48. Singh, S. P.; Pal, R. K. Ionizing Radiation Treatment to Improve Postharvest Life and Maintain Quality of Fresh Guava Fruit. *Radiat. Phys. Chem.* **2009,** *78* (2), 135–140.
49. Song, H. P.; Byun, M. W.; Jo, C.; Lee, C. H.; Kim, K. S.; Kim, D. H. Effects of Gamma Irradiation on the Microbiological, Nutritional, and Sensory Properties of Fresh Vegetable Juice. *Food Control* **2007,** *18* (1), 5–10.
50. Tetriana, D.; Sugoro, I. Application of Nuclear Technology in the Fields of Vaccines. *Alara J.* (PTKMR–Batan, Patir Batan–the Market Friday, Jakarta), **2007,** *1*, 12–18.
51. Venugopal, V.; Doke, S. N.; Thomas, P. Radiation Processing to Improve the Quality of Fishery Products. *Crit. Rev. Food Sci. Nutr.* **1999,** *39* (5), 391–440.
52. Wani, A. M.; Hussain, P. R.; Meena, R. S.; Dar, M. A. Effect of Gamma-Irradiation and Refrigerated Storage on the Improvement of Quality and Shelf-Life of Pear (*Pyruscommunis* L.). *Radiat. Phys. Chem.* **2008,** *77* (8), 983–989.
53. Xuetong, F. Irradiation of Fresh and Fresh Cut Fruits and Vegetables: Quality and Shelf-Life. *Food Irradiat. Res. Technol.* **2012,** *2012*, 271–293.
54. Yun, J.; Li, X.; Fan, X.; Tang, Y.; Xiao, Y.; Wan, S. Effect of Gamma Irradiation on Microbial Load, Physicochemical and Sensory Characteristics of Soybeans (*Glycine max L.*). *Radiat. Phys. Chem.* **2012,** *81* (8), 1198–1202.
55. Zaman, W.; Paul, D.; Alam, K.; Ibrahim, M.; Hassan, P. Shelf-Life Extension of Banana (*Musa sapientum*) by Gamma Radiation. *J. Bio-Sci.* **2007,** *15*, 47–53.
56. Zhang, K.; Deng, Y.; Fu, H.; Weng, Q. Effects of Co-60 Gamma-Irradiation and Refrigerated Storage on the Quality of *Shatang mandarin. Food Sci. Human Wellness* **2014,** *3* (1), 9–15.

CHAPTER 4

SCOPE OF NOVEL PULSED LIGHT TECHNOLOGY: SAFETY AND SHELF-LIFE OF MINIMALLY PROCESSED FRUITS AND VEGETABLES

JUNAID AHMAD PANDITH, SATYAM BANSAL,
SHIV KUMAR KATIYAR, ARSHIED MANZOOR,
PIR MOHAMMAD JUNAID, and SAGHIR AHMAD

ABSTRACT

Fruits and vegetables constitute a potential source of essential nutrients to prevent several chronic diseases such as cancers, cardiovascular diseases. Freshly cut fruits and vegetables are a healthy alternative to meet the requirements of this demand. Because of the fierce competition in the industry, processors have been forced to consider new nonthermal approaches for extending shelf-life, retaining nutrients and sensory attributes of these products. The utilization of pulsed light (PL) treatment represents advanced technology for increasing the shelf-life of freshly cut fruits and vegetables. Therefore, this chapter will thoroughly discuss the potential of pulsed light treatment as an emerging method to decontaminate minimally treated fruits and vegetables, extend their shelf-life, and the chief factors affecting pulsed light treatment. Moreover, this chapter will also focus on physiochemical changes (i.e., appearance, texture, firmness, retention of vitamin C content, total phenolic content, antioxidants potential, etc.) and the impact of PL on different minimally processed fruits and vegetables. Besides the present status of PL legislation, specifications, its consequences, and future perspective will be discussed.

Quality Control in Fruit and Vegetable Processing: Methods and Strategies. Megh R. Goyal, & Faizan Ahmad (Eds.)

4.1 INTRODUCTION

In this modern era, everyone wants to lead a healthy lifestyle, and to achieve this desire, fruits and vegetables play an indispensable role. However, due to the unavailability of sufficient time or a busy lifestyle, intake of fruits and vegetables in the human diet is below the recommended intake that is a major factor for leading an unhealthy life. Therefore, to overcome this problem minimally processed fresh-cut fruits and vegetables serve the best.[29]

The main goal of the development of fresh minimally processed vegetables and fruits is to supply consumers with fresh, convenient, and nutritious products of extended shelf-life and high-quality attributes. The rate of deterioration in fresh-cut produce is much higher as compared with the unprocessed raw material, majorly because of the injuries caused by minimal processing such as peeling, dicing, slicing, shredding. These processing procedures generally subsidize the storability of freshly cut fruits and vegetables and the effects can be observed by a succession of typical symptoms, including tissue tenderization, surface browning, nutritional content, reduction, the existence of undesired off-flavor, and degradation due to microbes during preservation periods, increase in phenolic activity, vitamin C loss, antioxidant reduction, and loss in texture.

The storage life of these ready-to-eat (RTE) fresh-cut fruits and vegetables is enhanced by various methods such as the use of edible coatings and films, irradiation, controlled atmospheric storage and modified atmospheric packaging, pulsed light, etc. Among various methods, pulsed light is one of the potential novel techniques that extend the shelf-life of freshly cut fruits and vegetables. The technique works on eradicating the microbes and ceasing the physiochemical changes on the superficial surfaces of freshly cut commodities such as fruits and vegetables up to an extent and hence elongating the shelf-life of freshly cut articles to a considerable period within the constraint of maintaining the cold chain during the entire process. Pulsed light treatments were given to cut fruits and vegetables to enhance their shelf-life. Some treatment has been applied on various fruits and vegetables that include carrot, tomato, apple, cantaloupe, avocado, green onion, cabbage, mango, melons, lettuce, yard-long bean, exotic aneaden fruits, etc. Table 4.2 provides a list of pulsed light-treated fruits and vegetables and the impact on physiochemical changes after the treatment.

This chapter will discuss in detail the pulsed light technology, equipment and generation of pulsed light, mechanism of action of pulsed light on microbes, and microbial and physiochemical changes in various lightly processed fruits and vegetables. Specifically, pulsed light treatment and its impact on fruits and vegetables has been discussed, such as:

- Andean fruits uvilla (*Physalis peruviana L.*),
- apple (*Malus domestica*),
- avocado (*Persea americana*),
- cabbage (*Brassica oleracea*),
- cantaloupe (*Cucumis melo L.*),
- carrot (*Daucus carota subsp sativus*),
- green onion (*Allium fistulosum*),
- Kensington Pride mangoes (*Mangifera indica L.*),
- melon (*Cucumis melo var. cantalupensis*),
- mortino (*Vaccinium floribundum* Kunth),
- naranjilla (*Solanum quitoense Lam.*),
- raspberry (*Rubus idaeus),*
- Red Oak Leaf lettuce (*Lactuca sativa*),
- Red pepper (*Capsicum annuum* L.),
- strawberry (*Fragaria ananassa),*
- tomato (*Lycopersicum esculentum),*
- Yard-long bean *(Vigna unguiculata),*

4.2 FUNDAMENTALS OF PULSED LIGHT TREATMENT

4.2.1 DEFINITION AND TERMINOLOGY

Pulsed light is a technique employed to preserve and maintain the safety and qualitative attributes of food products. It involves the application of pulses of wide range "white light" that are extreme and brief in length. The light spectrum involves light of wavelength varying from UV to IR region. The subject is given pulses of light ranging from 1 pulse to 20 pulses per second for a duration of 1 μs to 0.1 s and each pulse possesses an energy density of range from 0.01 to 50 Joules/cm^2. In general, the wavelength distribution is in a manner that approximately 70% of the electromagnetic radiations fall in the spectrum from 170 to 2600 nm.[12]

Pulsed light has got approval from US FDA for its use as the total aggregate therapy not exceeding 12.0 J/cm^2. In various studies it has been illustrated that the use of pulses of light is much more beneficial than continuous light as the power delivered by the pulse is more intense and concentrated. Thus pulse is more effective in controlling microbial contamination than continuous light. It can be better understood by correlating pulse with impulsive force as it is more damage-causing than a continuous force.[39]

4.2.2 TERMINOLOGIES USED

Fluence Rate: It can be explained as the quantity of power received by the sample from the lamp per differential surface area. Hence the rate of Fluence is calculated in Watt/meter^2 (W/m^2).

Dose/Fluence: The total energy radiations gained by the sample from the lamp per unit area mean while the therapy is going on are defined as Fluence/dose. Its unit is Joule/meter^2 (J/m^2). The PL treatment or PL inactivation effectiveness depends on the delivered fluence. The parameter impacting the fluence (on a specific separation from the radiating lamp) is the width of the pulse, duration of treatment (t), number of pulses (n), the rate at which pulses repeat themselves (f). The fluence is obtained using a mathematical expression as the product of the energy of each pulse, the total number of pulses, and the pulse frequency. The governing equation is as follows[39]:

$$F_o = f \times t \times e \tag{4.1}$$

where F_o = fluence (J/cm^2), f = pulse frequency (Hz), t = treatment time (s), and, e = energy of single pulse (Joules/cm^2/pulse).

Pulse Width is defined as the moment (fractions of second) during which energy is supplied by the pulse to the sample.

Exposure Time is the period measured in seconds for which the sample is treated with PL.

Peak Power is the ratio of the energy of the pulse and the duration of a pulse. It is measured in Watt (W).

Pulse-Repetition-Rate (PRR) is defined as the pulses per second (Hertz [Hz]) or frequently indicated as PPS (pulses per second). It is also termed frequency.[1]

4.3 TYPE AND SPECTRUM OF PULSED LIGHT

Pulsed light technology employs intense white light to decontaminate the desired surfaces. The white light used is quite similar to the light of the sun but having much greater intensities than PL.[39] The type and spectrum of pulsed light rely on several parameters such as the type of flash lamp used as different gases emit the different spectrum of light. For instance, the xenon-light source releases a wide spectrum of flash light generally in the range of 200–1100 nm with roughly 25% in the UV range.[32]

Broad-spectrum pulses of light consist of white light, which is the combination of UV light, IR light, and the visible spectrum of light with their wavelengths in the range of infrared (IR) rays (λ = 700–1100 nm), for visible light (λ = 400–700 nm), and UV rays (λ = 200–400 nm). Further, the UV light is again classified into three classes according to the wavelength as ultraviolet Type-A: 315–400 nm, ultraviolet Type-B: 280–315 nm, ultraviolet Type-C: 200–280 nm. Among these, ultraviolet-C is the most potential light for microbial decontamination due to its shorter wavelength.[39]

In 1996, pulsed light was granted approval by the FDA for its implementation in the production, processing, and handling of food materials with the limitation that that maximum UV dose or fluence of 12 J cm^{-2} can be employed, whereas emission spectra are to be kept in the range of 200 and 1100 nm and pulse duration of less than or equal to 2 milliseconds.[15]

The type of pulsed light is categorized by the intensity, that is, the fluence rate of the pulse light. As the fluence of pulsed light changes, the intensity varies. High-intensity pulsed light (HIPL) is used to decontaminate the surfaces.[38]

4.4 EQUIPMENT AND GENERATION OF PULSED LIGHT

A typical style pulsed light system works on the principle of energy multiplication by concentrating electrical energy into a capacitor in less than a second and utilizing this stored energy in a millionth of a second. Therefore, the energy requirement is not very high as the energy is increased many folds by capacitor circuitry. After this, the pulses are generated in the pulse generation section by the use of high-power switches that perform the on/of operations at a very rapid and in the desired fashion. Then these electrical pulses are converted into light flashes by flash lamps in the lamp unit where a variety of lamps are present to deliver the desired spectrum of pulsed light.

4.5 THE MAJOR COMPONENTS AND EQUIPMENT USED FOR THE GENERATION OF PULSED LIGHT

Power Generation Unit: In this, the generation of DC high voltage is done by using normal low-voltage AC from the regular power supply systems.

Capacitor Circuitry: High-voltage capacitors are joined in parallel which focuses energy in the charge cycle in less than a second and releases this energy in the discharge cycle in a millionth of a second, thus multiplying the electrical energy many times.

Pulse Generation and Configuration Setup: In this section, the generation of pulse takes place by the use of specific high-power operating switches that carry out on/off cycles in a flash changing the continuous flow of low-electrical power into a high power pulsed light as shown in Figure 4.1.

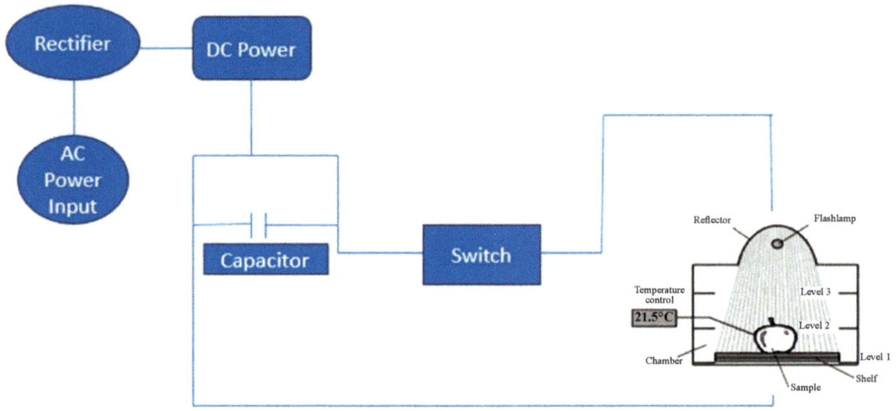

FIGURE 4.1 Pulsed light generation equipment.

Lamp Unit consists of batteries of flash lamps possessing gases that are excited because of high-electrical pulsed power enclosed into a glass tube. These gases are excited because of the strong electrical discharge from the pulse configuration device. In the excitation phase, atoms gain energy in the form of an electrical pulse, and in the de-excitation phase, the atom comes to the ground state releasing the energy which they have gained in the excitation phase in the form of light. Different gases emit a different wavelength of light and in this way, a desired spectrum of light is generated.

4.6 PULSED LIGHT TECHNOLOGY: PRESENT STATUS AND REGULATIONS

According to US FDA:

- The total successive additions of treatment concerning total fluence cannot be more than 12 J/cm^2
- The time of pulses is less than or equal to 2 ms

- Generally, frequencies of the pulse applied vary from 1 to 20 pulses per second
- Emission spectra range should be between 200 and 1100 nm

The application of pulsed light is a promising technology for minimal processing and shelf-life extension of fresh-cut commodities as it can inhibit microbial proliferation on account of the UV-C light. PL account for surface decontamination and retention of various physicochemical properties of the fresh-cut produce to a considerable extent. The pulsed light got its description as a cost-effective nonthermal technology and a method that does not give rise to undesirable residuals on treated surfaces of food.[39,56]

4.7 PULSED LIGHT ACTIVITY ON MICROBIAL INACTIVATION

Pulsed light technology involves the use of light pulses of intense high power. The main components of light used are the UV, the visible spectrum light, and the IR light and the combination leads to microbial inactivation. Photophysical, photochemical, and photothermal effects are typically responsible for microbial eradication. They can inactivate microorganisms separately or in combination with one another, and their effectiveness is directly proportional to the treatment method and target microorganisms.

In literature, it has been indicated that though photophysical and photothermal effects of pulsed light might have an impact on microbial inactivation, but the chief factor that leads to the destruction of microorganisms is the photochemical effect of pulsed light.[39]

The rich wide spectrum of UV radiations ascribes lethal power to pulsed light. UV light has a pivotal role in the cell inactivation of microorganisms. It was also discovered that the lethality of PL is affected a lot by the no-killing effect in case a filter is employed to eliminate UV radiations. Wavelengths below 320 nm are found to have maximum potential to inactivate the microbes.

The UV spectrum is branched into three wave sectors: short-wave UV-C (200–280 nm), medium-wave UV-B (280–320 nm), long-wave UV-A (320–400 nm).[1] Chiefly, the dreadful impact is possessed by the photochemical effect where UV-C light photons of wavelength noteworthy 253.7 nm are absorbed by the DNA of microbes.[39] Conjugated carbon–carbon present in proteins and nucleic acids absorb UV radiations, which bring about antimicrobial effect because it transforms structures of DNA and RNA, dimers of pyrimidine chiefly thymine dimers.[23]

A large number of thymine dimers, a low amount of cytosine dimers, and intermediate amounts of mixed dimers are typically generated through UV irradiation.[42]

During the process of cell division, the dimmers prevent the formation of new DNA chains which helps to prove lethality of the UV light by chlorogenic death of cell.[14] Through the application of UV-C light, photo products such as 5-thyminyl-5, 6-dihydrothymine, and cyclo-butane pyrimidine dimmers from bacterial spores may be formed.[57]

In various studies, it has been demonstrated that the enzymatic repair of the DNA damaged by pulsed light is also not possible. The photo-chemical effect is another aspect of photosensitization. Several photoactive compounds or photo-sensitizers are administered in the microbial cells targeted during the process of photosensitization and then it is followed by light flashes.[39]

According to the research studies, the sensitivity of microbes to PL is in the order: fungal spores < Gram-positive bacteria < Gram-negative bacteria.[55] Another study done by a researcher stated that similar trends in microbial sensitivity were observed.[7] However, contrarily a researcher reported a different pattern of microbial sensitivity in their analysis.[26] Spores are less susceptible than bacterial vegetative forms.

The decontamination because of the thermal effect is effective if the temperature of the surface reaches the pasteurization temperature as at this temperature the lethality for microbes is quite high. Nonetheless, in a study, it was observed that when pulsed light was focused on A. niger cells and the molds, they instantaneously got heated and caused cell death.[54] A possibility is also that light pulses can raise the surface temperature to destroy inhabiting bacteria.[19]

The reason behind cell death might be the alterations in the cell structure and physiology, which could be known as photothermal stress. The regional overheating caused by pulsed light may lead to the explosion of the cell causing rupture of cellular material. The scanning electron microscope (SEM) images of *A. niger* spores treated with PL helped in visualizing the release of overheated contents of spores.[39,61] Moreover, Figure 4.2 shows the *L. monocytogenes under TEM*: (A) Control, (B) treatment of 150 pulses for 30 s, (C) treatment of 900 pulses for 180 s with intense pulsed light (ILP) on the fluence of 1.75 Joules/cm^2/pulse, (D) treatment of 1000 s by UV-C at 254 nms. The bar in this figure relates to 200 nms.[17] Table 4.1 shows the impact of pulsed light treatment on various fruits and vegetables.

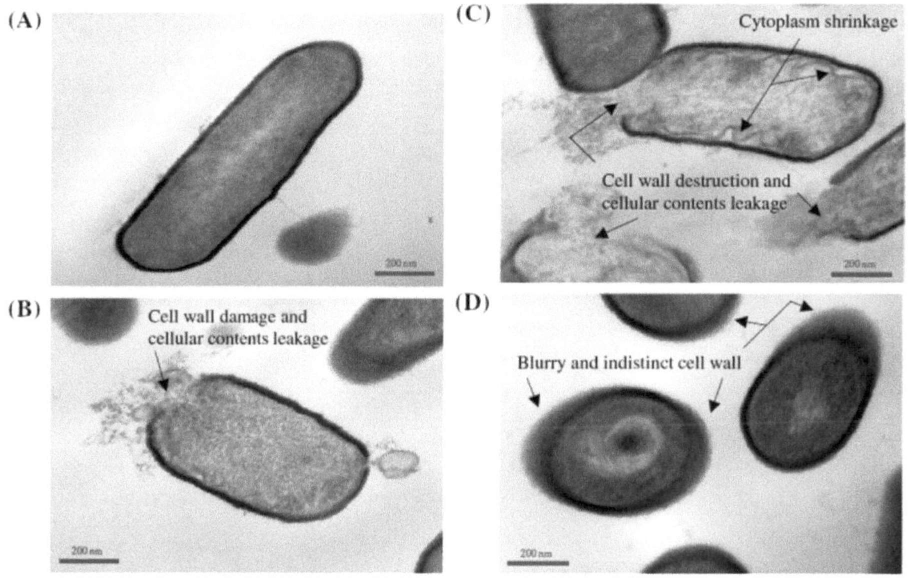

FIGURE 4.2 TEM depicting the action of UV light on microbes.
Source: Reprinted with permission from Ref. [17]. © 2013 Elsevier.

4.8 CRITICAL FACTORS AFFECTING EFFICACY OF PULSED LIGHT DECONTAMINATION

4.8.1 DESIGN AND PROCESSING FACTOR OF PULSED LIGHT

Each manufacturer designs and layouts their pulsed light system differently, but the basic components are the same. The basic elements are the source of energy, which provides power to the pulsed light system, the capacitor circuitry, which magnifies the power multiple times by using capacitors connected in parallel, the pulse production and synchronization system, which configures the pulse of the continuous current by using high power circuits. The pulses are then sent to the flash lamps, which transform electrical energy into light, and eventually, a triggering mechanism is set up to start all of the operations.[1]

The flash lamp is the most essential element because it emits pulsed light. An inert gas, like xenon or krypton, is used to fill the lamp. Xenon is chosen for most microbial inactivation purposes due to its greater conversion ability and because it is a gas of preference for most microbial inactivation applications. The key structural elements of the flash lamp are the casing, seals,

TABLE 4.1 Pulsed Light Treatment on Various Fruits and Vegetables.

Cut fruit and vegetable	Treatment		Microorganism	Reduction in microbial loads in log1010 CFU/g	References
	Fluency	Time and temperature			
Apple	11.9 J/cm^2	5°C for 7 days	*Escherichia (ATCC 11229)*	–0.26	[24]
			Listeria (ATCC 33090)	–1.5	
			Saccharomyces (KE 162)	–0.31	
			Aerobic mesophilic	4	
			Molds and yeasts	40	
Avocado	0.4 J/cm^2	5°C for 15 days	*L. innocua*	2.719	[53]
	30 pulses		*Escherichia*	2.12	
Cabbage	7 J/cm^2	45 sec	Mesophilic aerobic counts- sample wt-11 g	1.03	[25]
			Mesophilic aerobic counts- sample wt-11 g	1.64	
Cantaloupe	254 nm	10°C for 6 days	*Mesophilic aerobes*	Aprox 3	[37]
			Lactic acid bacteria		
			Pseudomonas spp	Aprox 1.7	
			Yeast and molds	Aprox 1.5	
				Below detection limits	
Green onion	(14.3 J/cm^2 for 15 sec.		*Escherichia coli (O157:H7)* inactivation on turned spot inoculated stems	4.9	[62]
			Escherichia coli (O157:H7) inactivation on turned spot inoculated leaves	4.5	
Mellon	1200 J/m^2	6°C for 10 days	Total viable count	2.16	[40]
			Enterobacteriaceae	2.61	

TABLE 4.1 *(Continued)*

Cut fruit and vegetable	Treatment		Microorganism	Reduction in microbial loads in log1010 CFU/g	References
	Fluency	Time and temperature			
Raspberry	34.2 J/cm^2	60 s	*E. coli* (O157:H7)	2.6	[13]
			Salmonella	2.9	
Strawberry	34.2 J/cm^2	60 s	*E. coli*(O157:H7)	2	[13]
			Salmonella	2.8	
Tomato	2.2 J/cm^2	20°C for 15 days	*S. cerevisiae*	2.3	[4]
			Aerobic mesophlic bacteria mold and yeast counts	2.9 1.9	

and electrodes, with the envelope serving as a jacket that both contains and covers the electrodes.

The lamp must be transparent to the lamp's emitted radiations, resistant to both the filling gas and air, capable of withstanding high temperatures and thermal shocks, and mechanically strong. Suprasil, transparent fused quartz with a thickness of around 1 mm, is used to produce the envelopes. Metallic electrodes hang down from the envelope's ends and are attached to a capacitor powered to a high voltage. The electrodes conduct electricity through the gas. The cathode decides the lamp's lifespan and is, therefore, an essential aspect.

The material used to construct the electrode is determined by the operating criteria. Since sputtering induced by hot spots produced while peak power is being supplied may trigger degradation of the cathode material, the cathode's purpose is to provide a dispersed and sufficient amount of electrons. As a result, the cathode's lifespan will be reduced. The anode must have enough mass/surface area so that it can withstand the power loading induced by the electric arc's electron bombardment. Flash lamp's enclosure must be sealed. Solder seals, rod seals, and ribbon seals are also widely employed seals.[1]

The gas inside the flash lamps is ionized as a result of the extremely high current released by the capacitor and pulse configuration system. When a molecule is ionized, the atoms move to higher transition energy, and the energy is consumed in the form of electrical energy. When the atom returns to its ground state, the absorbed energy is discharged in a mode of light and heat energy. Due to the possibility of overheating during this process, cooling systems must be provided to ensure a good lamp life and consistent working. Cooling fans may be employed to do this.[1]

Some other pulsed light sources such as a solid-state Marx-generator are used for pulsing an UV lamp to inactivate microbes, static discharge lamps with spectral outputs just as the flash lamps and sparker technology produces both a light and a sonic sound pulse.[52] Electricity to light conversion efficiency is about 50%. The 25% UV, 45% visible radiation, and 30% infrared make up the spectral range. The flashes occur at a rate of 1–20 flashes/s. The type of food preservation equipment relies on multiple factors, which include ozone formation, the surface area of the food component, the measurements of the individual therapy unit, and the optimal level of decontamination. If the food being treated is temperature-sensitive, a cooling unit facility might be needed. Depending on the use, pulsed light apparatus may be batch or continuous. Batch processing, for example, that produced by Xenon Corp. (Waltham, MA), involves placing the packets into the chamber along with lamps positioned as the chamber's walls. A single lamp mounted over the subject and a customizable tray to house these articles are the most basic prototypes.[39]

More sophisticated designs can include eight lamps in a cabinet, as well as a quartz stand for containing the article, and provide for 360° exposure and therapy.[52] With this configuration, an in-line treatment system is feasible, and continuous pulsed light tests have been carried out.[1,9,49]

4.9 PRODUCT PARAMETERS AND MICROBIAL PROPERTIES

Pulsed light is a multifaceted operation, with various modes of response depending on the dosage and the microbe to be inactivated. Microorganism shielding by a rugged apple superficial layer and microorganism internalization, for example, in pores of apple tissues, have a major effect on inactivation trends.[24] The product's reflection coefficient is also taken into account; it should be low for successful microbial inactivation. Pulsed light effectiveness is also influenced by optical properties and dietary quality. Solid foods and packaging stuff should be clear, flat, and free of defects such as roughness which also include pores, grooves that might protect microbes from the incident light. Also extremely smooth finished surfaces can reflect light, reducing the effectiveness of pulsed light[39] Table 4.2 shows physio-chemical changes after pulsed light treatment on various fruits/vegetables in the fresh-cut state.

4.10 PULSED LIGHT-ASSISTED MICROBIAL DECONTAMINATION IN MINIMALLY PROCESSED FRESH-CUT FRUITS

4.10.1 CANTALOUPE

The main deterioration species in freshly cut cantaloupe includes aerobic, mesophilic bacteria, Pseudomonas spp. and lactic acid bacteria (LAB) are one of the major deterioration species for fresh-cut cantaloupe. These microbes were chosen to see how UV-C treatments impacted the microbial content of freshly cut cantaloupes. Fruit handled under UV-C light has the minimum levels of aerobic mesophilic bacteria and lactic acid bacteria. Yeast and mold cultures in fresh-cut pieces of whole melon that have been exposed to UV light and freshly cut pieces that have been handled immediately post cutting, on the other hand, were below detection (less than 1 colony-forming unit [CFU]). Regardless of the procedure, as storage time improved, a

TABLE 4.2 Physiochemical Changes after Pulsed Light Treatment on Various Fruits/Vegetables in the Fresh-Cut State.

Fruits and vegetables	Pulsed light operation		Physiochemical changes after pulsed light treatment	References
	Fluency	Storage time temp.		
Apple	17.5 J/m^2	6°C for 7 days	During storage, the flavor profile of sliced apples degraded, weight loss happened, and slices became dehydrated and brown.	[29]
Avocado	0.4 J/cm^2	5°C for 15 days	Fresh-cut avocados lost their lightness over time and were marginally lighter in treated than untreated samples, despite little difference in a* value and a small improvement in b* values. In samples treated with 30 pulses, a decrease in firmness was observed over time.	[53]
Cabbage	7 J/cm^2	7°C for 9 days	Off odor increased after treatment, decrement in taste was observed, Overall visual quality retained, slight to no change was observed in sogginess, browning dryness was observed.	[25]
Cantaloupe	0.9 J/cm^2	4°C for 28 days	Lipase activity in UV-C-treated fruit declined significantly and was untraceable after 7 days of storage. Fruit sliced under UV-C radiation retains its firmness better. During the preparation of cut fruit, the procedure also decreased respiration.	[37, 31]
Lettuce	7.11 kJ/m^2	5°C for 10 days	UV-C processed commodity packets depleted O_2 faster after 7 days of storage at 5°C and a faster increase in CO_2 than the control in the same conditions, as well as tissue softening and browning occurred.	[6]
Mango	11.7 KJ/m^2	20°C for 12 days	In comparison to untreated control fruit, UV-C therapy slowed skin degreening, decreased endogenous ethylene development, decreased respiration rate, and reduced chlorophyll material. TPC (Absolute Phenolic Content) and total antioxidant activity were significantly higher in treated fruits toward the termination of the storage period, and their firmness was maintained.	[51]
Melon	1200 J/m^2	6°C for 10 days	During storage, there was no noticeable impact of UV-C light treatment on product color or firmness. Melon leakage was reduced when exposed to UV-C radiation, on the other hand.	[40]

TABLE 4.2 *(Continued)*

Fruits and vegetables	Pulsed light operation		Physicochemical changes after pulsed light treatment	References
	Fluency	Storage time temp.		
Mortino	12.5 kJ/m^2	6°C for 21 days	Delay in weight loss firmness, it maintained higher antioxidant capacity (AC) than control fruit.	[9]
Naranjilla	12.5 kJ/m^2	6°C for 21 days	Reduced weight loss and a significant delay in naranjilla softening.	[9]
Raspberry	29.7 J/cm^2	--	Little color improvements in treated samples.	[13]
Red pepper	10 kJ/m^2	0°C for 21 days	The treatments did not induce significant changes in antioxidant capability, but they did minimize chilling injury, weight loss, and color retention was there.	[8]
Straw berry	29.7 J/cm^2	–	The color of the treated sample was slightly better, but the difference was not significant.	[13]
Tomato	8 J/cm^2	5°C for 20 days	A slight increment in soluble solids and titrable acidity was observed, slight decrement in pH and firmness was observed.	[58]
Uvilla	12.5 kJ/ m^2	6°C for 21 days	Weight loss is reduced, but there is no impact on firmness loss. UV-C therapy has no adverse effects on the flavor of the fruit and improved the Soluble Solids Content (SSC) in uvilla.	[9]
Yard-long bean	9.0 J/cm^2	4 ± 1°C for 16 days	Dehydration and textural problems were encountered; it can be accounted for the photothermal effect of pulsed light.	[45]

community of aerobic, mesophilic bacteria, Pseudomonas spp. including lactic acid bacteria were raised in almost all articles. Treatments of fresh-cut pieces during storage were more successful in reducing spoilage microbes than processing the pieces of fruit under UV light.[37]

4.10.2 STRAWBERRY

Strawberry eating is beneficial to our health in a variety of ways, but fresh-cut strawberries are tough to obtain due to their gradual decline in consistency. Pulsed light therapy (4, 8, 12, and 16 Joules/cm^2) was applied to fresh-cut strawberries. At 5°C, improvements in consistency and antioxidant properties were observed for 14 days. Fresh-cut strawberries were analyzed for their phenolic content, anthocyanin content, and vitamin C content.[11]

4.10.2.1 FUNGAL DECAY INCIDENCE

By causing photochemical alterations in DNA/RNA components and/or localized heating, pulsed light treatment induces cell damage. PL treatments were found to decrease the occurrence of fungus on the whole strawberry in trials.[18] The employment of pulsed light therapy to freshly cut strawberries, on the other hand, resulted in a longer-term inhibition of fungal growth.[11] This might be attributed to a reduced prevalence of achenes above the freshly cut surfaces to also be handled, which could have a defensive shadowing effect against pulsed UV-light in the fresh whole fruit.[13,18,48]

4.10.3 RASPBERRY

Raspberries were found concerned in a minimum of five outbursts of *Cyclospora cayetanensis*.[48] At 72 and 59.2 J/cm^2, maximum degradations of *Escherichia coli* O157:H7 and Salmonella were found as 3.9 and 3.4 $\log_{10}$ colony forming units/g, respectively. At these UV doses, there was no visible harm to the fruits. Pulsed UV light has the ability for being implemented as a disinfection tool for raspberries, according to the findings of this report. On doses of 0.19, 0.33, and 0.40 Joules/cm^2 per pulse, which referred to 13, 8, and 5 cm from the quartz window, respectively, pulsed UV-light was used to treat samples inoculated from *Escherichia coli* O157:H7 and Salmonella. Raspberries were processed for 5, 10, 30, 45, and 60 s, resulting in 34.2,

59.4, and 72 Joules/cm^2 cumulative maximal wideband energy doses for 13, 8, and 5 cm, respectively.

At fluencies of 2.9 and 72 Joules/cm^2, *Escherichia coli* O157:H7 reductions of 0.4 and 3.9 $\log_{10}$ colony forming units/g were recorded, respectively. At 3 cm from the quartz window, lowering in CFU/g varied from 0.9 to 3.9 $\log_{10}$ colony forming units/g after 5 and 60 s, respectively. The temperature of the raspberry rose dramatically post being exposed to pulsed UV light at all wavelengths. After the 60-s procedure, maximum temperature rises of 60, 55, and 30°C (fruit temperatures of 80, 75, and 48°C) were noted at 3, 8, and 13 cm, respectively. It is particularly essential to mention that, in addition to photochemical inactivation, pulsed UV-light inactivation of pathogens has a thermal aspect. The optimal treatment place was 8 cm from the quartz window with a 60-s therapy period, resulting in a reduction of 3.4 log10 CFU/g.[13]

4.10.3.1 DECONTAMINATION OF RASPBERRIES USING PULSED UV-LIGHT

Shadowing or shielding effects are most likely to blame for variations in microorganism reductions on the surfaces of raspberries, as described by.[33] The openings between raspberries' drupelets will protect microorganisms from light, resulting in only incomplete disinfection. The shadowing issue could reduce the effectiveness of pulsed UV light for decontaminating raspberries. The untreated raspberries had L*, a*, and b* attributes of 29.20, +20.25, and +11.66, respectively, compared to 31.29, +23.16, and +12.99, respectively, for the processed raspberries. There were no significant variations.[13]

4.10.4 APPLE

The impact of pulsed light doses on fresh-cut apples was studied for the microbiological contamination of control and the treated samples. For Saccharomyces cerevisiae KE162, no counts were observed after a 100-s PL application at 5 cm from the sample; the reduction values obtained for Escherichia ATCC 11229 and Listeria ATCC 33090 were found to be 2.25 and 1.7 logs, respectively, after the treatment with pulsed light for fluency of (2.4 J/cm^2). Nine pulses of pulsed light having fluence of 17.5 kJ/m^2 were imparted on the fresh-cut apple; the increment in temperature of 2°C was observed while the final temperature not exceeding the 26° C on the surface of the apple was recorded. The photothermal effect of pulsed light may be the reason behind the same.[24]

4.10.4.1 MICROBIAL INACTIVATION

The control sample's microbial counts were found to be exceedingly low. A subsidized residence of mesophilic bacteria (2.26±0.12 log colony-forming unit/cm^2) was discovered in particular. The counts of yeasts, molds, and Enterobacteriaceae obtained were all below the traceable level (50 CFU/cm^2). The effect of pulsed light on inoculated microbes, such as *L. brevis* and *L. monocytogenes* on apple samples was investigated to confirm pulsed light's inactivation effectiveness. As a symbol of spoilage and pathogenic microorganisms, these microorganisms were selected. *L. brevis* and *L. monocytogenes* were substantially reduced by 3.0 and 2.7 log decrements, respectively, when exposed to low-intensity pulsed light (17.5 kJ/m^2), although more changes in light fluence were less efficient.[24]

4.10.5 MELON

Microbial counts, color, firmness, juice leakage, sensory properties, and selectiveness were all calculated in melon cubes kept at 6°C for 14 days. The sample exhibited to UV-C rays resulted in two-log decreases in the overall viable count and Enterobacteriaceae counts, which stayed two log units less during storage than the nontreated sample. During storage, there was no discernible impact of UV-C light therapy on product color or firmness. Exposure to UV-C radiation, on the other hand, reduced melon leakage, most likely because of the development of a thin dry layer above the sample surface. UV-C-treated articles often had a greater taste, making them slightly more appealing than untreated samples.[40]

4.10.5.1 MICROBIAL ANALYSIS

The magnitude for the initial microbial count quite resembles that of fresh-cut melon.[37] Total viable count and Enterobacteriaceae populations were (2.0±0.5) 10^5 and (2.0±0.1) 10^3 colonies forming unit/gram, respectively, whereas yeast and mold colonies were missing. After cutting operations, melon cubes were exposed to UV-C light, which culminated in a number of logarithmic reductions greater than 2 involving both total viable count and Enterobacteriaceae.

A fluence of 4100 J/m^2 was found to decrease the overall viable population of freshly cut samples by 1.5 logs.[22] Slightly fewer decrements in aerobic counts (1 log) were obtained by,[10] irradiating watermelon samples to an UV-C fluence of 1600 Joules/m^2. The UV-C doses were investigated further to determine if increased exposure could enhance bacteria elimination without detrimental effect on fresh-cut watermelon quality. Increased microbial reduction was obtained by exposing cubes to UV-C for 10 min (13.7 kJ m^{-2}), but this dose adversely affected pulp color. In particular, red color fading was noticed, and this was evidenced by higher L* and lower a*, b*, C* and hue angle units in the 13.7 kJ m^{-2} treatment than in 6.9 kJ m^{-2} or lower dose. This discovery may be elaborated by the fact that UV-C light transmittance by fruit tissue is extremely limited. UV-C radiation can only have an effect on an extremely fine layer of the sample surface.

Fruit-cutting actions are prominently associated with creating damaged surfaces that are anything but away from flat and smooth on a microscopic basis. It could clarify the phenomenon of dispersed surface roughness, which is thought to cast a shadow on possible microbial cells, minimizing the germicidal effect of UV-C radiation. In other terms, using UV-C light fluencies greater than those needed to subsidize the microbial community directly open to the light will not enhance disinfection effectiveness as the light will not influence shadowed microbial cells. Higher fluence therapies, on the other hand, are thought to cause more reliability disruption.[44]

Mild processing conditions were found to be successful for microbial disinfection of freshly cut melon in a study. Samples were processed under UV-C light (20 W/m^2) and afterward exposed to 1200 Joules/m^2 UV-C rays for growing periods at 6°C to mimic fresh-cut fruit preservation along with the refrigerated chain. UV-C radiation not only disinfected freshly cut sample cubes post-harvest, but also impacted microbial population development during storage. For a 14-day storage period, gross viable and Enterobacteriaceae counts in light-handled articles maintained 2 log units less than in control samples. This duration reflects a very long storage time for this sort of commodity, which is important. Yeasts and molds were contained in the UV-C treated melon cubes at amounts just below the method's detection limit and slightly lower than the control up to 7 days of the experiment, but at 14 days of samples storage, the control, and pulsed light-handled samples yielded identical counts. Eventually, all UV-C exposed and control samples showed a decline in the population of lactic acid bacteria during storage.[40]

4.11 PULSED LIGHT ASSISTED MICROBIAL DECONTAMINATION IN MINIMALLY PROCESSED FRESH-CUT VEGETABLES

4.11.1 TOMATO

It is one of the most widely consumed vegetables and is well known for its remarkable sensory and nutritional properties. In their fresh-cut state, tomatoes are high in folates, vitamin C, vitamin E, and other natural antioxidants, especially carotenoids and phenolic compounds. It suffers huge losses in terms of microbial and physiochemical changes. To overcome these losses, pulsed light, a nonthermal technique is employed. Pulsed light at fluences of 4, 6, or 8 J/cm^2 was employed for the samples stored at 5°C for 20 days, and fluencies of 2.68 and 5.36 Joules/cm^2 were used, and the results were studied in red-ripe tomatoes kept at 20°C for 15 days. Microbial loads on the pulsed light treated tomatoes were 2log CFU/g lower than the control samples.[58]

4.11.1.1 MICROBIAL ANALYSIS

The fresh-cut state of tomato is encountered by various microbes such as psychrophilic bacteria, molds, yeasts, aerobic mesophilic bacteria, etc. The microbial infection could be managed to some degree using pulsed light treatment, as psychrophilic bacterial counts on control samples (3.1±0.1 log CFU/g) were not prominently impaired by pulsed light therapy, while initial yeast and mold loads on fresh-cut tomatoes (4.9±0.1 Log colony forming units /g) were significantly decreased only after handling at fluences of 6 and 8 Joules/cm^2 to 4.2 and 3.8 Log CFU/g, respectively.[58]

Tomatoes treated with two pulses (5.36 J/cm^2) had the lowest counts of aerobic mesophilic bacteria (2.9 $\log_{10}$ CFU/g). Fluences of 0.3, 0.56, 0.71, 1.1, 1.65, and 2.2 Joules/cm^2 were used to assess the antimicrobial effectiveness of PL on tomato surfaces inoculated with *S. Cerevisiae*. When *S. Cerevisiae* populations were inoculated on tomato sections and treated with PL, they were drastically decreased. At 2.2 J/cm^2, maximum reductions of *S. Cerevisiae* of 2.3 $\log_{10}$ were obtained in pulsed light treated tomato.[58] Psychrophilic bacteria counts on unhandled articles expanded prominently to 7.0 ± 0.2 log CFU/g, while those on pulse light treated articles reduced up to 0.7 ± 1.8 log CFU/g compared to control samples at the termination of storage. The yeast and mold loads on fresh-cut tomatoes, on the other hand, were 8.0 ± 0.2 log CFU/g, with the articles subjected to pulsed light-fluences of 8 Joules/cm^2 displaying the lowest changes (up to 0.5 log CFU/g) when compared to those on unhandled tomato samples.[58]

4.11.2 LETTUCE

The irradiation of "Red Oak Leaf" lettuce, which is minimally handled by UV-C treatment, was performed to prevent microbial growth and increase the shelf-life. UV-C resistance of 20 bacterial strains from various genera commonly associated with fresh produce (Enterobacter, Erwinia, Escherichia, Leuconostoc, Pantoea, Pseudomonas, Rahnela, Salmonella, Serratia, and Yersinia) was initially tested in vitro. Most bacterial strains were inhibited by the lowest dose (30 J m^{-2}). The most resistant strains were Erwinia carotovora, Leuconostoc carnosum, Yersinia aldovae, and Salmonella typhimurium which needed an UV-C dose of 85 Joules/m^2 to fully arrest development. Both UV-C therapies improved the product's shelf-life. However, post 7 days of keeping at 5°C, the 7.11 kJoules/m^2 dosage caused tissue softening and browning. UV-C radiation is also useful in suppressing existing microflora, moreover expanding the shelf-life of "Red Oak Leaf" lettuce that has been minimally processed when applied at the proper dosage.[6]

4.11.2.1 MICROBIAL COUNTS

The microbial counts for "Red Oak Leaf" lettuce, which is minimally handled, were raised during storage and were lying in the presumed range for fresh-cut vegetable items, as explained by.[43] While different bacterial groups had different growth habits, they all reacted to UV-C radiation treatment in the same way. The bulk of the microorganisms studied grew slower when exposed to UV-C radiation. Complete aerobic and facultative aerobic mesophilic bacteria development was shown to be smaller in radiated samples than in control samples. The highest growth declines were noted among two and six days of storage for the greater radiation levels (2.37 and 7.11 kJoules/m^2). The same results were obtained when minimally handled "'Lollo Rosso' and 'Red Oak Leaf' lettuces were dealt with 0.4, 0.81, 2.44, 4.07, and 8.14 kJoules/m^2 on just one side of the tissue." The concentrations of mesophilic bacteria in squash slices were found to be greater in nonradiated controls compared to handled with 0.493 or 9.80 kJ/m^2 UV light in many studies. Based on overall microbial counts, UV-C radiation increased the shelf-life of freshly cut lettuce.[20]

The shelf-life of untreated lettuce was a minimum of two days less than that of UV-C treated lettuce. When higher doses (2.37 and 7.11 kJoules/m^2) of UV-C radiation were used, they were successful in reducing lactic acid bacteria. Among the unhandled control and the radiated articles, (1.18 and 7.11 kJoules/m^2) radiation fluencies, a discrepancy of around 1 log unit was recorded. When the growth of enteric bacteria was examined, similar findings were discovered. After two days of storage, all of the UV-C radiation doses tested were successful in reducing enteric growth. Nevertheless, no major variations were found between the treatments applied. Radiation treatments have had an impact on yeast development. On all days except the third and tenth, there was a substantial discrepancy among the control and radiated articles. Toward the end of the storage time (day 10) when yeast development was at its peak, no major variations between the control and radiated samples were discovered.[6]

4.11.3 YARD-LONG BEAN

Freshly cut yard-long beans were subjected to pulsed light treatment at various doses (1.8 Joules /cm^2, 5.4 Joules /cm^2, 9.0 Joules /cm^2,, and 12.6 Joules /cm^2). Over 14 days, the microbial stability of freshly cut yard-long beans processed at 4±1°C was tracked. The implementation of a PL dosage of 9.0 Joules/cm^2 improved the shelf-life of freshly cut samples while preserving consistency when kept at 4±1°C, according to various researches.[45]

4.11.3.1 MICROBIOLOGICAL SAFETY

The research studies show that the control sample suffered huge microbiological contamination compared to the pulsed light treated sample. After 7 days of preparation, untreated samples achieved their full aerobic count (7 log CFU/g), while pulsed light extended the shelf-life of the freshly cut sample for 14 days. Aside from this, yeast and molds are contained in abundance in freshly cut yard-long beans. Yeast counts surpassed 5 log CFU/g, microbial spoilage in fresh-cut vegetables was observed.[28,27] Yeast and mold counts were held below 5 log CFU/g in articles handled with 9.0 Joules/cm^2 and 12.6 Joules/cm^2 for up to 14 days. As a result, microbial replication is prevented.[45]

4.11.4 GREEN ONIONS

Pulsed light-assisted microbial decontamination of fresh-cut green onion has been studied. Various types of microbial loads were observed on fresh-cut green onion; the inactivation efficacy of *Escherichia coli* O157:H7 by pulsed light (PL) was studied. To represent two separate matrixes, green onions were split into two regions: stems and leaves. Decontamination of stems was found to be more difficult. To assess decontamination levels, two separate inoculation procedures were used. When the methods of spot and dip inoculation were compared, it was discovered that dipped inoculated green onions were more difficult to decontaminate. The findings revealed that the treatments included 5 s of dry pulsed light (articles were not submerged in water during pulsed light therapy) and 60 s of wet pulsed light (articles were soaked in water and shaken during pulsed light therapy); both were successful in inactivating the bacteria.[62]

4.11.4.1 MICROBIAL INACTIVATION

Dry PL was used to treat fresh-cut green onions for 5, 15, 30, and 60 s. Only 5 and 15 s were considered for further evaluation because substantial green onions processed for 30 and 60 s showed consistent decline, involving softer and shrunk tissue, darker texture, and a cooked scent, and because this treatment was only on the surface, samples were treated for better outcomes. Spot inoculation simulated contamination caused by unsanitary contact from workers' hands or equipment while harvesting, packaging, or value-addition manufacturing; also dip inoculation simulated contamination during cleaning, which is typically the primary phase in the packing unit. The dry treatment for the spot inoculated one was found to be superior, with log reductions ranging from 4.1 to 5.2 $\log_{10}$ CFU/g. The nonsignificant difference was found in 5 s PL with and without switching; moreover, 5 s and 15 s treatments indicate that 5 s dry pulsed light therapies would be adequate to inactivate *E. coli* on stems and leaves for spot inoculated (106 CFU/g) green onions. For dip inoculated green onions, the average log reduction was less than 1.2 $\log_{10}$ CFU/g. 15 s treated stems in both with and without turnings (1.1 $\log_{10}$ CFU/g) were considered to be the most efficient therapy. In the case of leaves, there were no major variations between the 5 s and 15 s PL therapies, as well as between switching and not switching at each treatment. The researchers concluded that PL should be employed to decontaminate green onions with *E. coli* O157:H7. The consistency of the wet pulsed light therapy was greater than the dry pulsed light treatment. For spot inoculated stems and leaves, the 5 s dry and 60 s wet pulsed light treatments demonstrated optimistic inactivation efficacy (N4 $\log_{10}$ CFU/g).[62]

4.11.5 CABBAGE

Surveilling headspace gas concentrations, microbial species, and sensory attributes were used to assess the effect of intense pulsed light (IPL) therapy on the shelf-life of cabbage, which was kept at 7°C in equilibrium modified atmosphere packages (EMAP).[25]

4.11.5.1 MICROBIAL ANALYSIS

For aerobic psychrotrophic count (APC), a 0.54 log reduction can be achieved in minimally processed (MP) white cabbage. Psychrotrophic and yeast counts show a consistent increase in both control and irradiated MP cabbage during storage. As time goes by, the value of ILP decontamination wears off; control and treated samples resulted in the same counts, and flashed samples had the highest psychrotrophic count from day 7 until the conclusion of the trial. Though its aerobic psychrotrophic count was reduced by 0.34–1.27 logs in the subsequent procedures, after 4 days at 5°C and 2 days at 15°C, the counts were higher in the treated samples. Some decontamination treatments can cause vegetable tissues to become softer, making them more susceptible to microbial degradation. Psychrotroph counts in treated samples surpassed the acceptable limit (108 CFU/g) on day 7, although controls were acceptable for the whole 9-day experiment; yeast counts never reached the permissible limits (105 CFU/g). As a response, ILP had little effect on the microbial shelf-life of minimally processed cabbage kept at EMAP 7°C. The appearance of off-odors reduced the shelf-life of the processed samples to just 7 days.[25]

4.12 PULSED LIGHT TREATMENT AND PHYSIOCHEMICAL CHANGES IN MINIMALLY PROCESSED FRESH-CUT FRUITS

4.12.1 ANEADEN FRUITS

Naranjilla, uvilla, and mortino were subjected to a pulsed light treatment of fluencies (0, 8, or 12.5 kJ/m^2) of UV-C light and evaluated weekly for the best-suited fluence of pulsed light to analyze deterioration index at the storage cycle for 21 days at 6°C. Reduction in deterioration index and retention of the quality was observed by the application of high doses of pulsed UV-C light. Short prestorage UV-C exposure may serve as a nonchemical method to supplement storage at a lower temperature, quality retention, and enhance the after-harvest life of Aneaden fruits naranjilla, uvilla, and mortino. The most prominent visual attribute losses that were observed are the pits on the peel of uvilla, softening and peel browning were prominent losses in naranjilla while the reduction in moisture content and losses in weight were the marginal losses in mortino.[9]

4.12.1.1 FIRMNESS

The UV-C treatment showed various effects on aneaden fruits as in *uvilla* the firmness was retained, for naranjilla softening was delayed. Firmness values of naranjillia on the last sampling day were similar to the control sample stored for 7 days.[9]

4.12.1.2 SOLUBLE SOLIDS CONTENT

Soluble solids content is arranged in the increasing order from the highest in uvillia (ca. 13.9–14.1%) then in mortino (9.8–10%) to the lowest in naranjillia (7.2–7.3%). A remarkable increment in SSC during storage was observed for uvillia while moderate changes were seen for naranjillia and mortino. Higher SSC was observed for the UV-C irradiated uvillia than the nonirradiated fruit post 14 and 21 days' storage.[28]

4.12.1.3 TITRATABLE ACIDITY

Acidity and ph were not prominently impacted by the application of the UV-C light. Roughly, it was evaluated as 328 for uvillia, 379 for naranjillia, and 190 for mornito in meq/kg.[9]

4.12.1.4 ANTIOXIDANT CAPACITY

The total antioxidant capacity in the control sample was found to be five and seven times increased in mortino (6800 mg/kg) as compared to uvilla

(1500 mg/kg) and naranjilla (870 mg/kg), respectively. Higher antioxidant capacity than control was observed in uvillia instantly post-treatment and during storage. Results suggest that UV-C treatment elicited antioxidant accumulation.[5] UV-C treated uvilla and mortino retained greater antioxidant capacity than control during long periods of storage. The activation of the biosynthesis pathway of the phenolic compound as a result of UV radiation may account for the triggered increase in antioxidant capacity of uvillia.[9]

4.12.1.5 QUALITY RETENTION, WEIGHT LOSS, AND SOFTENING

By the application of UV-C treatment, the quality retention for mortino was lower in the extent to the uvillia and naranjillia. Application of high fluence pulsed light can effectively reduce the loss of weight in mortino. Softening was also very effectively delayed by the UV-C treatment in naranjillia. Maintaining the tissue integrity as observed in mortino may account for the remarkable improvement in anthocyanin stability for a longer duration of storage. Overall, the findings suggest that short prestorage is beneficial. UV-C exposure can be a nonchemical method to improve reduced temperature handling, consistency preservation, and fresh-cut fruit shelf-life. Fruits from the Aneaden include naranjilla, uvilla, and mortino.[9]

4.12.2 CANTALOUPE

4.12.2.1 POD ACTIVITY

Fresh-cut cantaloupe melon underwent the pulsed light treatment. A significant increment in peroxidase (POD) activity was observed for the freshly cut fruit under UV light than the fruit treated with UV light post-cutting and control. The POD activities when compared between the control and the postcut pulsed light treated fruits showed that there was an increment in treated one. During storage, similar results were marked for both the sets of samples whereas samples processed under UV light showed slightly increased POD activity. Photo-oxidation products were quickly accumulated by the application of UV light; a defense mechanism of plants was triggered against oxidation in the response to UV light.[34]

4.12.2.2 ES ACTIVITY

Reduction in esterase (ES) activity was observed for the fruit treated with pulsed light post cut.[35] However, the deprivation of esters and the simultaneous development of phytoalexin compounds showed to be significant in early reaction phases concerned with the loss of freshness of some cut fruits during storage. Treatment of cantaloupe post cut stage with UV light showed the reduction in ES activity in the fresh-cut treated fruit.[36]

4.12.2.3 LIPASE ACTIVITY

Lower lipase activity was detected in fruits treated with pulsed light compared to the control sample. Lipase activity, on the other hand, decreased gradually in fruit treated with UV-C light, and after 7 days of storage at 10 °C, no lipase activity was detected.

4.12.2.4 SENSORY EVALUATION

Sensory analysis reveals that the development of off-flavor, that is, the development of rancid musty off-flavors was controlled by cutting the cantaloupe in UV light exposure; however, improvement in retention of preferable flavors in melon such as sweet aromatic and fruity melon aroma may not be observed as for post-cut UV irradiated fruit. The judging suit did not determine considerable variations for acceptable flavors among post-cut UV light treated and fruit cut under UV light. The score obtained for the samples cut under UV light was considerably reduced in rancid aroma compared to after-cut UV-treated and untreated control.[37]

4.12.2.5 HEADSPACE GAS

Headspace gas composition shows a reduction in the rate of respiration of fruit handled under UV light. The reduction in analogous respiration rates was found to be maintained during the storage of the fresh-cut fruit.[37]

4.12.2.6 FIRMNESS

Firmness retention for the pulsed light treated fresh-cut cantaloupe was observed compared to the control.[37]

4.12.3 MANGO

A tropical fruit well known for its immense nutritional and sensory characteristics has the potential to provide about half of the suggested daily consumption of vitamin C and carotenoids, etc. These qualities attracted the scientific community to research fresh-cut states. Analysis performed on physical factors such as firmness, color, etc., and chemical factors (such as phenolic activity, carotenoid, ascorbic acid contents, and enzyme activities (Phenylalanineammonialyase, PAL and Polyphenoloxydase, PPO)) were observed on fresh-cut Kent mangoes. The mangoes were irradiated with eight lamps to provide a total fluence of 8 J/cm^2.[16] Haden mangoes were also subjected to UV type-C irradiation power of 2.46 and 4.93 kJoules/m^2 after 18 days at 25°C. Total flavonoids (TF), lipoxygenase (LOX) enzymatic activity were recorded at 0, 0.5, 1, 8, 16, and 24 h and after that every 3 days. The sample decay and overall appearance (OA) were tested every 2 days by evaluating on a hedonic scale.[27]

4.12.3.1 FIRMNESS

Studies showed that in the control sample firmness decrease of around 1.5 N after 48 h of storage was observed and this decrement in firmness was continued at 6°C. In contrast, the pulsed light treated sample indicated positive fluctuation of firmness until 1.5 N toward the termination of storage for 2 to 3 days when treated with a total fluence of 8 J/cm^2. Hence, a cumulative fluency of 8 J/cm^2 inducted firmness estimation of irradiated sample cubes during the 7 days of storage, whereas control samples depicted tenderization properties.[16]

4.12.3.2 COLOR

The color was analyzed by L^* a^* b^* values by the conventional system and DE was calculated. An increment was observed in the values DE in the pulsed light treated sample compared to the initial sample, DE values increment was observed in the control sample. Overall the color retention for the pulsed light treated sample showed better color retention properties of fresh-cut mangoes.[16]

4.12.3.3 PPO ACTIVITY

PPO activity in control samples remained largely unchanged for 7 days of storage. After 7 days, PPO activity in PL samples improved to 58.16 UAE/min/mg proteins, reflecting a 50% improvement over the initial value. After 3 and 7 days of storage, PPO activities in PL samples were slightly higher than in controls. On storage for 2 days reduction in PPO activity was recorded in both control and treated samples.[16]

4.12.4 STRAWBERRIES

4.12.4.1 COLOR

Entire storage, the lightness parameter (L^*) of outer segments was preserved in all strawberries. The a^* parameter of outer areas, on the other hand, decreased in response to the procedure. Fresh-cut strawberries treated at low doses showed a small a^* decrease, while slices treated at 16 J/cm^2 showed a more dramatic decrease.[11]

4.12.4.2 FIRMNESS

Fresh-cut strawberries were determined to have no major firmness changes after being handled with PL. According to a study in whole strawberries using light and transmission electron microscopy, the tension induced by PL therapy is probable to trigger cell wall hardening, a significant reduction in the integrity of the walls of hypodermis cells, resulting in less softening of the sample.[18,48]

4.12.4.3 TOTAL PHENOLIC COMPOUNDS

PL treatments have little effect on the overall phenolic content during the storage time. Strawberry phenolic content was originally 1.083 g/kg. The overall phenolic content of strawberries illuminated with UV-C for 5 min at a fluency of 2.1 kJ/cm^2 contained the maximum total phenolic content, followed by 1 and 10 min of UV-C illumination. The overall phenolic content of the control fruit was the lowest. During the 15-day storage cycle, the overall phenolic content of strawberry fruit improved for both UV-C illuminated and the control sample. When compared to illuminated fruit, however, this rise was comparatively smaller in control fruit.[21]

4.12.4.4 VITAMIN C

There was no significant difference in vitamin C content among slices illuminated with doses less than 16 Joules/cm^2, which matches the results of[60] for whole strawberries. Fresh-cut strawberries treated at 4 J/cm^2, on the other hand, retained their vitamin C content during storage, while slices treated at 8 J/cm^2 were held for 6 days and then marginally reduced. Vitamin C levels in samples handled at 12 and 16 J/cm^2 depleted dramatically during storage. Temperature readings never approached 42°C, so the drop may be due to photochemical more willingly than thermal effects. The greater reduction in slices exposed to higher PL doses, on the other hand, could indicate oxidative injury on the tissues treatment by these doses, as ascorbic acid is extremely susceptible to oxidation while wounding.[18,51]

4.12.5 AVOCADO

The nutritional content of fresh avocado is very high as it contains unsaturated fats, which are thought potent to reduce the risk of cardiovascular and hepatic disease, in abundance. It is high in minerals like potassium and magnesium, as well as vitamins E, C, and B-complex. Avocado's nutritious composition renders it a perfect medium for the growth of microbes which leads to quality loss and spoilage.[53]

4.12.5.1 HEADSPACE GAS

Reduction in oxygen content was observed during the preliminary days of storage in both the treated and unhandled samples. From the third day past the oxygen content in the tray was maintained near zero. The dramatic fluctuations in oxygen levels may be said due to the minimal processing, which increased the respiration rate of the fruit and hence leads to the undesirable anaerobic respiration triggering the fermentative processes associated with the biochemical metabolism of fruits. Conversely, the rise in CO_2 level was noted in the treated samples compared to the control. The values reached as high as 20 kPa until maximum. It was observed that higher CO_2 levels lead to anoxia conditions. The higher CO_2 concentration was a result of the higher uptake of O_2. The pulsed light exposure showed the impact on the ethylene production as the storage period and pulses no. accounted for it. This shows that the more the number of pulses, the lesser the ethylene concentrations. It is critical to control ethylene production to prevent the trigger of stress response of fresh-cut fruits. Negligible impact on ethanol production was observed; the slight formation of alcohol can be accounted for the anaerobic conditions.[53]

4.12.5.2 pH

There was a significant difference in pH during the treatments, suggesting a drop in pH at the start of the experiment, which was significantly greater in treated samples than in unhandled samples. Fresh-cut avocados treated by 15 and 30 pulses yielded no major variations.[50]

4.12.5.3 FIRMNESS

In samples treated with 30 pulses, there was a reduction in firmness over time. The textural properties of fresh-cut avocado can be drastically altered by pulsed light. After 3 days of preparation, the firmness of treated fresh-cut avocados had significantly lowered. After 9 days of storage, it was discovered that both treated and untreated avocado cylinders had equal firmness values.[50]

4.12.5.4 COLOR

In both treated and untreated avocado samples, the L^* (lightness), a^* (green–red), and b^* (blue–yellow) attributes are calculated. The lightness of fresh-cut avocados was greatly influenced by treatments. There were no major variations among the control and the articles treated with 15 pulses at the outset of the trial, and freshly cut avocados handled with 30 pulses showed symptoms of browning. The extremely important indicator of avocado enzymatic browning is its lightness. Fresh-cut avocados' lightness reduced over time and was found slightly reduced in handled than unhandled samples. According to the results, there were no discrepancies in a^* attributes of samples implemented to varying numbers of pulses during storage, but b^* attributes were substantially varied among treatments. The average color alterations of freshly cut avocados were measured, with the freshly cut untreated article acting as a guide. As a result, the evidence suggests that ILP therapies have a detrimental impact on the product's color. The color loss may be explained by an increase in avocado respiratory activity following the treatments.[53]

4.12.6 APPLE

4.12.6.1 EFFECT ON COLOR

The fresh delicious apple disc has the color of creamy-white, well depicted by L^* attributes ≈ 70, $a^* \approx 6.3$, $b^* \approx 22$. Irradiation was done by pulsed light at the two sample locations from lamp viz. 5 and 10 cm from the lamp for 20 s, 60 s, and 100 s reflecting a reduction in luminosity (greater ΔL^*), moreover an increase in a^* value in comparison to the control samples through the entire storage period. This effect was more enhanced in apples with exposure at shorter distances. Sharp differences were evident on the zero-day and with storage time decrease in the overall significant differences was observed. Moreover, initially, b^* value increased with the increase in irradiation time. The variations in b^*, on the other hand, did not follow a consistent pattern during storage. The changes in the b^* value of apples processed at 5 cm were found to be close to that of the control. As compared to the control, the apple shown to a light pulse at 10 cm for 60 s and 100 s had a higher b^* value during storage, while b^* variants of samples irradiated for 20 s were lower compared to control. The b^* values of the apple samples that had been irradiated for 2 s did not improve dramatically after 7 days of storage.[24]

4.12.7 MELON

Physicochemical consistency parameters such as color, firmness, and juice leakage were also assessed in UV-C handled and the control samples were kept at 6°C for growing periods, and no major variations in color and firmness of freshly cut melon were found as a result of keeping the temperature at 6°C, by independent application of UV-C light. However, there were major variations in juice leakage. Till 3 days of storage, intracellular liquid leakage from UV-C irradiated articles was slightly smaller than that from control samples. Upon further storage, however, no variations in this parameter were found among the samples. The surface of sample cubes treated with UV-C proved to be drier than control samples. UV-C light promotes the dehydration of a thin superficial surface of samples without impacting their total firmness and color. This phenomenon can result in the creation of a thin superficial dried film that prevents leakage of juice within the initial days of storage. The defensive effect of this edible film will be lost after further storage and moisture equilibration among the internal tissues and the dry film. It was found that UV-C handled the sample as less juicy compared to freshly prepared melon. Terpenoids, such as geranyl acetone, β-ionone, β-cyclocitral, among others, could be considerably raised following their synthesis in direct reaction to UV light stress. On contrary, esters were observed to reduce only after thinly sliced laminar tissues were exposed to UV light for an extended period. Off flavor can result if endogenous terpenoid concentrations become extremely high.[40]

4.13 PULSED LIGHT TREATMENT AND PHYSIOCHEMICAL CHANGES IN MINIMALLY PROCESSED FRESH-CUT VEGETABLES

4.13.1 TOMATO

Tomato in its fresh-cut state suffers huge physiochemical losses. The main enzymes involved in tomato softening are polygalacturonase (PG) and pectin methylesterase (PME.); these enzymes are considered the triggering enzymes for cell wall softening and hydrolysis in tomatoes, hence causing loss of turgor and crispness in tomatoes. As a result, the effects of pulsed light with fluencies of 4, 6, or 8 Joules/cm^2 on freshly cut conditions for 20 days of keeping at 5°C are investigated, and the findings for weight reduction, PME, and PG activities are discussed.[58]

4.13.1.1 FIRMNESS

The impact of pulsed light was not so profuse on the firmness of fresh-cut tomatoes. Regardless of the procedure used, firmness values declined dramatically over the storage period. Faster firmness decrements were found as the fluence of pulsed light applied increased. Over 20 days of storage, control tomato slices lost 63% of their initial firmness, while those exposed to fluencies of 4 and 6 Joules/cm^2 lost even more (72 and 74%, respectively). Freshly cut tomatoes showed fluences of 8 Joules/cm^2 and just lost 55.7% of their original firmness during storage.[58]

4.13.1.2 WEIGHT LOSS

The weight loss in tomato slices is proportional to the fluency of the pulsed light used, with slices exposed to 8 Joules/cm^2 showing the largest reduction in weight (1.6%); it was found to be four times and twice that of freshly cut tomatoes handled with 4 and 6 Joules/cm^2, respectively. Weight loss increment was clear during days postprocessing (ca. day 4), beyond the first week of storage the increment in rates was dropped considerably. Freshly cut tomatoes exposed to pulsed light and processed for 20 days lost 9 to 10% of their weight, whereas untreated tomato slices lost a smaller but not insignificant amount of weight (7%).[58]

4.13.1.3 PECTIN METHYLESTERASE (PME) ACTIVITY

The effect of pulsed light fluencies on freshly cut tomatoes indicates that pectin methylesterase activity in control and pulsed light treated freshly cut tomatoes reduced dramatically over time. PL-treated tomato slices, on the other hand, maintained higher PME activity values over time than untreated samples. In fact, compared to those exposed to 4 and 6 J cm^{-2}, tomato samples exposed to the highest PL-treatments (8 J cm^{-2}) maintained higher PME activity.[58]

4.13.1.4 POLYGALACTURONASE ACTIVITIES

PG activity does not show the active impact of pulsed light on PG activity. It fell dramatically (by around 60%) in the days following processing, regardless of the treatment used. Beyond the initial depletion, residual PG activity values were maintained without significant modifications.[58]

4.13.1.5 SENSORY ANALYSIS

The sensory evaluation gave mixed results as no markable differences were observed until 5 days of storage; after the 5th day of storage control slices were preferred over the treated samples of fluencies 6, 8, and 4 J/cm^2 until 10th day of storage, and discrepancies in acceptability attributes for unhandled and pulsed light handled tomato samples were barely important, variations in sensory results between the samples tended to be less prominent. At that stage, judges favored sample slices subjected to increased fluences. Certain unfavorable quality characteristics, such as mealiness and dryness, were identified by several judges in several samples at random.[58]

4.13.2 CARROTS

The color, carotene, polyacetylene, and soluble sugar values of carrot slices were analyzed after they were afflicted to pulsed light (PL) therapies on either side "(total pulsed light fluencies of 2.26, 4.52, 5.41, 9.38, and 13.15 Joules/cm^2)." When carrot slices were exposed to fluences of 2.26 and 4.52 Joules/cm^2, higher concentrations of carotene were observed relative to unexposed carrots. Carrot slices were treated with PL to improve carotene content, polyacetylenes, and soluble sugars. When fluencies having a range from 2.26 to 5.41 Joules/cm^2 were added to carrot slices, the luminosity of the peripheral cortex improved.[2] Color changes in PL-treated carrot slices and the control slices were measured of the central stele (vascular tissue) and peripheral cortex layer; the results indicated that as the slices were shown to both 9.38 and 13.15 Joules/cm^2, the lightness for the peripheral cortical layer of carrot slices reduced dramatically when compared to untreated samples. The browning index (BI) findings in the carrot cortex are prominently higher compared to vascular tissue when elevated pulsed light doses are implemented. The carrot cortex mostly comprises parenchyma cells and it may contain carotenes and sugars, which has a profound impact on the color of the peripheral cortex layer. When processing samples at 13.15 J/cm^2, the primary a^* attributes (redness) of the central stele of sample slices decreased significantly, although there were no noticeable differences in the peripheral cortex layer. As increasing PL doses, a normal decline in b^* attributes (yellowness) was found in both the central stele and peripheral cortex layers.[2]

4.13.2.1 β-CAROTENE RETENTION IN PL-TREATED CARROT SLICES

Total carotenoids were 1.8±80.05 mg β-carotene equivalents/g DW in control carrot samples. The content of total carotenoids of sample slices prominently raised compared to unhandled slices, when irradiating at 2.26 (123.2%) or

4.52 (112.1%) Joules/cm^2. But no further increment in fluency above 4.52 Joules/cm^2 showed a negligible impact on the initial β-carotene content of carrot slices.[2]

4.13.2.2 POLY ACETYLENE RETENTION IN PL-TREATED CARROT SLICES

According to different researchers, the amounts of individual polyacetylenes were 54.02 μg/g DW for FaOH, 23.18 μg/g DW for FaDOH, and 12.10 μg/g DW for FaDOAc in control carrot pieces. FaOH and FaDOAc had the same convictions. Following the application of PL doses of 2.26 kJ/cm^2, a significant increase in FaDOH levels of 29.50 μg/g DW (127.1%) was observed. The same PL fluences resulted in the highest rate of FaOH and FaDOc preservation, with 94.7% (51.2 μg/g DW) and 84.7% (10.25 μg/g DW0), respectively.[2]

4.13.2.3 IMPACT OF PULSED LIGHT ON THE SOLUBLE SUGARS FROM CARROT SLICES

The sugar content was measured with the aid of NMR spectroscopy. In control and PL-treated carrot slices, fructose, sucrose, β-glucose, and α-glucose were found. Sucrose (187.9 mmol/g of dried weight [DW]) was the most concentrated sugar in untreated carrot slices, after that β- glucose (44.1 mmol/g of DW), fructose (16.5 mmol/g of DW), and α-glucose (26.9 mmol/g of DW). The primary application of pulsed light did not affect the sucrose and α-glucose content of carrot slices, but it did have an effect on β-glucose and fructose during treatment. When processing at 2.26 J/cm^2 PL doses, β-glucose concentrations were 19% raised in pulsed light treated slices than in control samples. When using 5.41 J/cm^2 concentrations, however, the maximum fructose levels (17.5 mmol/g of DW) were reported. The findings indicated that the increase in sugar extraction found in pulsed light-handled samples may be due to free sugars available to extract as an outcome of PL hydrolyzing more complex carbohydrates.[3]

4.13.3 RED PEPPER

Chilling injury (CI) is among the most important factors preventing refrigeration in subtropical and tropical horticultural commodities like pepper. A quick UV-C treatment before storage at low temperatures was found to be successful against CI. Red peppers (*Capsicum annuum* L.) are subjected to UV-C radiation (10 kJoules/m^2) for 21 days before being stored at 0°C. Chilling injury development, ascorbic (AA) and dehydroascorbic acids, and DPPH radical scavenging capacity are investigated during storage. The activities of superoxide dismutase (SOD), catalase (CAT), ascorbate peroxidase (APX), and guaiacol peroxidase (GPX) were also assessed. While the fruit was preserved for more than 14 days, the CI increased rapidly, but it was considerably reduced in UV-C-handled peppers. UV-C irradiation did not impact the color of the fruit, but it did minimize weight loss. While the control had lower AA and DPPH radical scavenging capacity, this happened at the termination of storage, when chilling injury was already preceded. SOD, CAT, and APX behavior, on the other hand, were increased in UV-C treated samples within the first 2 weeks of storage, when indications appeared. The findings indicate that UV-C application inhibits chilling injury and reductions in the weight of red pepper and that this may be due to improved antioxidant enzyme activity.[8]

4.13.3.1 UV-C DOSE OPTIMIZATION

Several studies have indicated a decrease in CI in red peppers handled with UV-C radiation. The dosage implemented (7 kJoules/m^2) was originally chosen to rely on decay control of fruit contained at 10°C, but it was later found that therapy was also effective in lowering chilling injury. Researchers first looked to see if a higher dosage would be any more effective. UV-C therapy of 10 and 20 kJ/m^2 was applied to red peppers, which were then stored at 0°C. The chilling injury was more common in control peppers than in fruit handled with 10 or 20 kJ/m^2 UV-C. Fruit showed to a UV-C dosage of 10 kJ/m^2 lost the least amount of weight (1.3%). The 10 kJ/m^2 therapy had a lower weight loss and a shorter exposure period to gain the same CI prevention gain.[27,51,]

4.13.3.2 IMPACT OF UV-C THERAPY ON CHILLING INJURY COLOR AND WEIGHT LOSS

The fruit was given a 10 kJ/m^2 treatment and held at 0°C. After 7 days, the control fruit started to exhibit Cl signs. After 21 days, UV-C applied fruit had a low-moderate chilling injury index (2.5). The UV-C therapy has no detrimental effects on the appearance of the fruit's surface and greatly decreased weight loss. It has to do with the decreased damage caused by UV-C penetration. The findings show that UV-C treatments of 10 kJ/m^2 will reduce pepper CI.[8] Ascorbic acid (AA), dehydroascorbic acid (DHA), and DPPH radical scavenging potential in various experiments have discovered that UV-B and UV-C radiation induces an increase in the deposition of AA and DHA. The AA concentration in both control and UV-C treated peppers was close to 800 mg/g at the start of the storage time in this sample.

AA content was improved marginally in both control and UV-C irradiated fruit during the first week of storage.[41] The 5% increase noted over the first 7 days of containing may be due to a minor advancement of fruit production, because the fruit was stored at 0°C. After 21 days of storage, no differences in AA were seen in UV-C irradiated fruits, while a decrease was observed in the untreated. Since ascorbic acid is particularly vulnerable to oxidation when injured, the increased decrement in untreated fruit may be attributed to the raised injury discovered. DHA made up less than 0.5% of the vit. C pool and did not differ substantially while containing. Even though AA content in control fruit declined over time, there was no concurrent accumulation of DHA, implying that further deterioration occurred quickly. At the early stages of CI growth, UV-C treatments haven't induced significant changes in AA or DPPH radical scavenging capability. When the disorder's manifestation was already advanced, UV-C treated peppers had a marginally higher AA and DPPH radical scavenging potential.[8]

4.13.4 LETTUCE

4.13.4.1 GAS COMPOSITION

The densities of oxygen and carbon dioxide in containers of less processed "Red Oak Leaf" lettuce kept in passive modified atmosphere packaging (MAP) varied over time and were influenced by the various treatments. In contrast to the control, UV-C treated commodity packages had a faster depletion of O_2 and a faster rise in CO_2. In addition, the gas chemistry changed as the radiation exposure increased. Higher UV-C radiation doses result in higher CO_2 levels and lower O_2 levels. UV type-C irradiation induces respiratory discomfort in freshly cut "Red Oak Leaf" lettuce and it also raises its respiration rate, as previously stated[6]

4.13.5 YARD-LONG BEAN

The fresh-cut Batonnet yard-long bean, when subjected to pulsed light treatment, has shown a reduction in microbial spoilage, as discussed earlier. The physical changes such as color and texture changes that occurred due to treatment are also discussed.[45]

4.13.5.1 TEXTURE

Texture analysis shows loss of turgor of fresh-cut pulsed light treated samples as the tissue tenderization of the beans occurs. Due to the photothermal effect of the pulsed light, the loss of moisture occurs as the surface temperature is increased and as a result, the loss of crispness is observed in the beans. Analysis shows the impact of fluency of the PL on the loss of texture as the fluency of the pulsed light increased the loss of texture is increased, As compared to untreated samples, fresh-cut samples handled by reduced doses (1.8, 5.4, and 9.0 Joules/cm^2) have greater texture retention[45]

4.13.5.2 COLOR

It is a critical factor for sensory and qualitative analysis. All treatments resulted in a decrease in L^* value. Browning is suggested by a reduction in the L^* value.[32] Physical damage and wounding during preparation and processing can cause this, which increases respiration and ethylene production, as well as related enhancements in rates of different biochemical reactions liable for color variations that were triggered. Overall the pulsed light treatment failed to preserve the color of the fresh-cut yard-long bean.[45]

4.14 FUTURE CHALLENGES AND SCOPE

The effectiveness of pulsed light on smooth surfaces such as packaging materials; another area of study is the production of pulsed light for solid foods having rugged surfaces, as well as granular substances such as grains, nuts, spices, and other granular materials. Further research into the applicability of photosensitization-based microbial load reduction in foods is necessary. A continuous and great amount of effort is required for the engineering design of the setup of the instrument for achieving minimal processing. This can be accomplished using conveyor mechanisms for solid article handling and continuous flow-through mechanisms for liquid objects. Product heating issues may be addressed using PL equipment with cooling systems, which are accounted for by the photothermal effect of pulsed light. Further research work is also required for the study and analysis of the toxic and unwanted byproducts formation during and post-treatment which could not be palatable.[39,56]

4.15 SUMMARY

The novel PL technique is a nonthermal, environmentally friendly technology that is promising for decontaminating food and surfaces in contact with foods, and packaging materials. PL has a bright future in the food industry, where it will improve the consistency of food by inactivating microorganisms like L. monocytogenes in a matter of seconds. PL, on the other hand, is one such technique that can counteract the harmful effects of typical thermal processing and the use of hazardous chemicals. Application of pulsed light for the sterilization of food products is quite difficult due to irregularities of food surfaces. This method has shown to be effective in minimizing peanut allergy symptoms. It should be looked at further to see how it can assist with other food allergies. Since this method is still new and has yet to be commercialized, it should be investigated for cost savings. In today's culture, almost all value-added goods have a consumer niche, so processing, marketing, and analysis on fresh-cut fruits and vegetables are growing. Researchers and industry must collaborate to address the obstacles that obstruct the year-round distribution of high-quality fresh-cut fruits across the country.

ACKNOWLEDGMENT

The authors wish to extend special thanks to the TEQIP-III project for providing funding for this work.

KEYWORDS

- **cut-vegetables**
- **decontamination**
- **fresh-cut fruits**
- **nonthermal processing**
- **physiochemical changes**
- **pulsed-light**
- **shelf-life**

REFERENCES

1. Abida, J.; Rayees, B.; Masoodi, F. A. Pulsed Light Technology: A Novel Method for Food Preservation. *Int. Food Res. J.* **2014**, *21* (3), 839–848.
2. Aguiló-Aguayo, I.; Gangopadhyay, N.; Lyng, J. G.; Brunton, N.; Rai, D. K. Impact of Pulsed Light on Color, Carotenoid, Polyacetylene and Sugar Content of Carrot Slices. *Innov. Food Sci. Emerg. Technol.* **2017**, *42*, 49–55.
3. Aguiló-Aguayo, I.; Hossain, M. B.; Brunton, N.; Lyng, J.; Valverde, J.; Rai, D. K. Pulsed Electric Fields Pre-Treatment of Carrot Purees to Enhance Their Polyacetylene and Sugar Contents. *Innov. Food Sci. Emerg. Technol.* **2014**, *23*, 79–86.
4. Aguiló-Aguayo, Ingrid, Charles, F.; Renard, C. M. G. C.; Page, D.; Carlin, F. Pulsed Light Effects on Surface Decontamination, Physical Qualities and Nutritional Composition of Tomato Fruit. *Postharv. Biol. Technol.* **2013**, *86*, 29–36.
5. Allende, A.; Marín, A.; Buendía, B.; Tomás-Barberán, F.; Gil, M. I. Impact of Combined Postharvest Treatments (UV-C Light, Gaseous O3, Superatmospheric O2 and High CO2) on Health Promoting Compounds and Shelf-Life of Strawberries. *Postharv. Biol. Technol.* **2007**, *46* (3), 201–211.
6. Allende, A.; McEvoy, J. L.; Luo, Y.; Artes, F.; Wang, C. Y. Effectiveness of Two-Sided UV-C Treatments in Inhibiting Natural Microflora and Extending the Shelf-Life of Minimally Processed Red Oak Leaf Lettuce. *Food Microbiol.* **2006**, *23* (3), 241–249.
7. Anderson, J. G.; Rowan, N. J.; MacGregor, S. J.; Fouracre, R. A.; Parish, O. Inactivation of Food-Borne Enteropathogenic Bacteria and Spoilage Fungi Using Pulsed-Light. *IEEE Trans. Plasma Sci.* **2000**, *28* (1), 83–88.
8. Andrade Cuvi, M. J.; Vicente, A. R.; Concellón, A.; Chaves, A. R. Changes in Red Pepper Antioxidants as Affected by UV-C Treatments and Storage at Chilling Temperatures. *LWT—Food Sci. Technol.*, **2011**, *44* (7), 1666–1671.
9. Andrade-Cuvi, M. J.; Moreno, C.; Zaro, M. J.; Vicente, A. R.; Concellón, A. Improvement of the Antioxidant Properties and Postharvest Life of Three Exotic Andean Fruits by UV-C Treatment. *J. Food Qual.* **2017**, 2017.

10. Artés-Hernández, F.; Robles, P. A.; Gómez, P. A.; Tomás-Callejas, A.; Artés, F. Low UV-C Illumination for Keeping Overall Quality of Fresh-Cut Watermelon. *Postharvest Biol. Technol.* **2010**, *55* (2), 114–120.
11. Avalos-Llano, K. R.; Martín-Belloso, O.; Soliva-Fortuny, R. Effect of Pulsed Light Treatments on Quality and Antioxidant Properties of Fresh-Cut Strawberries. *Food Chem.* **2018**, *264*, 393–400.
12. Barbosa-Canovas, G. V.; Schaffner, D. W.; Pierson, M. D.; Zhang, Q. H. Special Supplement Kinetics of Microbial Inactivation for Alternative Food Processing Technologies (Pulsed Light Technology). *J. Food Sci.* **2000**, *65*, 82–85.
13. Bialka, K. L.; Demirci, A. Efficacy of Pulsed UV-Light for the Decontamination of *Escherichia coli* O157:H7 and Salmonella spp. on Raspberries and Strawberries. *J. Food Sci.* **2008**, *73* (5), 201–207.
14. Bolton, J. R.; Linden, K. G. Standardization of Methods for Fluence (UV Dose) Determination in Bench-Scale UV Experiments. *J. Environ. Eng.* **2003**, *129* (3), 209–215.
15. *CFR (Code of Federal Regulations) Title 21.* https://www.accessdata.fda.gov/scripts/cdrh/cfdocs/cfcfr/CFRSearch.cfm?fr=179.41**;** Accessed on March 14, 2021.
16. Charles, F.; Vidal, V.; Olive, F.; Filgueiras, H.; Sallanon, H. Pulsed Light Treatment as New Method to Maintain Physical and Nutritional Quality of Fresh-Cut Mangoes. *Innov. Food Sci. Emerg. Technol.* **2013**, *18*, 190–195.
17. Cheigh, C. I.; Hwang, H. J.; Chung, M. S. Intense Pulsed Light and UV-C Treatments for Inactivating Listeria Monocytogenes on Solid Medium and Seafoods. *Food Res. Int.* **2013**, *54* (1), 745–752.
18. Duarte-Molina, F.; Gómez, P. L.; Castro, M. A.; Alzamora, S. M. Storage Quality of Strawberry Fruit Treated by Pulsed Light: Fungal Decay, Water Loss and Mechanical Properties. *Innov. Food Sci. Emerg. Technol.* **2016**, *34*, 267–274.
19. Dunn, I. J. E.; Costa, R. La; Clark, R. W.; Mar, D.; John, F.; Pearlman, J. S.; Painchaud, F.; Hofmann, G. A. *United States Patent (1) U. S. Patent. 19*; 1989; pp 23.
20. Erkan, M.; Wang, C. Y.; Krizek, D. T. (2001). UV-C Irradiation Reduces Microbial Populations and Deterioration in Cucurbita Pepo Fruit Tissue. *Environmental and Experimental Botany*, **2001**, *45* (1), 1–9.
21. Erkan, M.; Wang, S. Y.; Wang, C. Y. Effect of UV Treatment on Antioxidant Capacity, Antioxidant Enzyme Activity and Decay in Strawberry Fruit. *Postharv. Biol. Technol.* **2008**, *48* (2), 163–171.
22. Fonseca, J. M.; Rushing, J. W. Effect of Ultraviolet-C Light on Quality and Microbial Population of Fresh-Cut Watermelon. *Postharv. Biol. Technol.* **2006**, *40* (3), 256–261.
23. Giese, N.; Darby, J. Sensitivity of Microorganisms to Different Wavelengths of UV Light: Implications on Modeling of Medium Pressure UV Systems. *Water Res.* **2000**, *34* (16), 4007–4013.
24. Gómez, P. L.; Salvatori, D. M.; García-Loredo, A.; Alzamora, S. M. Pulsed Light Treatment of Cut Apple: Dose Effect on Color, Structure, and Microbiological Stability. *Food Bioprocess Technol.* **2012**, *5* (6), 2311–2322.
25. Gómez-López, V. M.; Devlieghere, F.; Bonduelle, V.; Debevere, J. Intense Light Pulses Decontamination of Minimally Processed Vegetables and Their Shelf-Life. *Int. J. Food Microbiol.* **2005b**, *103* (1), 79–89.
26. Gómez-López, V. M.; Devlieghere, F.; Bonduelle, V.; Debevere, J.Factors Affecting the Inactivation of Micro-Organisms By Intense Light Pulses. *J. Appl. Microbiol.* **2005a**, *99* (3), 460–470.

27. González-Aguilar, G. A.; Zavaleta-Gatica, R.; Tiznado-Hernández, M. E. Improving Postharvest Quality of Mango by UV-C Treatment. *Postharv. Biol. Technol.* **2007**, *45* (1), 108–116.
28. Gutierrez, M. S.; Trinchero, G. D.; Cerri, A. M.; Vilella, F.; Sozzi, G. O. Different Responses of Goldenberry Fruit Treated at Four Maturity Stages with the Ethylene Antagonist 1-Methylcyclopropene. *Postharv. Biol. Technol.* **2008**, *48* (2), 199–205.
29. Ignat, A.; Manzocco, L.; Maifreni, M.; Bartolomeoli, I.; Nicoli, M. C. Surface Decontamination of Fresh-Cut Apple by Pulsed Light: Effects on Structure, Color and Sensory Properties. *Postharv. Biol. Technol.* **2014**, *91*, 122–127.
30. Jacxsens, L.; Devlieghere, F.; Falcato, P.; Debevere, J. Behavior of Listeria Monocytogenes and Aeromonas spp. on Fresh-Cut Produce Packaged Under Equilibrium-Modified Atmosphere. *J. Food Protection* **1999**, *62* (10), 1128–1135.
31. Koh, P. C.; Noranizan, M. A.; Nur Hanani, Z. A.; Karim, R.; Rosli, S. Z. Application of Edible Coatings and Repetitive Pulsed Light for Shelf-Life Extension of Fresh-Cut Cantaloupe (*Cucumis melo* L.). *Postharv. Biol. Technol.* **2017**, *129*, 64–78.
32. Kramer, B.; Muranyi, P. Effect of Pulsed Light on Structural and Physiological Properties of *Listeria Innocua* and *Escherichia Coli*. *J. Appl. Microbiol.* **2014**, *116* (3), 596–611.
33. Lagunas-Solar, M. C.; Piña, C.; MacDonald, J. D.; Bolkan, L. Development of Pulsed UV Light Processes for Surface Fungal Disinfection of Fresh Fruits. *J. Food Protection* **2006**, *69* (2), 376–384.
34. Lamikanra, O.; Watson, M. A. Effects of Ascorbic Acid on Peroxidase and Polyphenoloxidase Activities in Fresh-Cut Cantaloupe Melon. *J. Food Sci.* **2001**, *66* (9), 1283–1286.
35. Lamikanra, O.; Watson, M. A. Storage Effects on Lipase Activity in Fresh-Cut Cantaloupe Melon. *J. Food Sci.* **2004**, *69* (2), 126–130.
36. Lamikanra, O.; Watson, M. A. Temperature and Storage Duration Effects on Esterase Activity in Fresh-Cut Cantaloupe Melon. *J. Food Sci.* **2003**, *68* (3), 790–793.
37. Lamikanra, O.; Kueneman, D.; Ukuku, D.; Bett-Garber, K. L. Effect of Processing Under Ultraviolet Light on the Shelf-Life of Fresh-Cut Cantaloupe Melon. *J. Food Sci.* **2005**, *70* (9), 110–114.
38. Luksiene, Z.; Buchovec, I.; Kairyte, K.; Paskeviciute, E.; Viskelis, P. High-Power Pulsed Light for Microbial Decontamination of Some Fruits and Vegetables with Different Surfaces. *J. Food Agric. Environ.* **2012**, *10* (3–4), 162–167.
39. Mandal, R.; Mohammadi, X.; Wiktor, A.; Singh, A.; Singh, A. P. Applications of Pulsed Light Decontamination Technology in Food Processing: An Overview. *Appl. Sci. (Switzerland)*, **2020**, *10* (10).
40. Manzocco, L.; Da Pieve, S.; Maifreni, M. Impact of UV-C Light On Safety and Quality of Fresh-Cut Melon. *Innov. Food Sci. Emerg. Technol.* **2011**, *12* (1), 13–17.
41. Martínez, S.; López, M.; González-Raurich, M.; Alvarez, A. B. The Effects of Ripening Stage and Processing Systems on Vitamin C Content in Sweet Peppers (Capsicum annuum L.). *Int. J. Food Sci. Nutr.* **2005**, *56* (1), 45–51.
42. Moseley, B. E. B. The Repair of Damaged DNA in Irradiated Bacteria. *Adv. Microb. Physiol.* **1968**, *2* (C), 173–194.
43. Nguyen-the, C.; Carlin, F. The Microbiology of Minimally Processed Fresh Fruits and Vegetables. *Crit. Rev. Food Sci. Nutr.* **1994**, *34* (4), 25–30.
44. Nigro, F.; Ippolito, A.; Lima, G. Use of UV-C Light To Reduce Botrytis Storage Rot of Table Grapes. *Postharv. Biol. Technol.* **1998**, *13* (3), 171–181.

45. Nor-Hasni, H.; Noranizan, M. A.; Roselina, K. Pulsed Light Treatment as an Alternative Technology To Extend Shelf-Life Of Fresh-Cut Yardlong bean (Vigna unguiculata). *Int. Food Res. J.* **2016**, *23* (6), 2588–2593.
46. Oms-Oliu, G.; Aguiló-Aguayo, I.; Martín-Belloso, O.; Soliva-Fortuny, R. Effects of Pulsed Light Treatments on Quality and Antioxidant Properties of Fresh-Cut Mushrooms (Agaricus bisporus). *Postharv. Biol. Technol.* **2010**, *56* (3), 216–222.
47. Oms-Oliu, G.; Martín-Belloso, O.; Soliva-Fortuny, R. Pulsed Light Treatments for Food Preservation: A Review. *Food Bioprocess Technol.* **2010**, *3* (1), 13–23.
48. *Outbreak of Cyclosporiasis*—Northern Virginia; Washington, DC; Baltimore, MD: Metropolitan Area, 1997. (n.d.). https://www.cdc.gov/mmwr/preview/mmwrhtml/00048551.htm (accessed on Mar 14, 2021).
49. Pataro, G.; Muñoz, A.; Palgan, I.; Noci, F.; Ferrari, G.; Lyng, J. G. Bacterial Inactivation in Fruit Juices Using a Continuous Flow Pulsed Light (PL) System. *Food Res. Int.* **2011**, *44* (6), 1642–1648.
50. Pintó, E.; Lentheric, I.; Vendrell, M.; Larrigaudière, C. Role of Fermentative and Antioxidant Metabolisms in the Induction of Core Browning in Controlled-Atmosphere Stored Pears. *J. Sci. Food Agric.* **2001**, *81* (3), 364–370.
51. Pristijono, P.; Golding, J. B.; Bowyer, M. C. Postharvest UV-C Treatment, Followed by Storage in a Continuous Low-Level Ethylene Atmosphere, Maintains The Quality of Mango Fruit Stored at 20°C. *Horticulturae* **2019**, *5* (1), 1–12.
52. Proctor, A. *Alternatives to Conventional Food...—Google Scholar* (n.d.), 2011. https://scholar.google.com/scholar?hl=en&as_sdt=0%2C5&q=Proctor%2C+A.+2011.+Alternatives+to+Conventional+Food+Processing.+In+Proctor&btnG= (accessed on Mar 14, 2021).
53. Ramos-Villarroel, A. Y.; Martín-Belloso, O.; Soliva-Fortuny, R. Bacterial Inactivation and Quality Changes in Fresh-Cut Avocado Treated with Intense Light Pulses. *Eur. Food Res. Technol.* **2011**, *233* (3), 395–402.
54. Richman, P. E. S. *United States Patent (19)*, 1984.
55. Rowan, N. J.; Gregor, S. J. M. A. C.; Anderson, J. G.; Fouracre, R. A.; Ilvaney, L. M. Pulsed-Light Inactivation of Food-Related Microorganisms. *Appl. Environ. Microbiol. J.* **1999**, *65* (3), 1312–1315.
56. Rowan, Neil J. Pulsed Light as an Emerging Technology to Cause Disruption for Food and Adjacent Industries—Quo vadis? *Trends Food Sci. Technol.* **2019**, *88* (Nov), 316–332.
57. Slieman, T. A.; Nicholson, W. L. Artificial and Solar UV Radiation Induces Strand Breaks and Cyclobutane Pyrimidine Dimers in Bacillus Subtilis Spore DNA. *Appl. Environ. Microbiol.* **2000**, *66* (1), 199–205.
58. Valdivia-Nájar, C. G.; Martín-Belloso, O.; Soliva-Fortuny, R. Impact of Pulsed Light Treatments and Storage Time on the Texture Quality of Fresh-Cut Tomatoes. *Innov. Food Sci. Emerg. Technol.* **2018**, *45*, 29–35.
59. Vicente, A. R.; Pineda, C.; Lemoine, L.; Civello, P. M.; Martinez, G. A.; Chaves, A. R. UV-C Treatments Reduce Decay, Retain Quality and Alleviate Chilling Injury in Pepper. *Postharv. Biol. Technol.* **2005**, *35* (1), 69–78.
60. Viskelis, P.; Luksiene, Z.; Buchovec, I. Impact of High-Power Pulsed Light on Microbial Contamination, Health Promoting Components and Shelf-Life of Strawberries. *Food Biotechnol.* **2013**, *51* (2), 284–292.

61. Wekhof, A.; Trompeter, F.-J.; Franken, O. Pulsed UV Disintegration (PUVD): A New Sterilisation Mechanism for Packaging and Broad Medical-Hospital Applications. *First Int. Conf. Ultraviolet Technol.* **2001**, *Table* 1, 1–15.
62. Xu, W.; Chen, H.; Huang, Y.; Wu, C. Decontamination of Escherichia coli O157: H7 on Green Onions Using Pulsed Light (PL) and PL-Surfactant-Sanitizer Combinations. *Int. J. Food Microbiol.* **2013**, *166* (1), 102–108.

CHAPTER 5

SANITIZERS FOR DECONTAMINATION OF FRESH PRODUCE: ALTERNATIVE TO CHLORINE

ZAHRA H. MOHAMMAD, SALAM A. IBRAHIM, and
FAIZAN AHMAD

ABSTRACT

Postharvest washing and disinfection of fruits and vegetables is the most critical step in minimizing the risk of microbial contamination and enhancing the quality of produce. A common commercial disinfectant is chlorine. However, chlorine leaves chemical residues on the surface of produce. Consequently, chemical and organic acids (OAs) are being used for alternative postharvest disinfection. Most of these methods are recognized as safe (GRAS) by the FDA and result in no harmful byproducts. Additionally, some of these disinfection methods have been more effective than the chlorine while enhancing the quality and extending the shelf life of produce.

5.1 INTRODUCTION

With increasing consumer awareness of healthy foods and their benefits, the demand for fresh fruits and vegetables (FAV) has increased dramatically.[12,39] However, FAV are highly perishable and deteriorate rapidly due to spoilage and pathogenic microorganisms. FAV provide ideal conditions for the growth of natural micro flora, yeast, and molds, including moisture, pH, and nutrient that play a vital role in accelerating microbial growth. In addition, inadequate storage and handling also encourage microbial growth.[37]

Quality Control in Fruit and Vegetable Processing: Methods and Strategies. Megh R. Goyal, & Faizan Ahmad (Eds.)

FAV typically become contaminated with pathogenic microorganisms at various points during harvesting, postharvesting, storage, and handling[18] and can thus become vehicles for transmitting pathogenic microorganisms to cause foodborne illnesses. For example, FAV have been associated with many foodborne outbreaks[12], such as, those related to pathogenic *E. coli*, *Salmonella*, *L. monocytogens*, and norovirus. The products mostly linked to those outbreaks were from leafy greens that had been consumed raw or are minimally processed.[12]

Postharvest washing and sanitizing have been useful in reducing the spoilage and pathogens from FAV, extending the shelf life and retaining the quality. The most common commercial sanitizing agent for FAV is chlorine applied in concentrations ranging from 50 to 200 ppm for ~5 min. Chlorine is a popular sanitizer because it is easy to use, inexpensive, available in three forms (powder, liquid, gas), and effective against certain microorganisms.[6,10,13,39] However, its effectiveness can be greatly diminished by the presence of organic materials, concentration, the application time limits, and harmful chemical byproducts, such as (HCL, H).[6,39] Therefore, the application of alternative chemical or nonchemical sanitizers is vital for the food industry to ensure safety, extend the shelf life, enhance affordability, and avoid potentially harmful chemical byproducts.

This chapter explores the effectiveness of chemical disinfectants and organic acids (OAs) that are currently used or that could be used as an alternative treatment to chlorine. In addition, authors have explored the antimicrobial mechanisms of these alternative applications to ensure the safety and quality of fresh FAV and at the same time remove harmful pesticide residues.

5.2 CHEMICAL DISINFECTANTS FOR FRUITS AND VEGETABLES: AN OVERVIEW

Due to the increasing consumer attention toward the nutritional and health benefits of fresh produce, the demand for this food commodity has increased dramatically in recent years. However, cases of foodborne illnesses continue to be connected with consumption of fresh produce, particularly leafy greens.[5] The common commercial technique for postharvest washing of produce employs chlorine solutions.[5,6] However, in commercial practice, chlorine has been associated with a number of byproducts and chemical residues, some of which are related to the possible formation of carcinogenic chlorinated compounds in water.[5]

Consequently, the produce industry is continually seeking effective alternative disinfection methods that are safer and more environment friendly. Such alternative disinfection methods for FAV may reduce the risk of foodborne illnesses from food and food contact at commercial levels with no concern regarding harmful chemical byproducts.

In addition, the new techniques should retain the product quality and prolong the shelf life while inhibiting spoilage and pathogenic microbial growth in all production and distribution chain steps. Chemical disinfection with calcium-based compounds, chlorine dioxide, hydrogen peroxide, and peroxidic acid are examples of chlorine alternative disinfectants. OAs, such as, lactic acid, acetic acid, and citric acid are being used as an alternative to chlorine. Some of these alternative antimicrobial agents have shown to be more effective than chlorine, and their applications are increasing as postharvest and handling sanitizers.[8]

For example, calcium lactate has been proven to maintain the quality of products and has also been tested as a disinfection agent and found to be effective in reducing natural microbial levels without leaving chemical residues.[6,30,40] Previously, calcium lactate (0.5–2%) has been used as a firming agent for the produce. It has also shown that calcium lactate reduced coliform, aerobic bacteria, yeast, and mold populations in fresh cut jackfruit.[6] Calcium lactate could therefore be an excellent alternative to chlorine because it does not leave chemical byproducts.

Another chemical that could replace chlorine for washing FAV is chlorine dioxide (ClO_2), a stable dissolved gas used as an alternative to chlorine for water treatment.[4] Chlorine dioxide has high water solubility and higher oxidant capacity, and lower reactivity with organic matter, making it a good option for decontamination of produce.[4] However, it is unstable, and explosive, with a concentration of over 10%. Therefore, it should be used with extreme caution.

Other alternative chemical agents that have been shown to effectively reduce different microorganisms on various produce include: hydrogen peroxide and peroxidic acid. Additionally, OA (such as lactic acid, acetic acid, and citric acid) are widely used as antimicrobials against numerous microorganisms. However, the antimicrobial efficacy of both chemical disinfectants and OA depends on many parameters, including susceptibility to specific microorganisms, fruit or vegetable topography, their chemical composition, and concentration and contact time. More detailed information about each chemical or OA treatment is discussed separately in this section. Figure 5.1 shows the antimicrobial activities of chemical and OA on microbial cells.

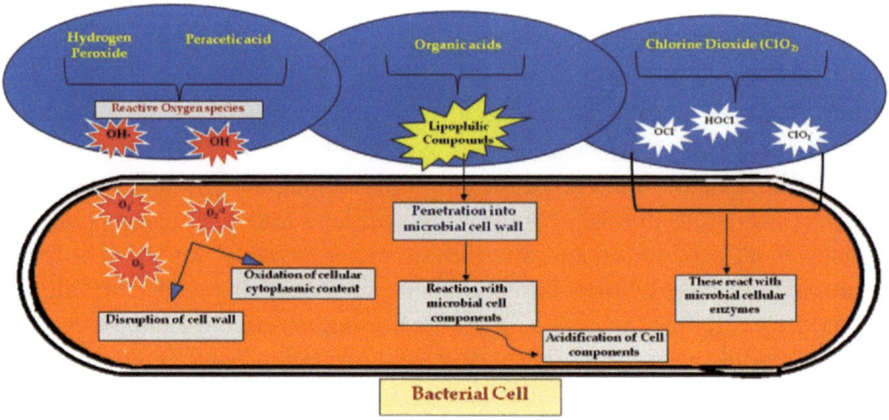

FIGURE 5.1 Mode of antimicrobial action of chemical and OA against bacterial cell.

5.3 ALTERNATIVE CHEMICAL DISINFECTANTS FOR FRESH FRUITS AND VEGETABLES

Chlorine is a common disinfectant for FAV and is used in different forms.[6,10,13,39] The mechanism of its germicidal or antimicrobial effects is by producing active hypochlorous (HOCl) acid, and increasing the level of solution pH. However, (HOCl) is the most active antimicrobial compound, and it is formed after chlorine (Cl) reacts with water. HOCl is a weak acid that can be further disassociated to hypochlorite ions.

Therefore, the primary mechanism of action begins, when HOCl penetrates the microbial cell wall and membrane and oxidizes various cellular compounds of microbial cell. This process causes the internal cellular pH to decrease, resulting in the inactivation of the cell function. Finally, the process leads the cell to pump H^+ ions out of the cell. After disassociation of HOCl, the resulting product OCl with a negative charge oxidizes the cell wall components. Overall, chlorine can create significant damage to the cell and inactivates the cell functions.[12] The following reaction describes the mechanism of chlorine:

$$Cl_2 + H_2O \rightarrow HOCl + H^+ + Cl$$

$$HOCl \rightarrow H^+ + OCl^- \quad (5.1)$$

However, the effectiveness of chlorine highly depends on pH. For example, at high pH, HOCl dissociates into OCl^- ions, but OCl^- is ineffective against pathogens. Additionally, byproducts and chemical residues are usually formed in the presence of natural organic substances.[6,39] Such chemical byproducts are formed during chlorination due to the reactions that occur with organic substances.

Therefore, the formation, amount, and type of chemical byproducts are highly dependent on the concentration of chlorine, reaction time, temperature, and pH.[6] Some of these byproducts and chemical residues are harmful to human health and may form carcinogenic chlorinated compounds in water.[5] Therefore, some chemical and OA are discussed, including calcium lactate, CIO_2, hydrogen peroxide, peroxidic acid, lactic acid, acetic acid, and citric acid.

5.3.1 CHLORINE DIOXIDE

Chlorine dioxide (ClO_2) is another alternative sanitizing agent for FAV. ClO_2 is a stable dissolved gas that has high water solubility and oxidant capacity for water treatment.[4] It can be a good sanitizing agent for fruits and vegetables decontamination because of its low sensitivity to organic matter.[4] However, careful attention should be given when using ClO_2 as a disinfectant as it is unstable and explosive with a concentration of >10%. Therefore, the concentration of using ClO_2 must not exceed 3 ppm; and after treatment with ClO_2, the produce should be rinsed with water.[35]

ClO_2 is a powerful oxidizing agent and is 2.5-folds more efficient than the chlorine as an antimicrobial.[33] The mechanism of antimicrobial activity of ClO_2 is based on its ability to destabilize cell membranes by reacting with amino acids, which cause the interruption of protein synthesis, oxidation of DNA, and damage to critical elements.[43] However, like any other chemical disinfectants, the microbial decontamination efficiency of ClO_2 depends on antimicrobial susceptibility of particular species, the composition of produce, concentration, holding time, and facility storage conditions.[12]

The use of ClO_2 (e.g., chloramines) can reduce the formation of chlorinated byproducts during primary disinfection. However, the production of chlorine dioxide, its decomposition, and reaction with nonorganic matters (NOM) can lead to the formation of byproducts. One of the byproducts is chlorite (ClO_2^-), a compound that is of health concern. Currently, the maximum allowable concentration of residual ClO_2 as established by the FDA is 3 ppm for postharvest washing of produce. This level of ClO_2 concentration ensures no production of harmful byproducts. However, a concentration of 5 ppm can be employed commercially since the concentration of ClO_2

decreases when it comes in contact with organic material.[12] On the other hand, Environmental Protection Agency (EPA) has set a different standard for ClO_2 effluent returning to the watershed. Currently, the EPA standard for ClO_2 maximum residual disinfectant level (MRDL) is 0.8 ppm.

Overall, ClO_2 confers more advantages for the disinfection of produce than does chlorine alone. For example, ClO_2 is not as affected by organic matter as chlorine; it is more selective at inactivation, does not generate chlorinated byproducts (such as trihalomethanes), is less corrosive compared to chlorine, and is not impacted by pH. However, there are still some limitations that make ClO_2 not the ideal alternative options to chlorine including an increase in salinity in soil due to the generation of byproduct that is sodium based and the limited concentration of effluent residual 0.8 ppm to control chlorate and chlorite disinfection byproducts in the field. In addition, the application of ClO_2 requires storage of two separate compounds that must be mixed prior to application (it has to be generated on site), and finally the cost.[32]

The production of ClO_2 gas is a result of the reaction of sodium chlorite with an acid. Depending on the manufacturing process, ClO_2 may have residual sodium, which can negatively impact fruits or vegetables by changing the sensorial and visual qualities. The formula below shows the production of ClO_2 without residual sodium:

$$\underset{\text{Sodium Chlorite acid}}{NaClO_2} + H^+ \leftrightarrow \underset{\text{Chlorous acid}}{HClO_2} \leftrightarrow \underset{\text{Chlorine Dioxide}}{ClO_2} \quad (5.2)$$

5.3.2 CALCIUM-BASED COMPOUNDS

Solutions containing calcium-based compounds are mainly used to control firmness and to extend the shelf life of fresh produce, especially delicate fruits.[6,39] Calcium lactate is one of the most important calcium-based compounds that have antimicrobial properties.[30] The antimicrobial properties of this compound are because of its capability to interrupt microbial transport phenomena.[40]

Calcium lactate Ca $[(CH)CHOHCOO]_2$ has two lactate anions and two calcium cations (Ca^+). It has been recognized as safe (GRAS) by the FDA and has been used as a firming agent.[30] Calcium lactate has shown to maintain the quality of products and has also been tested as a disinfection agent and found to be effective in reducing natural microbial levels without leaving chemical residues.[6,30,40]

Calcium lactate with the concentration of (0.5–2%) has been used as a firming agent for produce (such as strawberries, cantaloupes, and other fruits).[29] It has been observed that calcium lactate reduced the aerobic bacteria, coliform, yeast, and mold populations by approximately 85% to 99.6% from fresh cut jackfruit for up to 5 days of storage.[6] It has also been shown that calcium lactate possesses the same effectiveness as chlorine in reducing microbial counts on fresh cut lettuce and carrots.[39]

It has been suggested that the mechanism of the antimicrobial action of calcium lactate is based on the concentration of calcium oxide (alkaline pH from 12 to 13) in the solution.[6] It was also suggested that the production of oxygen species such as superoxide anions could be responsible for the antimicrobial activity.[6] Thus, calcium lactate or other calcium-based compounds could be a promising alternative to chlorine.

5.3.3 HYDROGEN PEROXIDE

Hydrogen peroxide (H_2O_2, also called hydrogen dioxide) is a colorless gas at room temperature and is generally recognized as safe (GRAS) for the application in foods.[13] H_2O_2 has bacteriostatic and bactericidal activity due to its high oxidation potential.[23] In addition, it has a broad spectrum of antimicrobial activity against many microorganisms including protozoal cysts and bacterial endospores, which give H_2O_2 an advantage over other disinfectant agents.[27,31] Therefore, H_2O_2 is widely used for FAV as disinfection and sanitizing agent to reduce microbial loads.[1,10] Another advantage of H_2O_2 is that in the presence of water, it decomposes to water and oxygen, making H_2O_2 eco-friendly with no toxic effects or harmful compounds.[10]

$$\underset{\text{Hydrogen Peroxide}}{2H_2O_2} \rightarrow \underset{\text{Water}}{2H_2O} + \underset{\text{Oxygen}}{O_2} \quad (5.3)$$

Moreover, H_2O_2 does not react with organic matter, does not release harmful byproducts, and any chemical residue that remains after treatment is easily removed by rinsing with water.[1,13,23] Additionally, H_2O_2 can be used as a liquid or as a gas to disinfect or sanitize FAV.[31] However, the effectiveness of H_2O_2 is highly influenced by its level of concentration. A concentration ranging from 1 to 5% has been used to disinfect and sanitize food and packaging materials.[10] For example, a concentration of 5% has shown to be effective and has resulted in 3 or more log reduction of microorganisms when used for washing apples at a temperature range from 50 to 60°C or at a temperature range of 70–80°C for cantaloupe.[23]

H_2O_2 also effectively reduced pathogenic microorganisms on a large variety of FAV including tomatoes, oranges, whole grapes, mushrooms, prunes, melons, bell peppers, and zucchini.[40] However, a concentration of 1% has been shown to be ineffective against some pathogenic microorganisms, depending on exposure time and temperature, and type of FAV.[23]

The mechanism of H_2O_2 antimicrobial activity against pathogens and spoilage microorganisms is based on its strong oxidizing ability and capacity to produce other cytotoxic oxidative species such as singlet oxygen species, hydroxyl radicals, and to damage membrane lipids and DNA.[13,27] As a result, H_2O_2 could be an affordable and environment friendly alternative disinfection agent for postharvest washing of produce due to its high antimicrobial capacity against a wide spectrum of spoilage and pathogenic microorganisms.[1,23]

The only limitation is that the higher concentration of H_2O_2 causes browning in some FAV, especially when the dipping method is used to wash the products.[23] Overall, H_2O_2 is an effective alternative to chlorine when combined with other treatment.

5.3.4 PERACETIC ACID (PAA)

Peracetic acid (PAA, also known as peroxyacetic acid) is produced by mixing hydrogen peroxide and acetic acid. PAA is widely used in postharvest washing. PAA was first used in 1950 as a disinfection agent for washing fruit and vegetable surfaces in the field and postharvest activity.[21] The FDA has approved the application of PAA as a sanitizer for food contact surfaces and washing of FAV, poultry, seafood, and meat.[14–16]

PAA has also been used as a disinfectant on food equipments, such as, dairy and cheese processing facilities.[23] Therefore, PAA is a promising alternative to chlorine for the food industry because it has strong oxidizing abilities and does not generate chemical residues or harmful by-products.[48] PAA could thus be an ideal alternative to chlorine.[22]

The mechanism of action of PAA is identical to that of chlorine. PAA and acetic acid are generated instead of hypochlorous acid, with the added benefit of hydrogen peroxide as an additional oxidizing compound, which decreases the pH of water as more PAA is added. Moreover, a concentration ranging from 0.03 to 2% solution is normally acceptable for PAA to produce a strong oxidation activity when used as an antimicrobial agent and disinfectant for FAV. PAA is stronger than hydrogen peroxide because PAA is not affected by catalase and peroxidase, the enzymes that break down hydrogen peroxide.[17] Moreover, the antimicrobial activity of PAA is not affected by the presence of organic matter or food residue when the solution retains the same acidic pH level.[23]

The mechanism of disinfectant activity of PAA is based on oxidation by the generation of reactive oxygen species.[28] The oxidation targets the bacterial cell membranes consisting of fatty, lipid-type chemistry for peroxides. Once inside the cell, PAA causes destruction of cell functionality by oxidizing proteins, DNA, enzymes, and metabolites, finally causing rapid death of the bacterium.[48]

The resultant products of the PAA antimicrobial action are acetic acid and peroxide, which are typically not harmful to humans. However, the antimicrobial efficiency of PAA depends on exposure time, concentration, and type of microorganisms.[48] The generation of PAA is shown below:

$$\underset{\text{Hydrogen Peroxide}}{H_2O_2} + \underset{\text{Acetic Acid.}}{CH_3CO_2H} \leftrightarrow \underset{\text{Peracetic Acid}}{CH_3CO_3H} + \underset{\text{Water}}{H_2O} \quad (5.4)$$

The primary advantage of PAA as an alternative disinfectant to chlorine is that PAA is more resistant to organic matter than chlorine, yet still loses its effectiveness in the presence of high levels of organic matter. Other advantage of PAA over chlorine is that PAA effectiveness does not depend on pH; thus, the application does not need pH control, and usually retains its activity at temperatures from 0 to 40°C and pH from 1 to 8. For these reasons, PAA has been widely used in food industry from the past several years.

5.3.5 ORGANIC ACIDS AND MODE OF ACTION AS DISINFECTANT: AN OVERVIEW

OAs (such as acetic acid, lactic acid, citric acid, and others) are natural products and do not produce toxic or harmful compounds.[5,9] As a result, OA are approved as generally recognized as safe (GRAS) by FDA.[26] The antimicrobial activity of each OA is different and is highly dependent on the acidity of each one.[6,9] The antimicrobial activities of OA are primarily biostatic but not biocidal, and relatively high concentrations are needed in order for them to be effective.[6]

OA are short chain OA that include one or more carboxyl groups. They are called OA because they include organic components that have acidic characteristics including aliphatic and aromatic compounds.[9] The acidity of OA is related to their carboxylic group –COOH.[26] Therefore, some OA that contain strong acids are not applicable for food matrices.[13]

OA possess strong antimicrobial activities against a broad spectrum of microorganisms.[21,41] Therefore, OA (such as lactic, acetic, and citric acids) can be used as antimicrobial agent to prevent food spoilage. They have been used as a food additive and preservative by being added to food products to prolong the shelf life. OA can also be used as an antioxidant to prevent visual changes (such as color and texture), or chemical changes (such as, flavor).[13] In recent years, the interest in using OA as antimicrobial agents for FAV has increased.

The mechanism of antimicrobial activity of OA is based on reducing the pH levels both outside and inside bacterial cells, which lead to acidification of the cells and disturbs membrane transport.[38] Briefly, the undissociated OA molecules can easily penetrate the microbial cell membrane and reduce the pH of the cell's interior, resulting in an acidic pH instead of a neutral pH. This process results in the inactivation of intracellular enzymes and dysfunction of cells' metabolic reactions, and ultimately, cell death.[47,21]

However, the use of OA as antimicrobial agents has some limitations, including low antimicrobial activity because the application of OA for FAV requires low concentrations and short exposure time. Otherwise, OA can adversely affect the color, flavor, and odor of food products at high concentrations or with long exposure time.[25,41] Thus, OA could be used as an alternative to chlorine if used in combination with other disinfection methods.[12]

5.4 SELECTED ORGANIC ACIDS AS AN ALTERNATIVE TO CHLORINE TO DISINFECT FRUITS AND VEGETABLES

5.4.1 ACETIC ACID

Acetic acid (CH3COOH) is an OA that has been widely used as a food preservative for decades.[25] Researchers have investigated the possible use of acetic acid as an antimicrobial agent against many pathogenic microorganisms, including *Salmonella* and *E. coli*. In addition, acetic acid has been used in the poultry industry as an antimicrobial against spoilage and pathogenic bacteria in order to improve the quality of poultry meat.[25]

Many research studies have shown the potential antimicrobial activity of acetic acid against pathogens in FAV through empirical experiments. Kang et al.[24] observed that the addition 3g/L of acetic acid into culture media resulted in the complete inhibition of the growth of *Colletotrichum gloeosporioides*. These authors also noticed that this fungus was able to grow again, when acetic acid was removed from the media.[24]

In the past two decades, the application of acidic acid and its effectiveness for controlling spoilage in fruits and vegetables has been investigated, and most of the reported results have been encouraging.[13] For example, acetic acid was effective against spoilage microorganisms on tomato when treated via dipping in a 4% acetic acid solution or exposed to vapor acetic acid at 40 μL/L.[2] In another study, acetic acid with a concentration of 4–6% was effective against different kinds of fungus, including *B. cinerea*, *M. fructicola*, and *P. expansum* that cause spoilage on some fruits, such as strawberries and apples.[42] However, acidic acid at high concentration resulted in adverse effects on the quality and sensory characteristics of fruits; and at low concentration may be ineffective at completely controlling decay microorganisms.[13]

The acetic acid at 10% concentration for washing cabbage for 20 min reduced about 65 to 79% of different pesticides, such as, chlorpyrifos, chlorothalonil, p,p-DDT, and cypermethrin.[46]

Radwan et al.[38] also observed that washing sweet pepper, hot pepper, and eggplant with 2% acetic acid for 1 min reduced 60–100% profenofos residues from these vegetables. These are some examples of useful studies that have demonstrated the effectiveness of acetic acid and its potential use as an alternative disinfectant to chlorine. However, product quality must still be considered as high concentrations always result in adverse effects. Thus, combining acetic acid with other disinfection methods could provide more effective and promising alternative disinfection for FAV.

5.4.2 LACTIC ACID

Lactic acid is a weak OA that has been commonly used in the food industry for food preservation and for sanitizing meat.[25] The efficacy of the antimicrobial activity of lactic acid is highly dependent on the pH of the washing solution. Studies have shown the potential of lactic acid as an antimicrobial agent to control spoilage and pathogenic microorganisms on FAV.

Neal et al.[34] investigated the efficacy of different disinfection methods in reducing *E. coli* O157:H7 and *Salmonella typhimurium* on fresh spinach leaves. These authors treated spinach leaves with calcium hypochlorite (200 mg/L), PAA (80 mg/L), ozonated water (1 mg/L), ClO_2 gas (2.1 mg/L), and 2% lactic acid and they found that lactic acid (2%) spraying for 15s at 55°C was the most effective treatment for reducing both pathogens compared to other approaches.

Another study reported a reduction range from 1.5 to 2.3 log CFU/g for *E. coli* O157: H7, *Salmonella Typhimurium*, and *L. monocytogenes* on sprouts dipped in lactic acid solution at a concentration of 0.5 to 2.0%.[45] Similarly, other researchers showed that lactic acid with a concentration of only 0.2% for 1 min reduced the population of *Y. enterocolitica* and *E. coli* O157: H7 on tomatoes by 5.2 and 2.5 log CFU/g, respectively.[11]

Therefore, it is evident that lactic acid could be an alternative disinfectant to chlorine as the results from the tomato study were comparable to those in which a chlorine concentration of 200 ppm reduced *E. coli* O157:H7 and *Y. enterocolitica* on tomatoes by 2.6 and 5.13 log CFU/g.[6] Moreover, another study demonstrated that lactic acid treatment for 15 s reduced pathogenic bacteria up to 3 log CFU/ cm^2 on cantaloupes and 3.6 log CFU/cm^2 on bell peppers.[3]

However, lactic acid was only able to reduce aerobic bacteria by 0.77 log CFU/g on broccoli sprouts treated with 2% for 2 min.[7] Thus, the concentration of 2% was not effective to reduce microorganisms on the sprouts. In addition, increasing the exposure time to 20 min did not solve the issue as such a high concentration significantly affected the quality of treated products, including color and visual appearance.[41] To this end, lactic acid could be an alternative replacement for chlorine, but the resultant product quality should be considered. The efficacy of lactic acid as a disinfectant for FAV could be enhanced, when it is used in combination with other treatments at low concentrations.

For example, one study showed that combining ascorbic acid at different concentrations and 0.2% lactic acid prohibited the growth of *E. coli* O157:H7 in growth media and carrot juice.[44] Similarly, the mixture of copper and lactic acid resulted in a significant reduction of *E. coli* O157:H7 up to 3.93 and 3.39 log on the surface of lettuce and tomato, respectively.[20,44]

5.4.3 CITRIC ACID

Citric acid is widely used as a preservative and flavoring agent in the food industry.[32] It is also used as an antimicrobial agent against spoilage and pathogenic microorganisms. The mechanism of the antimicrobial activity of citric acid is based on the reduction of pH outside and inside microbial cells. For example, the pH of internal microbial cells is typically maintained close to neutral.

Citric acid or other OA change the external pH; consequently, the internal pH is also dropped and lead to the alteration of microbial cell membrane structure, such as proteins and phospholipids. Ultimately, the permeability of the membrane and cell metabolites are affected, leading to microbial death.[12]

Park et al.[36] investigated the effects of different OA against pathogenic *E. coli* O157:H7, *Salmonella Typhimurium*, and *L. monocytogenes* on apples and lettuce, and on the quality of the produce. Among the different OA used, lactic acid and citric acid at a concentration of 2% for 10 min were most effective against pathogens, *E. coli* O157:H7, *Salmonella Typhimurium*, and *L. monocytogenes* that were reduced by 2.56 and 3.42 log CFU/fruit.[36]

The results from another study indicated that pathogenic *Staphylococcus aureus* was reduced by 2.7 to 3.9 log CFU/g on tabbouleh salad treated with citric acid and acetic acid at a concentration of 03–1.4% after 7 days of storage at 21°C.[12] Hence, citric acid can be used to reduce microorganism load on FAV,[41] and could be a potential alternative to chlorine as a disinfection agent. However, there are still some limitations that need to be considered when using citric acid as an antimicrobial agent at higher concentrations, as its application can negatively impact the color and sensory characteristics, and overall product appearance.

5.5 SUMMARY

Disinfection of FAV after harvest and during handling and storage is essential for controlling spoilage and pathogenic microorganisms and extending shelf life of the product. Chlorine is a typical disinfectant that is used for sanitizing FAV. However, there is continuous health and environmental concern on using chlorine to disinfect FAV due to its capability to produce unhealthy byproducts and possible carcinogenic compounds. A variety of alternative disinfection methods have been investigated or have already been applied. Some of these approaches and their applicability as alternatives to chlorine have been described in this chapter. In addition, authors addressed the advantages and limitations of chemical and OA techniques as potential alternatives to chlorine.

The chemical disinfection agents included: chlorine dioxide (ClO_2), calcium-based compounds, hydrogen peroxide, peroxidic acid, and OA (such as: lactic acid, acetic acid, and citric acid). Most of these agents are recognized as GRAS as they generally have a lower tendency to react with organic matter, do not produce harmful byproducts, and are cost-effective. However, such products are less effective at low concentrations; therefore, higher concentrations are typically needed to effectively control the growth of microorganisms on FAV.

Moreover, the application of these chemical agents at higher concentrations can negatively affect color, sensory, and the general appearance of FAV.

Although some of these disinfection methods have proven to be as effective and could thus be an excellent replacement, yet the efficacy of these methods is highly affected by concentration, exposure time, temperature, pH, the structure of the fruits or vegetables, and the presence of organic matter. Thus, the selection of a suitable disinfection method ultimately depends on the type of FAV, the targeted microorganisms, and facility condition conditions.

KEYWORDS

- **antimicrobial**
- **chemical disinfectant**
- **chlorine**
- **disinfestations**
- **fruit and vegetables**
- **novel disinfection**
- **organic acids**

REFERENCES

1. Abadias, M.; Alegre, I.; Usall, J.; Torres, R.; Viñas, I. Evaluation of Alternative Sanitizers to Chlorine Disinfection for Reducing Foodborne Pathogens in Fresh-Cut Apple. *Postharv. Bio. Technol.* **2011,** *59* (3), 289–297.
2. Alawlaqi, M. M.; Alharbi, A. A. Impact of Acetic Acid on Controlling Tomato Fruit Decay. *Life Sci. J.* **2014,** *11*, 114–119.
3. Alvarado-Casillas, S.; Ibarra-Sánchez, S.; Rodríguez-García, O.; Martínez-Gonzáles, N.; Castillo, A. Comparison of Rinsing and Sanitizing Procedures for Reducing Bacterial Pathogens on Fresh Cantaloupe and Bell Peppers. *J. Food Prot.* **2007,** *70* (3), 655–660.
4. Artés-Hernández, F.; Gómez, P.; Aguayo, E. Technological Innovations to Preserve Quality and Safety of Fresh-Cut Horticultural Products. *Environ. Friend. Safe Technol. Qual. Fruit Veg.* **2010,** ***2010***, 190–198.
5. Artés-Hernández, F.; Martínez-Hernández, G. B.; Aguayo, E.; Gómez, P. A.; Artés, F. Fresh-Cut Fruit and Vegetables: Emerging Eco-Friendly Techniques for Sanitation and Preserving Safety. *Postharv. Handl.* **2017,** ***2017***, 7–45.
6. Bhilwadikar, T.; Pounraj, S.; Manivannan, S.; Rastogi, N. K.; Negi, P. S. Decontamination of Microorganisms and Pesticides from Fresh Fruits and Vegetables: A Comprehensive Review from Common Household Processes to Modern Techniques. *Comprehensive Rev. Food Sci. Food Safet.* **2019,** *18* (4), 1003–1038.

7. Chen, L.; Zhang, H.; Liu, Q.; Pang, X.; Zhao, X.; Yang, H. Sanitizing Efficacy of Lactic Acid Combined with Low-Concentration Sodium Hypochlorite on *Listeriainnocua* in Organic Broccoli Sprouts. *Int. J. Food Microbiol.* **2019,** *295*, 41–48.
8. Chhetri, R. K; Thornberg, D.; Berner, J.; Gramstad, R.; Ojstedt, U.; Sharma, A. K.; Anderson, H. R. Chemical Disinfection of Combined Sewer Overflow Waters Using Performic Acid or Peracetic Acids. *Sci. Total Environ.* **2014,** *490*, 1065–1072.
9. Coban, H. B. Organic Acids as Antimicrobial Food Agents: Applications and Microbial Productions. *Bioprocess Biosyst. Eng.* **2020,** *43* (4), 569-591.
10. De Siqueira Oliveira, L.; Eça, K. S.; de Aquino, A. C.; Vasconcelos, L. B. Hydrogen Peroxide (H_2O_2) for Postharvest Fruit and Vegetable Disinfection. In *Postharvest Disinfection of Fruits and Vegetables*; Siddiqui, M. W., Ed.; Academic Press: London, 2018; pp 91–99.
11. del Carmen Velázquez, L.; Barbini, N. B.; Escudero, M. E.; Estrada, C. L.; de Guzmán, A. M. S. Evaluation of Chlorine, Benzalkonium Chloride and Lactic Acid as Sanitizers for Reducing *Escherichiacoli* O157: H7 and *Yersiniaenterocolitica* on Fresh Vegetables. *Food Control* **2009,** *20* (3), 262–268.
12. Deng, L. Z.; Mujumdar, A. S.; Pan, Z.; Vidyarthi, S. K.; Xu, J.; Zielinska, M.; Xiao, H. W. Emerging Chemical and Physical Disinfection Technologies of Fruits and Vegetables: A Comprehensive Review. *Crit. Rev. Food Sci. Nutr.* **2020,** *60* (15), 2481–2508.
13. Feliziani, E.; Lichter, A.; Smilanick, J. L.; Ippolito, A. Disinfecting Agents for Controlling Fruit and Vegetable Diseases After Harvest. *Postharv. Biol. Technol.* **2016,** *122*, 53–69.
14. Food and Drug Administration (FDA). Peroxyacids. In *Food and Drug Administration Code of Federal Regulations*: 21 CFR Part 173; 2020; Section A. 173.370. https://www.accessdata.fda.gov/scripts/cdrh/cfdocs/cfcfr/CFRSearch.cfm?fr=173.370 (accessed Feb 17, 2021).
15. Food and Drug Administration (FDA). Sanitizing solutions. In *Food and Drug Administration Code of Federal Regulations*: 21 CFR Part 178.1010; 2020; Section B. https://www.accessdata.fda.gov/scripts/cdrh/cfdocs/cfcfr/CFRSearch.cfm?fr=178.1010 (accessed Feb 15, 2021).
16. Food and Drug Administration (FDA). Chemicals Used in Washing or to Assist in the Peeling of Fruits and Vegetables. In *Food and Drug Administration Code of Federal Regulations*: 21 CFR Part 173.315; 2020; Section C. https://www.accessdata.fda.gov/scripts/cdrh/cfdocs/cfcfr/CFRSearch.cfm?fr=173.315 (accessed Feb 17, 2021).
17. Gad, S. C. Peracetic Acid A2. In *Encyclopedia of Toxicology*; Wexler, P., Ed., 3rd ed.; Academic Press: Oxford, 2014; pp 788–790.
18. García, S.; Heredia, N. Microbiological Safety of Fruit and Vegetables in the Field, During Harvest, and Packaging: A Global Issue. In *Global Food Security and Wellness*; Springer: New York, 2017; pp 27–48.
19. Greenspan, F. P.; Margulies, P. H. Treatment of Raw Plant Tissue. *US Patent 2512-640,* Buffalo Electro-Chemical Co., 1950; p 19.
20. Gyawali, R.; Ibrahim, S. A.; Abu Hasfa, S. H.; Smqadri, S. Q.; Haik, Y. Antimicrobial Activity of Copper Alone and in Combination with Lactic Acid Against *Escherichiacoli* O157: H7 in Laboratory Medium and on the Surface of Lettuce and Tomatoes. *J. Pathog.* **2011,** *2011* (October 23), E-pub ID:650968. DOI: 10.4061/2011/650968.
21. Gyawali, R.; Ibrahim, S. A. Synergistic Effect of Copper and Lactic Acid Against *Salmonella* and *Escherichia coli* O157:H7: A Review. *Emirates J. Food Agric.* **2012,** ***2012***, 1–11.

22. Ho, K. L. G.; Luzuriaga, D. A.; Rodde, K. M.; Tang, S.; Phan, C. Efficacy of a Novel Sanitizer Composed of Lactic Acid and Peroxyacetic Acid Against Single Strains of Nonpathogenic *Escherichia coli* K-12, *Listeriainnocua*, and *Lactobacillus plantarum* in Aqueous Solution and on Surfaces of Romaine Lettuce and Spinach. *J. Food Protect.* **2011,** *74*, 1468–1474.
23. Joshi, K.; Mahendran, R.; Alagundaram, K.; Norton, T.; Tiwari, B. K. Novel Disinfectants for Fresh Produce. *Trends Food Sci. Technol.* **2013,** *34* (1), 54–61.
24. Kang, H. C.; Park, Y. H.; Go, S. J. Growth Inhibition of a Phytopathogenic Fungus, Colletotrichum Species by Acetic Acid. *Microbiol. Res.* **2003,** *158* (4), 321–326.
25. Khan, I.; Tango, C. N.; Miskeen, S.; Lee, B. H.; Oh, D. H. Hurdle Technology: A Novel Approach for Enhanced Food Quality and Safety–a Review. *Food Control* **2017,** *73*, 1426–1444.
26. Lianou, A.; Koutsoumanis, K. P.; Sofos, J. N. Organic Acids and Other Chemical Treatments for Microbial Decontamination of Food. In *Microbial Decontamination in the Food Industry;* Demirci, A., Ngadi, M. O., Eds.; Woodhead Publishing: Cambridge, 2012; pp 592–664.
27. Linley, E.; Denyer, S. P.; McDonnell, G.; Simons, C.; Maillard, J. Y. Use of Hydrogen Peroxide as Bio-Side: New Consideration of Its Mechanisms of Biocidal Action. *J. Antimicrob. Chemother.* **2012,** *67*, 1589–1596.
28. Luukkonen, T.; Pehkonen, S. O. Peracids in Water Treatment: A Critical Review. *Crit. Rev. Environ. Sci.* **2017,** *47*, 1–39.
29. Main, G. L.; Morris, J. R.; Wehunt, E. J. Effect of Preprocessing Treatments on the Firmness and Quality Characteristics of Whole and Sliced Strawberries After Freezing and Thermal Processing. *J. Food Sci.* **1986,** *51* (2), 391–394.
30. Martin-Diana, A. B.; Rico, D.; Barry-Ryan, C.; Frias, J. M.; Mulcahy, J.; Henehan, G. T. M. Comparison of Calcium Lactate with Chlorine as a Washing Treatment for Fresh-Cut Lettuce and Carrots; Quality and Nutritional Parameters. *J. Sci. Food Agric.* **2005,** *85* (13), 2260–2268.
31. McDonnell, G. The Use of Hydrogen Peroxide for Disinfection and Sterilization Application. *Patai's Chem. Funct. Groups* **2009,** *2009*, 1–34.
32. Meireles, A.; Giaouris, E.; Simões, M. Alternative Disinfection Methods to Chlorine for Use in the Fresh-Cut Industry. *Food Res. Int.* **2016,** *82*, 71–85.
33. Nam, H.; Seo, H. S.; Bang, J.; Kim, H.; Beuchat, L. R.; Ryu, J. H. Efficacy of Gaseous Chlorine Dioxide in Inactivating *Bacilluscereus* Spores Attached to and in a Biofilm on Stainless Steel. *Int. J. Food Microbiol.* **2014,** *188*, 122–127.
34. Neal, J. A.; Marquez-Gonzalez, M.; Cabrera-Diaz, E.; Lucia, L. M.; O'Bryan, C. A.; Crandall, P. G.; Ricke, S. C.; Castillo, A. Comparison of Multiple Chemical Sanitizers for Reducing *Salmonella* and *Escherichiacoli* O157: H7 on Spinach (*Spinacia oleracea*) leaves. *Food Res. Int.* **2012,** *45* (2), 1123–1128.
35. Ölmez, H.; Kretzschmar, U. Potential Alternative Disinfection Methods for Organic Fresh-Cut Industry for Minimizing Water Consumption and Environmental Impact. *LWT-Food Sci. Technol.* **2009,** *42* (3), 686–693.
36. Park, S. H.; Choi, M. R.; Park, J. W.; Park, K. H.; Chung, M. S.; Ryu, S.; Kang, D. H. Use of Organic Acids to Inactivate *Escherichiacoli* O157: H7, *Salmonellatyphimurium*, and *Listeriamonocytogenes* on Organic Fresh Apples and Lettuce. *J. Food Sci.* **2011,** *76* (6), M293–M298.

37. Qadri, O. S.; Yousuf, B; Srivastava, A. K. Fresh-Cut Fruits and Vegetables: Critical Factors Influencing Microbiology and Novel Approaches to Prevent Microbial Risks—A Review. *Cogent Food Agric.* **2015,** *1* (1), E-article **1121606,** 11.
38. Radwan, M. A.; Abu-Elamayem, M. M.; Shiboob, M. H.; Abdel-Aal, A. Residual Behavior of Profenofos on Some Field-Grown Vegetables and its Removal Using Various Washing Solutions and Household Processing. *Food Chem. Toxicol.* **2005,** *43* (4), 553–557.
39. Ramos, B.; Miller, F. A.; Brandão, T. R.; Teixeira, P.; Silva, C. L. Fresh Fruits and Vegetables: An Overview on Applied Methodologies to Improve Its Quality and Safety. *Innov. Food Sci. Emerg. Technol.* **2013,** *20,* 1–15.
40. Rico, D.; Martin-Diana, A. B.; Barat, J. M.; Barry-Ryan, C. Extending and Measuring the Quality of Fresh-Cut Fruit and Vegetables: A Review. *Trends Food Sci. Technol.* **2007,** *18* (7), 373–386.
41. Sagong, H. G.; Lee, S. Y.; Chang, P. S.; Heu, S.; Ryu, S.; Choi, Y. J.; Kang, D. H. Combined Effect of Ultrasound and Organic Acids to Reduce *Escherichia coli* O157: H7, *Salmonella Typhimurium*, and *Listeria Monocytogenes* on Organic Fresh Lettuce. *Int. J. Food Microbiol.* **2011,** *145* (1), 287–292.
42. Sholberg, P.; Haag, P.; Hocking, R.; Bedford, K. The Use of Vinegar Vapor to Reduce Postharvest Decay of Harvested Fruit. *Hort. Sci.* **2000,** *35* (5), 898–903.
43. Sun, X.; Baldwin, E.; Bai, J. Applications of Gaseous Chlorine Dioxide on Postharvest Handling and Storage of Fruits and Vegetables–a Review. *Food Control* **2019,** *95,* 18–26.
44. Tajkarimi, M.; Ibrahim, S. A. Antimicrobial Activity of Ascorbic Acid Alone or in Combination with Lactic Acid on *Escherichiacoli* O157: H7 in Laboratory Medium and Carrot Juice. *Food Control* **2011,** *22* (6), 801–804.
45. Wang, C.; Wang, S.; Chang, T.; Shi, L.; Yang, H.; Shao, Y.; Feng, W.; Cui, M. Efficacy of Lactic Acid in Reducing Foodborne Pathogens in Minimally Processed Lotus Sprouts. *Food Control* **2013,** *30* (2), 721–726.
46. Zhang, Z. Y.; Liu, X. J.; Hong, X. Y. Effects of Home Preparation on Pesticide Residues in Cabbage. *Food Control* **2007,** *18* (12), 1484–1487.
47. Zhou, G. H.; Xu, X. L.; Liu, Y. Preservation Technologies for Fresh Meat—A Review. *Meat Sci.* **2010,** *86* (1), 119–28.
48. Zoellner, C.; Aguayo-Acosta, A.; Siddiqui, M. W.; Dávila, J. E. Peracetic Acid in Disinfection of Fruits and Vegetables. In *Postharvest Disinfection of Fruits and Vegetables*; Siddiqui, M. W., Ed.; Academic Press: London, 2018; pp 53–66.

PART II

PRESERVATION STRATEGIES FOR FRESH PRODUCE

CHAPTER 6

PRESERVATION STRATEGIES FOR FRUITS AND VEGETABLES: PAST, PRESENT, AND FUTURE SCOPE

PIR MOHAMMAD JUNAID, FAIZAN AHMAD, ISHFAQ HAMID DAR, AABID HUSSAIN BHAT, SHAZIYA MANZOOR, JUNAID AHMAD PANDITH, and FAISAL KHAN

ABSTRACT

To ensure year-round availability with special concern to perishability and to reserve the surplus harvest, various preservation methods have emerged. Mankind, from very beginning, has exploited various preservation techniques starting from sun drying and cellar storage to pickling. The advancement in science and technology has directed this art of preservation to new heights with precise advancements. Techniques such as inhibition of microbial growth preservation by specific chemicals, lowering of water activity, restricted use of the amounts of atmospheric gases and use of heat, and so on, are employed in the preservation of fruits and vegetables. The use of ultrasound, high-pressure processing, pulsed electric field, irradiation methods, non-thermal plasma, controlled atmospheric storage, freeze drying, vacuum packing, and hurdle technology are some remarkable breakthroughs in the preservation of fruits and vegetables. The reduction in the postharvest loss, the selection and use of preservation techniques, and the trends in the preservation of the fruits and vegetables will be discussed in this chapter.

Quality Control in Fruit and Vegetable Processing: Methods and Strategies. Megh R. Goyal, & Faizan Ahmad (Eds.)

6.1 INTRODUCTION

The core motto of food preservation is the utilization of the surplus and highly perishable food for the future procurement with the desired quality. The process of food preservation is diverse and involves mandatory unit operations. Food preservation techniques differ with the availability of the food, nature of the food, and the value of the food that is to be preserved. The advancement in technology had profound domination in the food preservation technology. The use of cellar storage systems to the modern modified atmospheric systems had shifted the outlook of food availability and quality. Among the food diversities available, the most important, but at the same time, the most vulnerable to postharvest loss are fruits and vegetables (FAV).

It is estimated that the postharvest loss of fruits and vegetable in India has reached almost 40% of its total production. The World Food Conference held in Rome (1974) drew attention to the concept of postharvest food loss reduction as a significant means to increase food availability. The expansion and the progression in the agriculture is no doubt blooming because of the introduction of new techniques and hybrid varieties; however, at the same time, the number in postharvest losses is also a grave concern.

The rise in the loss of agricultural produce affects lives, economy, social wellbeing, and environmental interests. To curb the postharvest losses in fruits and vegetables, researchers have been designing novel methods.[51] In the ancient days, the methods of food preservation were either based on the fruit or vegetable type or on the environmental conditions of that area. Although the techniques were not so much sophisticated, they were efficient in achieving the goal. Scientific progression in agriculture and food sector has now introduced up-to-date technologies in the preservation of fruits and vegetables.

The incorporation of food preservation techniques such as active packaging, modified packaging, high-pressure processing (HPP), PEF, microware, edible coatings, low-temperature preservation, and advance drying methods have taken the preservation of fruits and vegetables to the next level.[7,33]

This chapter will focus on (1) the perspective of preservation of fruits and vegetables with different approaches; (2) preservation methods; (3) the evolution of preservation; and (4) understanding the landmarks accomplished and the targets, which are still to be achieved in the preservation methods of fruits and vegetables.

6.2 WHY FRUITS AND VEGETABLES?

The technology of fruit and vegetables has made a widespread interest in the field of food, health, and allied sectors. The exact definition of fruits and vegetables is quite impossible to understand although we had a botanical definition for fruit. The classification based on the botanical system and on the basis of food composition doesn't provide adequate satisfaction. For a layman, farmers, and the health professionals, the definition is however greatly influenced by cultural norms based on food preparation and selection.[40]

The use of fruits and vegetables as the prime source in the diet has been recommended by numerous organizations, such as the Food and Agricultural Organization (FAO), World Health Organization (WHO), European Food Authority (EFSA), and United States Department of Agriculture (USDA).[51,57] Therefore, it is necessary that the fruits and vegetables are handled, prepared, and stored in suitable conditions so as to take full benefit of their nutrients. To reduce the bacterial load and to make them free from contaminants, it is necessary to subject them to primary unit operations followed by proper preservation means.[28,58] The fruits and vegetables can reduce the chances of non-communicable diseases like stroke, cardiovascular diseases, clean the free radical accumulation in body, and help in gradual nutrient absorption and less chances of constipation.

6.3 PRESERVATION OF FRUITS AND VEGETABLES

Generally, fruits, vegetables, and seasonal perishable agricultural commodities are highly susceptible to spoilage. For their year-round availability, it is essential to preserve them in suitable conditions. The preservation of fruits and vegetables is a day-old technique for the procurement of the surplus produce during the needy hours. The preservation of fruits and vegetables has modified from time to time with advancements in knowledge, available fruits and vegetables, type, and the climatic conditions.

The preservation of fruits and vegetables is done either by lowering the water activity or by increasing the antimicrobial medium or sometimes modifying the environment or the produce or sometime restricting the microbial growth. Based on these, we can have a range of preservation methods from drying, dehydration, curing, modified atmospheric packaging, and low temperature preservation.[27,49,63] Therefore, the preservation of fruits and vegetables is a technique of handling or treating of F&V to control its spoilage by either restricting the growth of microorganisms, decrease the respiration rate and to maintain its shelf life, nutritive value, and flavor or its conversion to the valuable food products like jam, jelly, etc.

6.4 IMPORTANCE OF FRUITS AND VEGETABLES

Fruits and vegetables are rich sources of carbohydrate, poor source of protein and fats, and have a neutral to high acidic value, with highest amount of water activity, which makes them suitable for microbial activity, thus making them a perishable commodity. Fruits and vegetables are one of the important sources of dietary nutrients for body metabolism to help in the muscle build-up and required energy yield.

The fruits and vegetables take an important position among all the available food varieties due to their beneficial aspects.[5,63] The presence of profusion quantity of vitamins, minerals, essential dietary fiber, and other bioactive compounds makes them essential portion of balanced diet for healthy well-being and disease-fighting properties. The most important feature of most fruits and vegetables is that they can be consumed, unprocessed, and don't require food processing unit operations, and thus are cost-effective. This use of unprocessed food has triggered a trend in the food sector as consumption of unprocessed or minimally processed foods for the health-promoting benefits.[51]

One of the best examples of minimally processed foods are fresh-cut fruits and vegetables, which are suitably peeled and cut into pieces before packaging and are said to be highly nutritious with a pleasant flavor.[44] Unfortunately, the producers are not aware about the importance and functional aspects of the fruit and vegetable sector. The manual and day-old techniques of handling such kind of commodity increase the chances of contamination, loss of nutritional value, and decreased shelf life. To overcome such problems, the postharvest handing of fruits and vegetables should be proper and well-suitable. The shift of preservation methods from the conventional to the modern scientific moods should be a priority.

6.5 POSTHARVEST OF FRUITS AND VEGETABLES

Fruits and vegetables are more susceptible to postharvest damage due to physiological, microbial, and physical factors. The inadequate handling, cultural practice, unfavorable climate, and poor storage conditions result in the delay and decay of the product quality and life. The natural biochemical reactions in the fruits and vegetables naturally deteriorate the eating, nutritional, and keeping quality and thus add up to the physiological loss. The biological loss generally happens due to the presence of insects, yeasts, molds, rodents, birds, and animals either by direct or by cross-contamination of the infected ones. These types of infections are much susceptible to spreading during storage or transportation. The microbial infection may be of water origin or soil or sometimes airborne, which includes coliforms, micrococci, coryniforms, lactic acid bacteria, and pseudomonas.[40]

Other types of losses during harvesting, transportation, handling, and packaging are due to fragile tissue damage of fruits and vegetables leading to bruising, abrasion, cuts, and breakage. These types of damages either give rise to microbial spoilage or damage to entire structure of produce with time. It is estimated that nearly 50–80% of the particular commodity reaches to the consumers and the rest is discarded and damaged during the entire postharvest processing.[15] These postharvest losses affect directly the producers and indirectly the economic conditions of the country. Therefore, postharvest losses adversely influence the socio-economic conditions and lead to unemployment, economic breakdown, famer suicide, and other social and cultural issues.[24] The inadequate and improper food supply chain and government policies need to be revived to minimize the postharvest loss at the producer end.[15]

6.6 PRESERVATION METHODS

Preservation of the food especially fruits and vegetables is the art to make the food or perishable fruits and vegetables available throughout the year and in the needy time. Increasing of shelf life of the fruits and vegetables requires some processing parameters and vary with the nature of fruit and vegetable. The fruits and vegetables are produced in abundance during a particular time in year and needs to be immediately consumed or preserved due to their susceptible nature.

The process of delaying the rotten process or the quality degradation is done by means of different techniques. As the fruits and vegetables are rich source of essential nutrients and the quality gets decreased upon inadequate long storage, the preservation systems are designed as such to increase the shelf stability and maintain the quality of the produce.[24] In order to preserve the fruits and vegetables, it is required to prevent microbial growth by lowering of enzymatic and physiological changes. Improved postharvest value envelops the adequate visual, favor, textural, and nutritive nature of food. This chapter will discuss the various techniques and methods for preservation and quality maintenance of fruits and vegetables.

6.7 TRADITIONAL METHODS OF PRESERVATION

Man from the very beginning had not only been a food hunter but also a food gatherer as well as the farmer. This art of farming has led him to reserve the surplus commodity for the future needs. For the preservation of surplus commodity, that is, fruits and vegetables, he established various preservation techniques and methods with prosperous results. This art of preservation was not scientifically advanced and was based on the naturally available methods like sun drying.[1,49] The preservation of fruits and vegetables started from sun drying, curing, smoking, cellar storage, and low-temperature preservation.

6.7.1 SUN DRYING

Drying is a prehistoric method of preservation available to mankind. In our day-to-day life, we come across a vast variety of dried fruits and vegetables. The drying technique generally accompanies the decrease in the water percentage of the produce to restrict or hinder the microbial and enzymatic spoilage of fruits and vegetables. The term dehydration is sometimes used for drying which is but the advanced and efficient form of drying.

In the early days of food preservation, this technique was widely used and in the developing countries like India this technique is still in practice. The sun drying of fruits and vegetables is easy, cost free, and available almost throughout the year. The cold or hilly areas where the intensity of sunlight is least, the produce is dried in the fixed time periods in a year. In this technique, the fruit or vegetable is either sliced or kept as such depending on the type of commodity and spread on the sheets of cloth, wood, flour, or in trays directed to the sun light.[1,40,45] The fruits and vegetables to be dried are then turned in regular intervals for even drying.

This technique is time consuming and sometimes inefficient. In the times when there were no alternate methods of preservation, this was the only and popular means of preservation. By this technique, not only fruits and vegetables were dried but the food like meat, fish, and vinegar (dried in cotton) were used. This technique has been modernized nowadays and many other assemblies have been adopted for its modification. The forced convective air dryers have been used for the purpose of quick and efficient drying by the solar dryers.

6.7.2 CELLAR STORAGE

The cellar storage technique also known as root cellar or earth cellar is the preservation of fruits and vegetables accomplish by storing them underground or partially underground for weeks to months depending on the produce and its harvest conditions. In the earlier times, it had gained much importance in the procurement of crops during winter conditions. Although in the present times, with the advancement of storage systems, this technology seems to be useless, but it has significantly served the humankind through decades as the chief means of preservation.

This method is still in practice in the underdeveloped countries in the rural areas where the means of technology is little or absent. Generally, potatoes, carrots, turnips, beets, onions, cabbage, apples, pears, apricots, etc., were preserved for few months through this technology.[27,49] The foods like fresh meat, jam, milk, and cream are also preserved by this technique. The crops subjected to cellar storage remain cool during the summer time and during winters they are in unfrozen state. The controlled temperature and the steady humidity keep fruits and vegetables fresh and in preserved state.

6.7.3 SALTING

The method of preservation food especially fruits and vegetables by the use of edible salt dates back to the ancient times. The use of this method was generally employed to the meat, fish, and vegetables. This method of salting sometimes known as curing is significant till nowadays for its wide application in the preservation technology.[3,27,55] Vegetables like beans, cabbage, etc. are covered with the salt and preserved by this method. The salting method develops the concentration gradient across the food material, and the osmosis the water is squeezed out of the food material which helps in the lowering of water activity. This lowering of water percentage inside the body of food material results in decreased enzymatic and microbial actions. The development of hypertonic environment across the food material restricts the growth of pathogenic microorganisms, commonly bacteria and fungi. This phenomenon by the addition of salt finally helps in the shellfire extension of the food material (vegetables).

6.7.4 PICKLING

Pickling is one among the oldest methods of food preservation especially fruits and vegetables. In the process of pickling, a new product formation takes place by the anaerobic fermentation in brine or vinegar. The product so formed is known as pickle having salty, sour taste. In the ancient times, the surplus vegetables were mixed with the spices and put it the jars and earthen vessels for couple of months to facilitate the fermentation process. This results in the production of organic acids (preservative agents) by bacteria like *lacto bacillales* to produce lactic acid which decreases the pH considerably.[49,60]

Some of the notable pickles are kimchi sauerkraut, surstromming, and nukazuke. Fruits like mango, amla, peaches, plums, etc., are also used for pickle making. The use of edible oil in addition to vinegar is used in pickle making in various south Asian countries. For the pickle making, different recipes are used depending upon the nature of crop and type of pickle making. Herbs, species, mixed varieties of fruits, and vegetables are sometimes used in the pickle formation. The flavor and the nature of pickle depend on the nature of food used in pickling process, temperature of fermentation, salinity or acidity of solution, and oxygen elimination. The *Leuconostocmes enteroides* bacteria dominate the pickle formation with release of alcohol, acid mixes, and aromatic compounds upon the low temperature and salt concentration. But, on higher salt concentration and temperature, the production of lactic acids occurs due to presence of bacteria *Lactobacillus plantarum.*

6.8 ADVANCED PRESERVATION METHODS

As the people are well aware of the health benefits of fruits and vegetables, their consumption has considerably increased, so as their farming. The micro-level evaluation of fruits and vegetables has enabled the present generation to look and demand for the specific type of nutrients. This health-conscious consumption of food has paved the way forward for new varieties, preservation methods, and amount of consumption.

The WHO has therefore recommended the consumable amounts of fruits and vegetables of around 400 g for an adult per day. As per the dietary guidelines for Americans, it is recommended to consume seven to 10 portions of fruits and vegetables equal to half plate of fruits and vegetables per day.[6] Therefore, the need of hour is to make the year-round availability of fruits and vegetables for the maintenance of heath and habit. The ongoing research in food science has thus combined the modern technologies and exploited them in food preservation technology, which are been discussed here.

6.8.1 USE OF HEAT

The use of heat is the most and prime technique of food preservation. The microorganisms which are responsible for growth and spoilage of fruits and vegetables are inactivated by the application of heat. The use of mild heat in fruits and vegetables also lower the enzymatic activity of the produce thus helping in the extended shelf life. Heat can also destroy the anti-nutritional compounds in some foods (e.g., trypsin inhibitors in legumes) and gelatinize the starch improvement of proteins. The use of heat in case of modern controlled drying process makes fruits and vegetables fit for consumption for years.

Some of the techniques used are solar drying, oven drying, freeze drying, and osmotic dehydration.[62] Solar drying is actually the traditional sun drying but in an improved form having a compact chamber called as dehydrator. The fruits and vegetables are dried in this chamber and consisting of foil surface that increases the temperature in addition to the ventilation. Another method for preservation of fruits and vegetables is the material is subjected to freeze drying and is the suitable method for perishables.[25,43,49] This process involves the material to be frozen first and then the surrounding pressure of that material is significantly reduced, which facilitates the sublimation of frozen water from solid to gaseous phase. In the commercial scale of drying of fruits and vegetables, ovens are used. These are dehydrators consisting of heating chamber with air flow, low humidity, etc., although it does not have built-in fan system like in that of modern dehydrators.

6.8.2 OSMOTIC DEHYDRATION

This is one of the common methods of fruit and vegetable preservation and has nowadays appearing as a trending technique. It is basically a pre-treatment process that employs the diffusion phenomena of water from food substance when immersed in a hypertonic solution.[27,40,50] The solute substance such as glucose, sodium chloride, corn syrup, fructose, and other osmotic agents are used in osmotic dehydration.

Other methods are air drying and freeze drying. In this process, the solute substances migrate inside the food material and the moisture gets drained out by compensating the solute gradient outside the food. This process is generally slow in nature. The water removal is facilitated by the creation of osmotic pressure gradient across the fruit or vegetable surface. Osmotic dehydration makes a little processing in fruits and vegetables due to their much similarity between the natural and dehydrated products.

6.8.3 MODIFIED ATMOSPHERE PRESERVATION (MAP)

The preservation of fruits and vegetables with consequent increase in the shelf life has been a concern because of the fact that continuing respiration and other biochemical processes going on in the tissues. Therefore, it is necessary to control these processes in order to ascertain the quality and nutritional value of the product with a long shelf life.

Modified and controlled atmosphere technologies utilize this principle for the preservation of fruits and vegetables. In case of modified atmosphere technologies, the gas composition around the fruits and vegetables is modified or replaced with some inert gas. This results in the decrease in the rate of respiration and other biochemical processes within the product matrix, thereby increasing the shelf life of the fruits and vegetables.[37,67] There may occur changes in the composition of gas surrounding the product due to the respiration process. This is overcome in the controlled atmosphere technologies where the composition of gases is changed and continuously monitored so as to keep it invariable at the desired level.[66]

6.8.3.1 MODIFIED ATMOSPHERE TECHNOLOGIES

The modified atmosphere technologies have been successfully used in storage and packaging systems for vegetables and fruits. The shelf life and quality of fruits and vegetables is attributed to the different physicochemical and sensory characteristics along with other safety aspects. Primarily, changes in such properties which may include browning, moisture loss and firmness, softening of tissues, etc., have a significant contribution to the quality loss in fruits and vegetables, and hence rendering them unsuitable or harmful for consumption.[14]

These deteriorative and undesirable changes in the fruits and vegetables are as a result of different biochemical processes which include oxidation of phenols, degradation of enzymes and other volatile chemicals, loss of nutritional components, or contamination by spoilage microorganisms. This is accompanied by changes in the physicochemical characteristics like browning, loss of flavor and texture, etc. and can be seen in a number of fruits and vegetables like mango, apple, peach, plum, onion, eggplant, potato, etc.[4,12] Among different preservation technologies used in fruits and vegetables, modified atmosphere technology is classified under physical methods. The preservation is achieved by alteration in the gas composition, which is replaced with an inert gas inside the package or storage structure.[38]

The selection of gas and its concentration depends upon the various factors like characteristics of the food product, condition of surrounding air and also the type of packaging along with other factors. The combined effect of these factors helps in determining the optimum gas composition which can extend the shelf life of fruits and vegetables for considerable time periods.[47] The other parameters taken into consideration while designing the appropriate modified atmosphere preservation system include rate of respiration and transpiration, permeability, environmental and storage conditions, rate of heat and mass transfer, etc.[11]

The most commonly used gas mixture for preservation of fresh-cut fruits includes CO_2, O_2, and N_2 because of the fact that these gases help in preserving the quality and are also directly involved in the general metabolism of fruits and vegetables. These impact the metabolism rates in fruits and vegetables thereby hampering the growth of spoilage microorganisms and also preserving the quality by retarding the ripening in climacteric fruits.[61] O_2 when present in low concentration (<5%) retards the growth of aerobic microbes while as when present in higher concentration hampers the growth of anaerobes. Determination of optimum concentration to be used is therefore given good thought and analysis needs to be done in order to achieve the desired preservation.[26] High concentration of O_2 has also been shown to be very helpful in controlling the enzymatic activity and growth of some microorganisms.[26,36]

To preserve high quality in fruits and vegetables, other gas compositions can be used depending on the nature of the food product like nitrogen for preserving the physical structure and appearance, CO_2 for preserving the flavor and aroma etc..[49] Nitrogen is used to replace the O_2 thereby decreasing the rate of respiration and also with high levels of O_2 to prevent the growth of microbes. In case of vegetables, with respiration rate an important parameter in determining the self-life, low levels of O_2, and high levels of CO_2 are effectively used to extend the shelf life [12, 59].[12,59] CO_2 is highly beneficial in the retarding the growth of bacteria and fungi. Its use in modified atmosphere technologies for preservation of fruits and vegetables has been demonstrated to inhibit the growth of bacteria such as *Erwinia* and *Xanthomonas*, and fungi, such as: *Alternaria, Botrytis, Fusarium, Aspergillus, Penicillium, Rhizopus,* etc..[34]

Recently, there has been increased focus on use of unconventional atmospheres in the modified atmosphere technologies for the preservation of fruits and vegetables. The different gases used include noble gases like helium, argon and other compounds like nitrous oxide, N_2O. In comparison to conventional modified atmosphere technologies, these provide better-quality preservation along with longer shelf life. They also help in minimizing the postharvest losses in fruits and preservation of different bioactive compounds present without any undesirable changes in the sensory attributes.[47]

The preservation mechanism of some unconventional gases has not been fully explored yet. Argon, which is used in the modified atmosphere packaging, has been proposed to act on receptor sites of enzymes, thereby resulting in their deactivation and reduction of microbial growth.[10] N_2O assists in the reduction in rate of respiration in fruits and the subsequent retardation in maturation.[16] The preservation technologies using modified atmosphere are used in extending the shelf life of different fresh-cut vegetables and fruits. However, there is a need of extensive research in order to commercialize the different gas combinations and consequent possibility to increase shelf life of various agricultural products.

6.8.3.2 CONTROLLED ATMOSPHERE TECHNOLOGIES

The controlled atmosphere technology for preservation is similar to the modified atmosphere technology except that there is precise control of the concentration of gases in the storage space. There is continuous monitoring and rectification in the concentration of gases. Such a system essentially comprises of air tight storage containers with O_2 and CO_2 analyzers along with suitable systems for removal and addition of gases. The O_2 to CO_2 ratio is very significant and is determined specific to different food products.[13] There has been a considerable development of controlled atmosphere technology over the years. Earlier, it consisted of an airtight system with a provision of constant temperature maintained by mechanical refrigeration along with control of atmosphere by the products respiration which increased the CO_2 concentration while reducing the O_2 level. Afterwards, it was further improved by the addition of injection systems and catalytic generators to have an active control on the control of gas concentrations. Now, it has been modernized by the use of dynamic control systems and bio-sensing of the physiology of fruits and vegetables to set the concentration of air according to the metabolic condition of the stored fruit or vegetable.[13]

The controlled atmosphere storage is one of the most important and effective preservation methods for fruits and vegetables. It decreases the rate of respiration in the fruits and vegetables, inhibits the growth of spoilage microorganisms and also slows down the enzymatic degradation.[18,20] The controlled atmosphere has minimal effects on the nutritional characteristics of fruits and vegetables. It has been revealed that there was significant reduction in the loss of bioactive compounds including vitamin E, vitamin C, etc. from fresh-cut fruits.[64] There may be development of some unwanted changes in the texture and off-flavors due to accumulation of some volatile compounds.[53]

6.8.4 PRESERVATION BY HIGH-PRESSURE PROCESSING (HPP)

The demand for more convenient and minimally processed foods has increased significantly over the time. This has further necessitated the need for achieving optimum food preservation with minimum changes in the nutritional and organoleptic properties of food products. In this line, many emerging preservation technologies have been devised as alternative to the conventional preservation methods which involved thermal treatment and resulted in the loss of nutritional value of food products.

High-pressure processing (HPP) is considered as a promising technology for ensuring safety along with minimizing the effects on quality and nutritional parameters. HPP treatment when compared to the thermal methods of preservation prevents the loss of quality. Application of high-pressure technology yields no untreated parts as they pressure is distributed uniformly in the food product. It is also very beneficial in the modification of various physicochemical properties in foods and enhancement of different processes like extraction, thawing, etc..[42,48]

There has been comprehensive research in the application of high-pressure technology since long. Its capability to reduce the microbial activity and deactivation of enzymes while preserving the nutritional quality has increased the prospects of this technology in preservation systems. High-pressure processing for preservation has been successfully used in many food products including fruits, vegetables, meat, beverages, etc. HPP has been successfully demonstrated to result in inhibiting the growth of microorganisms while preserving the quality. It also brings in the modification of bioactives and development of novel foods with different functional properties.[53]

The two major principles governing the HPP preservation technology are: Le Chatelier's Principle and the Isostatic Law. The former explains the changes in molecular configurations, pH, phase changes and other chemical reactions, which are affected by the increase in pressure. The latter principle explains the instant and uniform transmission of pressure throughout the food matrix.[42] The process of HPP preservation is carried out in a high pressure equipment, which essentially consists of a pressure chamber, high pressure pumps, temperature and pressure monitoring, and controlling devices and a product handling system.

The products are subjected to HPP in the pressure chamber, which is previously filled with food product and any suitable pressure-transmitting medium. The pressure is generated either by filling the chamber with the pressure-transmitting medium or by reducing the volume of the pressure chamber using piston system. The pressure-transmitting media commonly used include water and other food-grade liquids like glycol, ethanol, castor oil, etc. The selection of suitable pressure-transmitting medium is done after considering its viscosity, corrosive nature, and also the temperature and pressure ranges for the process.[52,65]

For most preservation purposes, the working pressure is generally between 50 and 1000 MPa. No additional energies are required to keep the food product at the given pressure. However, there is an increase in the temperature with the increase in pressure due to adiabatic heating. This increase in temperature, also referred as compression heating is significant in case foods with high moisture like fruits, milk, etc..[9] In case of foods with high fat content, compression heating is also higher, which depends upon the pressurizing rate along with other factors.[52] The application of HPP technology to improve the stability and safety of food products, which is evaluated primarily on the basis of its effect on inhibition of the spoilage and pathogenic microbes, has been extensively studied. The HPP treatment has been shown to cause sub-lethal injuries to the microbes which have a considerable effect on the preservation.[39]

This technology has been effective in inactivation of microorganisms due to varying degree of damage caused to the cell constituents mainly cell membrane. The high-pressure results in the disruption of cell transport due to changes in the membrane transport proteins and also the phospholipid bilayer.[53] However, the complete destruction of bacterial spores presents a challenge as they are highly resistant to extremes of temperature, pressure, radiation, etc..[35] The efficacy of HPP for preservation in fruits and vegetables is mainly due to change in pH induced by the changes in pressure.[56] The effect of HPP on inactivation of microorganisms varies with the type of strain, processing conditions, among other factors.

6.8.5 PULSED ELECTRIC FIELD

The use of PEF has emerged as an alternative technology to the traditional thermal methods of preservation. The delaying of the microbial proliferation and the reduction in biochemical reaction of fruits and vegetables facilitated by drying on account of Pulse electric field is one of the evolving approaches on the way to the preservation technology. The conversion of raw agricultural commodity to a dried solid form by dehydration is an effective method in the preservation and transportation of the produce. The development of high electric field of the order of 10–80 KV/cm between the two electrodes by the application of high-intensity electric pulses ensures the microbial destruction at relatively low temperatures.

The use of PEF for rapid electric breakdown of the cell membrane and intensification of pores has led to the extraction, dehydration, tenderization, sterilization, aging, and byproduct valorization approaches in food processing. PEF technology is used to completely inactivate the enzymes, to preserve the harvested fruits and vegetables. Various studies suggest the role of PEF drying in the extension and preservation of bioactive compounds (polyphenols, carotenoids) of different fruits such as apples and carrots.[23] The PEF technology can be successfully used for the biosynthesis stimulation of bioactive compounds from various fruits and vegetables to promote the preservation and health benefits.

6.8.6 USE OF LIGHT EMITTING DIODES

The application of ultraviolet (UV) light as food-processing method carries on growing widely because of its dry, non-chemically intrusive and non-thermal nature. Ultraviolet light method is still utilized in commercial processing of numerous solid, liquids food items, various meats items, as well as bakery items.[17] UV light of 200–400 nm splits in germicidal UV three zones A: B: C lights (320–400 nm, 280–320 nm, and 200–280 nm).[17] Traditional types of prolonged monochrome UV light tend to be high-voltage arc-discharge mercury and mercury-amalgam lamps, usually known as low pressure (LP) lamps, which discharge UV-C light at 253.7 nm.

Presently, an improvement in light-emitting diode (LED) method has permitted the advancement of UV light sources that will outwit various drawbacks while providing several other benefits. Light-emitting diode (semiconductor) creates illumination by electroluminescence with a p-side and an n-side, forming a p–n junction by electronic overlap. The flowing of electrons and holes from one side to other in diode is facilitated by the electric Potential across the ends. The interaction of electron–hole to create electroluminescence takes triggers the electrons to slide to a lesser energy state, thus discharging a photon leading to the discharge of light of a unique frequency[19] giving rise to the generation of monochrome illumination, having fine bandwidth of wavelengths exactly like distinctive shades to a person's eye.

Light of different wavelengths (red and IR light from Gallium arsenide, blue lights from gallium nitride and yellow, green, red and orange from indium gallium aluminum phosphide) is obtained from different semiconducting materials.[19] All kinds of sugars, amino acids, multivitamins, as well as organic acids within fruit and veggies have always been valuable compounds when it comes to human health. Lower standard will impact the content of Vitamin C inside plants by governing the function of Vitamin C synthesis and breakdown enzymes, and it also governs carbohydrate build-up as well as amino acids metabolism in plants.

The Red light boosts the build-up of carbohydrate, and the blue light stimulates the development of protein. Flavonoids, phenolic compounds, anthocyanins, and other compounds present in fruit and veggies not merely have a crucial effect on the flavor, hues, as well as economical worth of veggies and fruits, but additionally own natural free radical cleansing activity, which enable it to considerably suppress free radicals in human. The use of blue LED light can assist the increase in anthocyanin content, and the mixture of blue and red light helps in the flavonoids content build-up. Therefore, the LED illumination can be effectively used for the enhancement in the nutritional quality of vegetables and fruits. The strategy to cultivate the food products in closed domain (control of water, atmosphere nutrients, and light) by the use of Upright farming allows the development of pesticide free crops. Also, the use of different type LED lighting plans (red/blue or red/white both with and without far red) in such business ranches with the main focus to minimize postharvest losses.

Generally, the temperature, relative humidity, and gaseous control are the main conditions to control postharvest quality.[30] Also, the succinct dose of specific light can be helpful in the keeping quality of fruits and vegetables then when put in darkness. However, sufficient literature is still underway to see the exact influence of light on postharvest behavior of vegetables and fruits.[8] Accordingly, the better comprehension of the connection among light and postharvest quality might bring about helpful postharvest applications.

The various studies demonstrate that some attributes such as nutrient content, appearance can be profoundly increased by the application of pre-harvest continuous-light strategy.[41] The LEDs can act as triggers and modifiers in enzymatic and stilbene synthesis in some fruits[2] and can enhance the creation of supplementary metabolites, could be an impulse to these stresses or an effect due to sun-screening to shield plants from ionizing radiations.[46] LEDs can be developed as a milestone in the crop protection, increasing various health benefits, deactivating the enzymes, food pathogens, and in the shelf-life increase in fruits and vegetables by exploitation of comparable and compatible LED wavelengths combination.

6.8.7 EDIBLE PACKAGING

The aspect of the shelf-life increase in the fruit and vegetable sector is their proper packaging. Different types of food packaging are available in the food sector ranging from cardboard, glass to polymeric packaging, yet the packaging of fruits and vegetables needs to be well perceived considering their delicate and perishable nature. The abundance of natural renewable sources used as biodegradable packaging reshaped the packaging technology and ensured the safe and appropriate utilization of biological resources.

The use of natural materials like cotton, wool, and leather, silk for packaging due to their structural and functional properties has inspired in creating new forms of packaging systems. The development and utilization of edible films and coatings as eco-friendly packaging technology has offered the extensive increase in the shelf life of various food products. The use of natural biopolymers such as polysaccharides, proteins, and lipids or the combination of these has potential of development and effective utilization as edible packaging.[22]

The use of polysaccharides in the development of edible packaging has shown to be effective due to their functional and unique structural properties. These polymers are mostly derived from plant, animal and marine sources. The increased quality by decreasing the desiccation of the fruits and vegetables to the prevention of oxidation of various food products edible coatings and films is the promising breakthrough. The use of edible coating and films although seems to be a new technology but its usage has been found very early in the Chinese traditional way of preserving fruits by coating methods.[21]

In the early 1900s, the dry fruits and nuts were coated with sucrose for prevention of oxidative rancidity, but with the concerns of appearance (gloss, color), delaying in ripening, delaying pf dehydration, the emulsion-based waxes were introduced. With the advancement and compatibility of the products, various biopolymers are now introduced to reduce the postharvest shrinkage and other factors in the produce. The attention toward the edible packaging sectors is increasing due to its biodegradable, environment friendly, and, simply, availability.

In fruits and vegetables, the delaying of ripening, increasing shelf life, facilitated by the decrease in the respiration rate have manifold scope. One of the major benefits of edible coating and films is their edibility, increase in the sensory attributes, and their extra nutritional benefits. The edible coating and films can be added with the bioactive substances (antimicrobials, antioxidants, and nutraceuticals) to increase the functionality in fruits and vegetables. The goals like improvement in barrier, nutritional, sensory, and carrier properties are to some extent improved by the use of edible films and coating in the field of fruit and vegetable preservation.[33,46,54]

6.8.8 ULTRASOUND TECHNOLOGY

The use of novel techniques such as ultrasound is gaining popularity in the food sectors due to its application in the extraction, processing, and food preservation. The various frequencies range of the ultrasound makes it suitable to different functions. The use of frequency more than 100 kHz with intensity less than 1 Wcm^2 is used for quality analysis and is known as low energy ultrasound. The use of frequency in the range of 20 and 500 kHz with more than 1 $W\text{-}cm^2$ intensity causes food disintegration and is known as high-energy ultrasound.[32,61] The power, frequency, treatment temperature, and time of ultrasound exposure assist in the effectiveness of the ultrasound. This non-thermal method is able to reduce the microbial concentration through the process of cavitation.[51] This ultrasound technique is also used for the destruction of various pathogenic microbes like *E. coli, Listeria monocytogenes* in different fruit juices.

The various comparative studies suggest that the use of ultrasound can maintain the various physiological parameters (Vitamin C, pH, acidity, and TSS) of fruits in comparison with the untreated fruits with 40 Hz frequency at constant power of 350 W.[31,61] Therefore, the ultrasound is the one of the emerging non-thermal method of decontamination of microbial load from fruits and vegetables to increase the shelf life and quality parameters.[29] The ultrasound can also be used in bringing the structural change of various enzymes such as pectin-methyl esterase and other enzymes present in fruits and vegetables. The change in secondary and tertiary protein structure by the breakage of intermolecular bonds within the molecule by cavitation helps in the restriction of enzyme activity. This restriction of enzymatic activity ceases the metabolic processes by enzyme inactivation and restriction, and thus help in the increase of postharvest storage.

This technology being environment friendly, energy saving is efficient in the increases of produce shelf life with reduction in cost compared to other processing methods. The fruit and vegetable surfaces can be cleaned to enable the removal of pesticide residues and chemical contaminants for ensuring good quality and shelf life. Some studies suggest the application of ultrasound is helpful in improving the mechanical characteristics of the fruit and vegetables by the cell wall polysaccharide stabilization.

6.9 SUMMARY

Food preservation has been pivotal to our society since the beginning. Preservation has come from simple processes such as salting to more complex preserving methods, such as, irradiation and chemical additives. While the most-occupied preservation technologies have a long history of use, there is currently a real need for improved techniques, to meet the developing needs of consumers. Some improvements are being derived from the use of established techniques in new combinations or under improved control and other improvements are being derived essentially from the development of new techniques. These are finding, at first, new and attractive, but niche, markets. It is expected that these will expand as experience in the new techniques is gained. If the resistance of bacterial spores to some of the new techniques could be overcome, and in manner that was widely proven and accepted to be safe, then the potentially markets could be immeasurably larger. Particular attraction of the new techniques is that they act by inactivation rather than by inhibition. With respect to the reduction of the incidences of food poisoning, the introduction of effective inactivation techniques that lead to the elimination of the pathogens must be the ultimate target of the primary food producers, processors, distributors, and retailers. Looking into the future, high-pressure preservation seems to be the next logical step. In an increasingly health-conscious world, an uncontaminated world, well-preserved food source is the ultimate goal, and hopefully the costs of production will decrease for all of us to enjoy HPP foods.

KEYWORDS

- **chemical additives**
- **food producers**
- **fruits**
- **hygiene**
- **irradiation**
- **preservation strategies**
- **vegetables**

REFERENCES

1. Ahmed, N.; Singh, J.; Chauhan, H.; Anjum, P. G. A.; Kour, H. Different Drying Methods: Their Applications and Recent Advances. *Int. J. Food Nutr. Safet.* **2013,** *4* (1), 34–42.
2. Ahn, S. Y.; Kim, S. A.; Choi, S. J.; Yun, H. K. Comparison of Accumulation of Stilbene Compounds and Stilbene Related Gene Expression in Two Grape Berries Irradiated with Different Light Sources. *Hort. Environ. Biotechnol.* **2015,** *56* (1), 36–43.
3. Alzamora, S. M.; Guerrero, S. N.; López-Malo, A.; Welti-Chanes, J.; Palou, E.; Argaiz, A.; Tapia, M. S. Combined Preservation Techniques for Fresh Fruit. In Improving the Safety of Fresh Fruit and Vegetables; Woodhead Publishing: London, 2005; pp 599–630.
4. Amaro, A. L.; Almeida, D. P. F.; Malcata, F. X.; Beaulieu, J. C.; Stein, R. E. Volatile Composition and Quality of Fresh-Cut *Cucumis melo* as Affected by Modified and Controlled Atmospheres. VI *Int. Postharv. Symp.* **2009,** *877*, 603–609.
5. Baert, L.; Mattison, K.; Loisy-Hamon, F.; Harlow, J.; Martyres, A.; Lebeau, B.; Stals, A.; Van Coillie, E.; Herman, L.; Uyttendaele, M. Norovirus Prevalence in Belgian, Canadian and French Fresh Produce: A Threat to Human Health? *Int. J. Food Microbiol.* **2011,** *151* (3), 261–269.
6. Barrett, D. M.; Lloyd, B. Advanced Preservation Methods and Nutrient Retention in Fruits and Vegetables. *J. Sci. Food Agric.* **2012,** *92* (1), 7–22.
7. Bibi, N.; Khattak, A. B.; Mehmood, Z. Quality Improvement and Shelf-Life Extension of Persimmon Fruit (Diospyros kaki). *J. Food Eng.* **2007,** *79* (4), 1359–1363.
8. Braidot, E.; Petrussa, E.; Peresson, C.; Patui, S.; Bertolini, A.; Tubaro, F.; Wählby, U.; Coan, M.; Vianello, A.; Zancani, M. Low-Intensity Light Cycles Improve the Quality of Lamb's Lettuce (*Valerian ellaolitoria* L.) During Storage at Low Temperature. *Postharv. Biol. Technol.* **2014,** *90*, 15–23.

9. Buzrul, S.; Alpas, H.; Largeteau, A.; Bozoglu, F.; Demazeau, G. Compression Heating of Selected Pressure Transmitting Fluids and Liquid Foods During High Hydrostatic Pressure Treatment. *J. Food Eng.* **2008,** *85* (3), 466–472.
10. Char, C.; Silveira, A. C.; Inestroza-Lizardo, C.; Hinojosa, A.; Machuca, A.; Escalona, V. H. Effect of Noble Gas-Enriched Atmospheres on the Overall Quality of Ready-to-Eat Arugula Salads. *Postharv. Biol. Technol.* **2012,** *73*, 50–55.
11. Chitravathi, K.; Chauhan, O. P.; Raju, P. S.; Madhukar, N. Efficacy of Aqueous Ozone and Chlorine in Combination with Passive Modified Atmosphere Packaging on the Postharvest Shelf-Life Extension of Green Chillies (*Capsicum annuum* L.). *Food Bioprocess Technol.* **2015,** *8* (6), 386–1392.
12. Costa, C.; Lucera, A.; Conte, A.; Mastromatteo, M.; Speranza, B.; Antonacci, A.; Del Nobile, M. A. Effects of Passive and Active Modified Atmosphere Packaging Conditions on Ready-to-Eat Table Grape. *J. Food Eng.* **2011,** *102* (2), 115–121.
13. de Siqueira Oliveira, L.; Eça, K. S.; de Aquino, A. C.; da Silva, L. M. R. Modified and Controlled Atmosphere Packaging. In *Fresh-Cut Fruits and Vegetables*; Academic Press: New York, 2020; pp 151–164.
14. Finnegan, E.; O'Beirne, D. Characterizing Deterioration Patterns in Fresh-Cut Fruit Using Principal Component Analysis, Part II: Effects of Ripeness Stage, Seasonality, Processing and Packaging. *Postharv. Biol. Technol.* **2015,** *100*, 91–98.
15. Goleman, A.; Boyatzis, D.; Mckee, R. Postharvest Losses in Selected Fruits and Vegetables in India. *J. Chem. Inf. Model.* **2019,** *53* (9), 1689–1699.
16. Gouble, B.; Fath, D.; Soudain, P. Nitrous Oxide Inhibition of Ethylene Production in Ripening and Senescing Climacteric Fruits. *Postharv. Biol. Technol.* **1995,** *5* (4), 311–321.
17. Green, A. UV Light-Based Interventions to Inactivate Pathogens on Fresh Produce Through the Use of UV-LEDs and Water-Assisted Processing. PhD Dissertation; Department of Food Science at University of Guelph, Toronto, 2018; p 231.
18. Guerreiro, A. C.; Gago, C. M.; Faleiro, M. L.; Miguel, M. G.; Antunes, M. D. The Effect of Edible Coatings on the Nutritional Quality of Fresh-Cut Apple Through Shelf-Life. *LWT*, **2017,** *75*, 210–219.
19. Gupta, S. D.; Jatothu, B. Fundamentals and Applications of Light-Emitting Diodes (LEDs) in In Vitro Plant Growth and Morphogenesis. *Plant Biotechnol. Rep.* **2013,** *7* (3), 211–220.
20. Gutiérrez, T. J., Ed. *Polymers for Agri-Food Applications*; Springer International Publishing: Switzerland, 2019; p 622.
21. Han, J. H. Edible Films and Coatings: A Review. *Innov. Food Packaging* **2014,** *2014*, 213–255.
22. Hassan, B.; Chatha, S. A. S.; Hussain, A. I.; Zia, K. M.; Akhtar, N. Recent Advances on Polysaccharides, Lipids and Protein Based Edible Films and Coatings: A Review. *Int. J. Biol. Macromol.* **2018,** *109*, 1095–1107.
23. Hazervazifeh, A.; Nikbakht, A. M.; Moghaddam, P. A. Novel Hybridized Drying Methods for Processing of Apple Fruit: Energy Conservation Approach. *Energy* **2016,** *103*, 679–687.
24. Hegazy, R. Postharvest Situation and Losses in India: Figshare. Journal Contribution. https://doi.org/10.6084/m9.figshare.3206851.v1. (accessed Aug 1, 2021).
25. Hsieh, C. F.; Liu, W.; Whaley, J. K.; Shi, Y. C. Structure, Properties, and Potential Applications of Waxy Tapioca Starches–A Review. *Trends Food Sci. Technol.* **2019,** *83*, 225–234.

26. Jacxsens, L.; Devlieghere, F.; Van der Steen, C.; Debevere, J. Effect of High Oxygen Modified Atmosphere Packaging on Microbial Growth and Sensorial Qualities of Fresh-Cut Produce. *Int. J. Food Microbiol.* **2001,** *71* (2–3), 197–210.
27. James, I. F. Preservation of Fruit and Vegetables. Agrodok Series 3; Wageningen: Agromisa Foundation; **2003,** Series 3; p 87.
28. Jeddi, M. Z.; Yunesian, M.; Gorji, M. E. H.; Noori, N.; Pourmand, M. R.; Khaniki, G. R. J. Microbial Evaluation of Fresh, Minimally-Processed Vegetables and Bagged Sprouts from Chain Supermarkets. *J. Health Population Nutr.* **2014,** *32* (3), 391–399.
29. Jiang, Q.; Zhang, M.; Xu, B. Application of Ultrasonic Technology in Post Harvested Fruits and Vegetables Storage: A Review. *Ultrasonics Sonochem.* **2020,**105261.
30. Kader, A. A. Increasing Food Availability by Reducing Postharvest Losses of Fresh Produce. In *V International Postharvest Symposium* 682; *Acta Horticulture*; 2004 (June); pp 2169–2176.
31. Khademi, O.; Ashtari, M.; Razavi, F. Effects of Salicylic Acid and Ultrasound Treatments on Chilling Injury Control and Quality Preservation in Banana Fruit During Cold Storage. *Sci. Hort.* **2019,** *249*, 334–339.
32. Khandpur, P.; Gogate, P. R. Effect of Novel Ultrasound—Based Processing on the Nutrition Quality of Different Fruit and Vegetable Juices. *Ultrasonics Sonochem.* **2015,** *27*, 125–136.
33. Kore, V. T.; Tawade, S. S.; Kabir, J. Application of Edible Coatings on Fruits and Vegetables Application of Edible Coatings on Fruits and Vegetables. *Imperial J. Interdiscipl. Res.* (Online), **2016,** *3* (1), 591–603.
34. Latou, E.; Mexis, S. F.; Badeka, A. V.; Kontakos, S.; Kontominas, M. G. Combined Effect of Chitosan and Modified Atmosphere Packaging for Shelf-Life Extension of Chicken Breast Fillets. *LWT-Food Sci. Technol.* **2014,** *55* (1), 263–268.
35. Liu, F.; Wang, Y.; Li, R.; Bi, X.; Liao, X. Effects of High Hydrostatic Pressure and High Temperature Short Time on Antioxidant Activity, Antioxidant Compounds and Color of Mango Nectars. *Innov. Food Sci. Emerg. Technol.* **2014,** *21*, 35–43.
36. López-Gálvez, F.; Ragaert, P.; Haque, M. A.; Eriksson, M.; van Labeke, M. C.; Devlieghere, F. High Oxygen Atmospheres Can Induce Russet Spotting Development in Minimally Processed Iceberg Lettuce. *Postharv. Biol. Technol.* **2015,** *100*, 168–175.
37. Ma, L.; Zhang, M.; Bhandari, B.; Gao, Z. Recent Developments in Novel Shelf-Life Extension Technologies of Fresh-Cut Fruits and Vegetables. *Trends Food Sci. Technol.* **2017,** *64*, 23–38.
38. Mangaraj, S.; Goswami, T. K.; Mahajan, P. V. Applications of Plastic Films for Modified Atmosphere Packaging of Fruits and Vegetables: A Review. *Food Eng. Rev.* **2009,** *1* (2), 133–158.
39. McClements, J. M. J.; Patterson, M. F.; Linton, M. The Effect of Growth Stage and Growth Temperature on High Hydrostatic Pressure Inactivation of Some Psychotropic Bacteria in Milk. *J. Food Protection* **2001,** *64* (4), 514–522.
40. 40. Mrema, G. C.; Kienzle, J. and Mpagalile, J. Current Status and Future Prospects of Agricultural Mechanization in Sub-Saharan Africa (SSA). *Agric. Mech. Asia Africa Latin Am.* **2018,** *49* (2), 13–30.
41. Nicole, C. C. S.; Mooren, J.; Pereira Terra, A. T.; Larsen, D. H.; Woltering, E. J.; Marcelis, L. F. M.; Verdonk, J.; Schouten, R.; Troost, F. Effects of LED Lighting Recipes on Postharvest Quality of Leafy Vegetables Grown in a Vertical Farm. In *VI International Conference Postharvest Unlimited* #1256; *Acta Horticulture*, **2019,** *2019* (Nov.), 481–488.

42. Norton, T.; Delgado, A.; Hogan, E.; Grace, P.; Sun, D. W. Simulation of High-Pressure Freezing Processes by Enthalpy Method. *J. Food Eng.* **2009,** *91* (2), 260–268.
43. Oladipupo, S. O.; Osunyemi, O. S.; Jayi, O. M. Synergistic Effect of Packaging Materials and Methanolic Extract of *Myrianthus arboreus* on the Control of *Callosobruchus maculatus*. *J. Food Technol. Pres.* **2018,** *2*, 6–10.
44. Oliveira, M.; Abadias, M.; Usall, J.; Torres, R.; Teixidó, N.; Viñas, I. Application of Modified Atmosphere Packaging as a Safety Approach to Fresh-Cut Fruits and Vegetables–A Review. *Trends Food Sci. Technol.* **2015,** *46* (1), 13–26.
45. Onwude, D. I.; Hashim, N.; Janius, R.; Abdan, K.; Chen, G.; Oladejo, A. O. Non-Thermal Hybrid Drying of Fruits and Vegetables: A Review of Current Technologies. *Innov. Food Sci. Emerg. Technol.* **2017,** *43*, 223–238.
46. Park, S. U.; Ahn, D. J.; Jeon, H. J.; Kwon, T. R.; Lim, H. S.; Choi, B. S.; Baek, K. H.; Bae, H. Increase in the Contents of Ginsenosides in Raw Ginseng Roots in Response to Exposure to 450 and 470 nm Light from Light-Emitting Diodes. *J. Ginseng Res.* **2012,** *36* (2), 198–203.
47. Pinela, J.; Barreira, J. C.; Barros, L.; Antonio, A. L.; Carvalho, A. M.; Oliveira, M. B. P.; Ferreira, I. C. Postharvest Quality Changes in Fresh-Cut Watercress Stored Under Conventional and Inert Gas-Enriched Modified Atmosphere Packaging. *Postharv. Biol. Technol.* **2016,** *112*, 55–63.
48. Prasad, N. K.; Yang, B.; Zhao, M.; Wang, B. S.; Chen, F.; Jiang, Y. Effects of High-Pressure Treatment on the Extraction Yield, Phenolic Content and Antioxidant Activity of Litchi (*Litchi chinensis* Sonn.) Fruit Pericarp. *Int. J. Food Sci. Technol.* **2009,** *44* (5), 960–966.
49. Rahman, M. S., Ed. *Handbook of Food Preservation*; CRC Press, 2007.
50. Ramaswamy, H. S. *Postharvest Technologies of Fruits and Vegetables*; DEStech Publications, Inc.: Lancaster, Pennsylvania, 2014; p 364.
51. Ramos, B.; Miller, F. A.; Brandão, T. R. S.; Teixeira, P.; Silva, C. L. M. Fresh Fruits and Vegetables—An Overview on Applied Methodologies to Improve Its Quality and Safety. *Innov. Food Sci. Emerg. Technol.* **2013,** *20*, 1.
52. Rastogi, N. K.; Raghavarao, K. S. M. S.; Balasubramaniam, V. M.; Niranjan, K.; Knorr, D. Opportunities and Challenges in High Pressure Processing of Foods. *Crit. Rev. Food Sci. Nutr.* **2007,** *47* (1), 69–112.
53. Rivalain, N.; Roquain, J.; Demazeau, G. Development of High Hydrostatic Pressure in Biosciences: Pressure Effect on Biological Structures and Potential Applications in Biotechnologies. *Biotechnol. Adv.* **2010,** *28* (6), 659–672.
54. Rojas-Graü, M. A.; Soliva-Fortuny, R.; Martín-Belloso, O. Edible Coatings to Incorporate Active Ingredients to Fresh-Cut Fruits: A Review. *Trends Food Sci. Technol.* **2009,** *20* (10), 438–447.
55. Rosenthal, A.; Torrezan, R.; Schmidt, F. L.; Narain, N. Preservation and Processing of Tropical and Subtropical Fruits. In: *Postharvest Biology and Technology of Tropical and Subtropical Fruits: Cocona to Mango*; Woodhead Publishing: London, 2011; pp 419–485.
56. Samaranayake, C. P.; Sastry, S. K. In Situ Measurement of pH Under High Pressure. *J. Phys. Chem. B* **2010,** *114* (42), 13326–13332.
57. Sandhu, P. P. Status and Recent Trends in Fresh-Cut Fruits and Vegetables. In: *Fresh-Cut Fruits and Vegetables*: *Technologies and Mechanisms for Safety Control*, Vol. 17; Academic Press: New York, 2019; pp 17–49.

58. Schuh, V.; Schuh, J.; Fronza, N.; Foralosso, F. B.; Verruck, S.; Vargas Junior, A.; Silveira, S. M. D. Evaluation of the Microbiological Quality of Minimally Processed Vegetables. *Food Sci. Technol.* **2020,** *40* (2), 290–295.
59. Sen, C.; Mishra, H. N.; Srivastav, P. P. Modified Atmosphere Packaging and Active Packaging of Banana: A Review on Control of Ripening and Extension of Shelf-Life. *J. Stored Prod. Postharv. Res.* **2012,** *3* (9), 122–132.
60. Sharif, Z. I. M.; Mustapha, F. A.; Jai, J.; Yusof, N. M.; Zaki, N. A. M. Review on Methods for Preservation and Natural Preservatives for Extending the Food Longevity. *Chem. Eng. Res. Bull.* **2017,** *2017*, 145–153.
61. Sousa-Gallagher, M. J.; Tank, A.; Sousa, R. Emerging Technologies to Extend the Shelf-Life and Stability of Fruits and Vegetables. *Stab. Shelf-Life Food* **2016,** *2016*, 399–430.
62. Stitt, Maud E. Preservation of Fruits and Vegetables by Drying; 1943; SDSU Extension Leaflets 76; pp 5. https://openprairie.sdstate.edu/extension_leaflets/76 (accessed Aug 1, 2021).
63. Thompson, F. E.; Willis, G. B.; Thompson, O. M.; Yaroch, A. L. The Meaning of Fruits and Vegetables. *Public Health Nutr.* **2011,** *11* (7), 1222–1228.
64. Wright, K. P.; Kader, A. A. Effect of Slicing and Controlled-Atmosphere Storage on the Ascorbate Content and Quality of Strawberries and Persimmons. *Postharv. Biol. Technol.* **1997,** *10* (1), 39–48.
65. Yaldagard, M.; Mortazavi, S. A.; Tabatabaie, F. The principles of Ultra-High Pressure Technology and Its Application in Food Processing/Preservation: A Review of Microbiological and Quality Aspects. *Afr. J. Biotechnol.* **2008,** *7* (16), 2739–2767.
66. Yang, Q.; Zhang, X.; Wang, F.; Zhao, Q. Effect of Pressurized Argon Combined with Controlled Atmosphere on the Postharvest Quality and Browning of Sweet Cherries. *Postharv. Biol. Technol.* **2019,** *147*, 59–67.
67. Yousuf, B.; Qadri, O. S.; Srivastava, A. K. Recent Developments in Shelf-Life Extension of Fresh-Cut Fruits and Vegetables by Application of Different Edible Coatings: A Review. *LWT,* **2018,** *89*, 198–209.

CHAPTER 7

OSMOTIC DEHYDRATION TECHNOLOGY FOR PRESERVATION OF FRUITS AND VEGETABLES

ARSHIED MANZOOR, BISMA JAN, QURAT UL EAIN HYDER RIZVI, PIR MOHAMMAD JUNAID, JUNAID AHMAD PANDITH, ISHFAQ HAMID DAR, SHAYEEB AHMAD BHAT, and SAGHIR AHMAD

ABSTRACT

Dehydration is the earliest form of conventional preservation method to provide long-lasting durability of foods. Preserving the freshness of fruits and vegetables (FAV) through the process of osmotic dehydration (OD) has led to several research studies in recent years, and this method consists of impregnation of vegetables or fruits in hypertonic solutions (mostly sugar and salt solution), thereby developing intermediate moisture products through moisture content reduction and solute gain. This technique is commonly applied before the actual drying process. Also, the OD can extend the shelf life of food products by slowing down the deteriorating physicochemical and biological activities. It operates at ambient temperature thus preventing the phase change and consumes less energy than other drying processes, besides improving the sensory and nutritional characteristics of foods. This process also helps in the retention of natural color, inhibition of enzymatic browning, and retention of volatile compounds in foods.

Quality Control in Fruit and Vegetable Processing: Methods and Strategies. Megh R. Goyal, & Faizan Ahmad (Eds.)

7.1 INTRODUCTION

Fruits and vegetables (FAV) are source of natural antioxidants and dietary fiber that have attracted researchers for their exploration and preservation.[29] In India, fresh produce contributes to about 30% of its gross domestic product (GDP) and it is also the second-largest producer of horticultural produce. However, almost 40% of the total production is wasted due to inadequate storage and proper retailing potentials. This has paved the way for the food industries and food technologists to search for strategies that can cater to these issues and help to extend the shelf life, while maintaining their quality in the safest possible way.

The osmotic dehydration (OD) process is the difference in osmotic pressure to direct the transfer of water from food to the solution at low temperature and low-energy consumption. It signifies that the water is moved to a higher concentration of solute from a lower solute concentration through complex cellular structure acting as a semipermeable membrane.[49,54] OD is also a preferred pretreatment due to its inability to remove moisture to a level necessary to arrest the microbial growth. In this process of OD, the food material in different forms (peeled, cut or even sliced) are placed within the medium containing sugars (sucrose, glucose, etc.) with a concentration on the higher side.

Fruit slices or cubes are dehydrated osmotically in the sugar syrup while brine/sodium chloride is preferred for the vegetables. The process of osmosis has been practiced for a long for the water removal from food materials thereby preserving them. However, in the last few decades, researchers have paid attention to this "dewatering-impregnation" process, which has brought an immense improvement and upgradation in techniques on osmotic dehydration.[9] OD has proved as an important tool in the minimal processing of FAV.

OD finds its applications in horticultural products highlighting FAV for water removal and soluble solid content increment.[21] The process is accompanied by loss of weight in the food material causing shrinkage due to the exchange of water and natural substances in the food–water system. Moreover, OD is devoid of significant damage to color, texture or flavor of FAV, where the water activity is reduced to 0.9, thereby maintaining its sensorial and textural integrity.[39]

Retention of color and flavor is also favored due to mild temperature in the OD process, mostly below 50°C. Also, enzymatic and oxidative discoloration have not been reported in OD as the product needs to be immersed in an osmotic solution that makes the process oxygen-free, hence it avoids blanching of the food product or use of SO_2. Longer duration of the process favors the bulk intake of solute into food material that was low initially, conferring its dependence on time duration. These positive impacts have made OD as an energy-efficient alternative besides widening its applications to many FAV (apricot, banana, mango, potato, apple guava, grapes, cherry, etc.).

This chapter highlights (1) the mechanism of osmatic dehydration (OD), (2) the factors influencing OD, and (3) the various processes in combination with OD to enhance the process of OD coupled with improving the quality of FAV during storage. Moreover, this chapter will emphasize on the applications of OD in food processing with recent advances in the process.

7.2 MECHANISM OF OSMOTIC DEHYDRATION (OD)

The OD process is carried out by the immersion of a fruit or a vegetable (whole or sliced) in hypertonic (with high osmotic pressure) osmotic solutions containing salt or sugars, such as sucrose, glucose, thus resulting in nutritious, safe, economical, and also concentrated food products with increased shelf life (Figure 7.1). The principle of operation for this process is a concentration gradient between the intracellular fluid in the food and the solution containing the food, and the process continues till equilibrium is established through the semipermeable membrane.[9] The complex internal structure of the food system does not allow the semipermeable membrane to be perfect. Therefore, allowing movement of solids into the food system portrays OD as the process of water and solute diffusion simultaneously.[33]

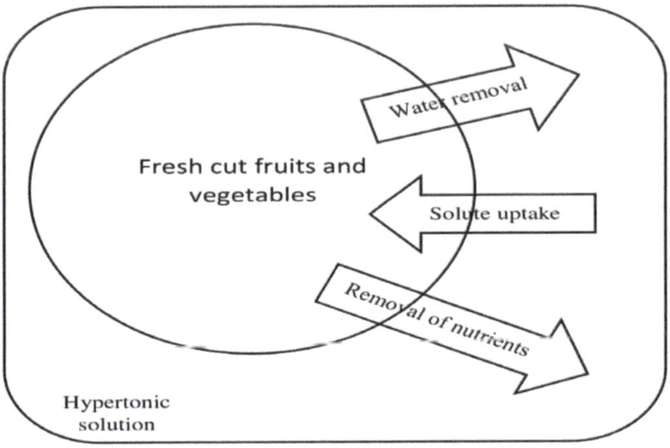

FIGURE 7.1 Different transfer processes in the osmotic dehydration technology.

During the osmotic dehydration process, water diffuses initially at a faster rate followed by a slow movement out of the food material, and solute present in the osmotic solution diffuses initially at a slower rate followed by faster rate into the immersed food product. At this point occurs a third process comprising of loss or leaching of sugars, acids, minerals, vitamins into the immersion medium side by side that is responsible for the alteration of the product's organoleptic and nutritional characteristics. Advantages and disadvantages of OD process are listed in Table 7.1.[16,18,36] Figure 7.2 describes the osmotic dehydration process.

TABLE 7.1 Advantages and Disadvantages of Osmotic Dehydration Process.

Advantages	Disadvantages
Minimizing the color and flavor loss credited to the low-temperature treatment	After OD process, the food product needs to be rinsed quickly to remove the coating of sugar adhered to it
Energy-efficiency is due to the absence of phase change.	When used in combination with other techniques, OD proves out to be more expensive
A minimum load on the dryer used is credited to the partial removal of water.	It takes a long time to result in the minimum water activity
Efficient flavor retention is credited to the use of sugar or sugar syrup as the osmotic solution.	Water activity in the food products after dehydrating osmotically consists of higher water activity
No need to use CO_2 or undergo blanching due to the full immersion of fruit/vegetable pieces into the osmotic solution	OD results in the alteration of food products concerned to its taste credited to the acidity level reduction
Leads to a better texture of the food product when reconstituted	OD fails to minimize the moisture up to a level necessary to arrest the microbial growth, hence preferred as a pretreatment only
Food product quality is improved credited to the solid density increment due to solid uptake	–
The food products can be stored for longer durations	–
Cost (packaging and distribution costs) reduction	–

7.3 OSMOTIC DEHYDRATION: APPLICATIONS IN PROCESSING AND PRESERVATION OF FOOD ITEMS

In foods, fruits, and vegetables, it is a well-established fact that several techniques for the OD process will decide their acceptance by consumer based on the extent of alternation in properties (both sensory and nutrition based) of foods.[22,34,50] OD helps to produce food products quite similar to natural or dehydrated products in terms of sensory aspects, hence, it is considered a simple and gentle method of processing for quality enhancement. OD of apple slices in 60–70° Brix sugar solution resulted in a 50% reduction in weight with superior quality and eliminated the need to use sulfur dioxide for color retention. The credit of determination of various activities (physical/chemical/biological) in foods is due to the OD process causing partial removal of water thereby forming intermediate moisture foods with the least water activity, which possibly reduces the chances of food deterioration.[35]

OD is used to develop new and high-quality products through the incorporation of various techniques resulting in the stabilization of texture and pigments by maximizing the sugar to acid ratio. It reduces enzymatic browning and prevents any damage to the color and flavor of food materials. Azoubell and Francinaide[6] suggested that the optimum conditions for 25% of maximum water removal and 6% of minimum solid uptake in mango fruits used 44% w/w osmotic solution at a temperature 38°C after an immersion time of 80 min. A combination of different osmotic agents can help in the evolution of flavor and sensory enhancement.

Osorio et al.[32] observed that ascorbic acid and lycopene content were significantly influenced in addition to anthocyanin and flavor elution into the osmotic solution coupled with decreased water activity using a combination of 70% sucrose, 70% sucrose + 65% glucose and ethanol for the OD of Tamarillo and Andes berry. Similarly, using a combination of sucrose and salt (1:1.5) solution significantly reduced the weight of tomato rings that avoided the need for sulfur dioxide for color retention compared with the OD of banana in 100% sucrose solution.[3] For removing the water to a level guaranteeing food safety, the food needs to be further dried as the OD does not provide the bone-dry products. However, high-temperature usage is not preferred for the OD process, as it renders the alteration of tissue structure and impregnation phenomenon.

7.4 FACTORS AFFECTING THE OSMOTIC DEHYDRATION

7.4.1 PRETREATMENTS

To prevent the food/biological materials from the detrimental effects caused through various drying techniques, researchers have incorporated different treatments before the OD process. Some of these prominent pretreatments include blanching, alkaline dipping, high hydrostatic pressure (HHP), sulfating, and freezing.[1] These pretreatments help to enhance the sensorial and nutritional quality according to Tadesse et al.[46] through the combination of blanching and osmotic pretreatments in the sliced carrots. Color, a primary attribute which decides the consumer acceptance of FAV, can be preserved through pretreatments with SO_2 or blanching to reduce their discoloration.

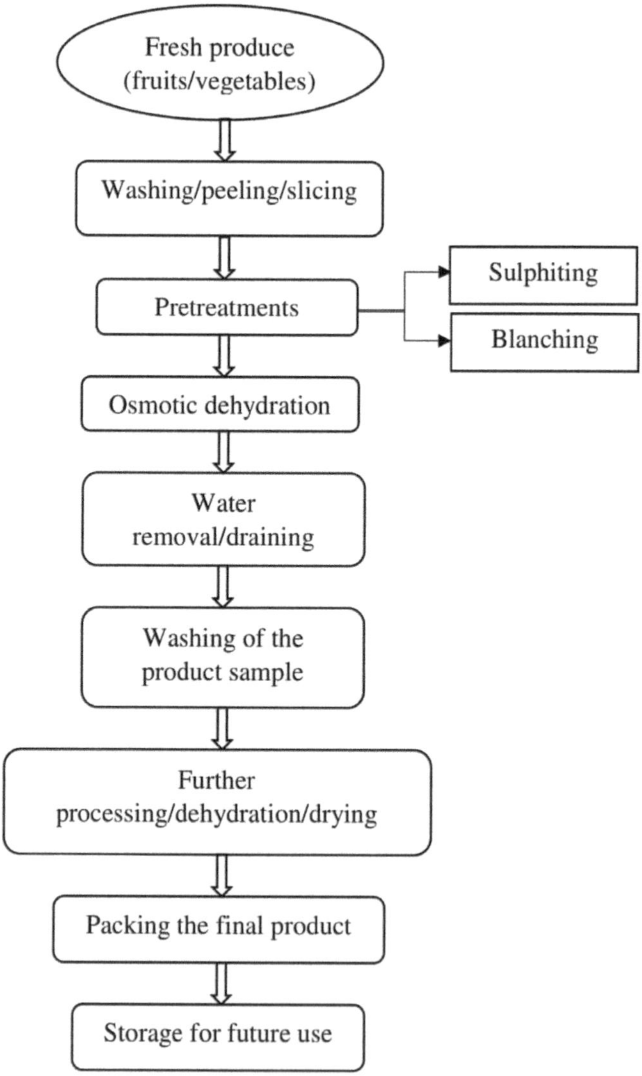

FIGURE 7.2 Flowchart of osmotic dehydration of FAV.

Another pretreatment method to prevent enzymatic browning includes immersion of FAV in citric acid solution (1%) followed by drying or osmotic dehydration. Dipping in ascorbic acid (0.4%) for 30 min helped to obtain high acceptability in papaya and mango slices before osmosis. Dipping food materials in 1% citric acid prior to OD also helps to postpone the enzymatic browning.[45]

7.4.2 CONCENTRATION AND TYPES OF OSMOTIC AGENTS

Various properties of osmotic agents (molecular weight, state of existence, solubility, and other physicochemical properties) can influence the process of OD.[23] Water removal kinetics along with equilibrium water content and solid gain is influenced by the osmotic solution concentration and characteristics for the sample immersion during OD process. Moreover, the osmotic solute and their concentration create a gradient resulting in the development of driving force necessary for the transfer of solute and water counter across the semipermeable membrane and determines the final product acceptability. Hence, the compatibility of the various osmotic solutes (glucose, sodium chloride, sucrose, glycerol, sorbitol, corn syrup, glucose syrup, and fructo-oligosaccharide) with the food material is an important aspect to be focused.[50]

Moreover, the osmotic agents in the process of OD must have an acceptable taste coupled with harmless behavior. The molecular weight of the osmotic agent is the function of the solute penetration; for example, low molecular weight solutes can penetrate at a faster rate into the cells/ tissues of FAV.[34] For this challenge, the effective diffusion and mass flow rate was highest in apricot when immersed in the solution containing sucrose while the diffusion was lowest in sorbitol containing osmotic solution,[17] such as glucose, fructose, maltodextrin, sucrose, and sorbitol. The authors also reported the occurrence of highest solid gains in the solution containing fructose and lowest in that containing maltodextrin. Osmotic agents in combination can increase the water removal efficiency of the OD process, according to Lenart and Flink,[25] who observed an enormous increase in water loss from potato by immersing it in a solution containing salt and sucrose together. However, the use of a combination of various osmotic solutions seems to be more complex and may have negative impact on the sensorial perception hence affecting their consumer acceptance.

Apart from the type of solvent, the rate of mass transfer in OD depends on the concentration of the solvents in use, based on the fact that water loss and solute intake in FAV are enhanced by increasing the osmotic solution concentration until equilibrium has reached.[24,34,50] For example, studies on OD of watermelon slabs, apricot, and pomegranate arils in sucrose solution of 40, 50, and 60°Brix reported that the highest transfer of water was from the sample to the solution and solids in the reverse direction in the samples immersed in 60°Brix solutions.[13,17,30] After a longer duration of OD process, there occurs formation of a hard outer surface of the food samples due to a phenomenon called "case hardening" that reduces the mass flow rate to a larger extent.[34]

7.4.3 DURATION OF THE OD PROCESS

The process duration also affects the osmotic dehydration related to water loss and solid gain parameters. Generally, greater is the time of immersion of the product, the higher is the moisture loss and vice versa up to a certain limit called equilibrium, after which both occur at the same rate.[17,30,34] However, a significant reduction in water loss and solid gain occurs after 2 h of product immersion, during which the highest transfer rate occurs. The transfers are at a faster rate during the initial time of product immersion that determines the fate of the OD process significantly afterward. This fact labels the initial period of immersion as a crucial one.

7.4.4 THE TEMPERATURE OF THE OSMOTIC SOLUTION

Heating represents an important step in food processing and enzymatic deactivation produced externally followed by passing into the food product.[19] OD kinetics has been affected by the temperature markedly, which is a prime factor in the OD process.[50] Both the solid gain as well as water loss are increased upto 50°C, after which it decreases sharply near 60°C and remain nearly constant afterward. A temperature rise decreases the osmotic solution viscosity coupled with the plasticizing effect on the cell membranes that have been reported as the main factors responsible for the increase in the water loss during OD.

However, a less significant effect is evident on the solid uptake by increasing the temperature.[22] This is reported within a 30–60°C temperature range for FAV. However, Lazarides[23] witnessed >50% increase in solid gain during the process of OD of apples at a temperature of 30–50°C due to the increase in cell permeability caused by the plasticizing effect. FAV are affected adversely, when they are processed osmotically above a temperature range of 50°C, for example, blue berries can be damaged if the temperature is above 60°C during OD.[37] An increase in temperature during the OD process can significantly affect the chlorophyll and ascorbic acid retention.

7.4.5 AGITATION DURING THE OD PROCESS

Agitation reduces the mass transfer resistance resulting in increment in the osmotic dehydration rate. The role of agitation is quite beneficial due to the fact that it minimizes the mass transfer rates during OD due to the prevention of osmotic solution and food product contact by high concentrated solutions.[34] In general, agitation does not affect the solid gain directly during the OD process because of the limitless nature of the external transfer of osmotic solutes. Nevertheless, higher agitation rates for a longer duration can result in the reduction of the solid gain rate due to the creation of a thin subsurface layer by the slow solute diffusion rate and the change in solute concentration gradient.[50]

Agitation reduces the chances for the formation of dilute solution film on the sample surface.[2] However, agitation helps the process of water loss during OD during the beginning of the process and mostly at low temperature. Agitation is reported to have the least or no effect on the solids gain if the OD occurs for a shorter period. However, a drastic or steep decrease in the solute gain has been reported for the longer duration of OD process. Vigorous agitation may render the FAV to be damaged leading to reduced consumer acceptance. Agitation can be performed by the use of a simple centrifugal pump that causes syrup circulation inside the solution. Gupta et al.[15] reported that the water loss and solid gain rates were increased to a large extent by increasing the agitation coupled with an increase in temperature and immersion time during the Osmo-convective dehydration of honey-ginger candy.[15] Therefore, agitation helps to decrease the contact time between food material and the osmotic solution thereby increasing the rate of mass transfer in OD.

7.4.6 SAMPLE/SOLUTION RATIO

The ratio of sample-to-solution results in an increase in the water and solute transfer rates during the OD process, which is decreased afterward.[36] In FAV, the sample/solution ratio is an important determining factor for the extent of the osmotic dehydration process. Several researchers have worked on the OD of FAV using different sample/solution ratio ranging from 1:1 to 1:5.[15]

However, larger ratios (1:60, 1:30) have been used to prevent the possibility of significant dilution of the medium that can cause a reduction in the driving force necessary for OD.[8] The most common combinations in practical applications are 1:2 or 1:3.[49] Hence, it is suggested that one should select the sample-to-solution ratio wisely keeping in mind the dependence of water removal on the driving force created by this ratio.

7.4.7 PRODUCT-RELATED FACTORS

The shape of the food material undergoing the process of OD also affects the water loss and solid gain. For example, the rate of moisture loss and solid gain is increased due to the increased surface area coming in contact with the immersion medium.[50] Moreover, the involved diffusion length of water and solutes depends on the sample geometry that will finally determine the rate of osmotic dehydration. For example, a bigger sized sample under consideration has an enlarged path compared with its surface area responsible for the exchange of solids.

Therefore, type of geometry restricts the entry of the solids into the sample and the rate of OD is decreased.[41] Also, the change in the permeability of the semipermeable membrane to full permeable from an initial partial nature is responsible for the tissue architecture change. The pore size and their distribution along the material surface determine the water removal rate along with the material interaction with the surrounding liquid. Cell shrinkage due to water loss can cause cell damage as the contact of the cell wall with the cell membrane gets reduced. However, the cell shrinkage is affected by the product porosity, the rate of mass transfer, and rehydration ratio of the dehydrated material.[7]

7.5 RECENT DEVELOPMENTS IN OSMOTIC DEHYDRATION TECHNOLOGY

Exposure of food materials to prominent techniques (nonthermal), such as ultrasound, vacuum, pulsed electric field, gamma-irradiation before undergoing OD can increase the mass flow rate, which is otherwise reduced by the resistance offered by cell membranes.

7.5.1 MICROWAVE (MW)-ASSISTED OD

Microwave heating is linked to various food applications in the food processing industries among which cooking, drying, baking, sterilization, pasteurization, and preservation are prominent ones. Microwaves ((electromagnetic waves) possess the frequency in the range of 300 MHz to 300 GHz. In the field of food applications, only two frequencies such as 915 and 2450 MHz are allowed for microwave heating. Microwave heating can achieve high and uniform heating rates, thereby reducing the cooking time, and is easy to operate with safe-handling.

Compared with the conventional methods, microwave heating causes least alteration of flavor and sensory qualities and has therefore been practiced for decades. Cooking spaghetti using microwaves resulted in enhanced softer and gelatinized product.[10] Moreover, the products were observed to show higher gel degrees and better color retention with less thermal effects compared with the spaghetti cooked by traditional methods.

Microwave cooking prevents the loss of B-vitamins (riboflavin, thiamin, niacin, and pyridoxine) compared with the traditional boiling and autoclaving and also reduced the cooking time of chickpea. It plays a vital role in omega-3 fatty acid retention and cholesterol containment up to an optimum level in tuna fish.[44] Also, there are reports of promotion of starch gelatinization and phenolic compounds retainment in potato samples that were processed using microwaves thus helping in the shelf-life improvement due to the antioxidant activity of phenolic compounds.

Microwave heating source results in products with improved quality and can meet the current consumer choice compared with the OD process alone. A different mechanism of internal heat generation by the microwave field is followed in the food materials undergoing microwave-assisted osmotic dehydration process. Heat generation is actually through water ionic polarization, dipole rotation coupled with other dielectric properties in microwave field that results in the water transfer to the surface by creating internal pressure gradient.[12] Water molecules present in the food materials absorb the energy of microwaves and undergo rotation at a high speed, which ultimately results in heat generation.

The process of heat generation is quite complex because of various factors related to the microwave field. There are numerous advantages of using microwaves for the osmotic dehydration of food materials, such as selective heating, energy-efficient, faster heating, and lesser time for the start-up and more precise control. Microwaves have been used for a long time in Europe for food pasteurization of products, such as potato chips and pasta.[26] Moreover, the drying time is reduced with an increase in the microwave power. However, higher power may affect the food quality.

7.5.2 PULSED VACUUM (PV)-ASSISTED OD

Sometimes the OD process is assisted with vacuum pulse for shorter periods at the beginning to improve the solid gain and water removal through the hydrodynamics mechanism imposed due to capillary flow and pressure changes.[14] A pulsed vacuum application resulted in better water loss when applied to guava slices and solid gain was also reduced to an appreciable level.[11]

An important application of pulsed vacuum is the production of semidehydrated food products through osmotic dehydration, for example, osmotic dehydration of pumpkin previously blanched. The pumpkin samples were improved based on water loss, solid gain, water activity and color characteristics. Using pulsed vacuum in combination with OD can reduce the drying cost by decreasing the drying periods to an appreciable level. Also, it leads to minimum alteration in the sensory properties of FAV other than flavor. Moreover, applying a pulsed vacuum decreases the fruit hardness, the retention of vitamin C and volume ratio were increased in cherry tomatoes exposed to the pulsed vacuum before undergoing osmotic dehydration.[5]

Pulsed vacuum OD represents the widely used technique for improving elasticity, shrinkage and maximum stress in biological products. Carotenoids are handsomely preserved by vacuum application.[28] However, it leads to the loss of phenolic acid and betalain in eggplant and carrot. Besides, higher temperature usage during pulsed vacuum resulted in the color deterioration and skin browning of fruits. Hence, the higher temperature must be avoided as in the case of lemon slices.[52] The procedure followed in the pulsed vacuum osmotic dehydration of food materials involves the immersion of food material in osmotic solution first followed by the application of working pressure and maintaining that pressure for an extended period. Finally, the process ends by holding the food material under normal atmospheric pressure again.

7.5.3 HIGH HYDROSTATIC PRESSURE (HHP)-ASSISTED OD

High hydrostatic pressure (HHP) is introduced to assist the osmotic dehydration of food materials in comprehensive water removal, thereby producing the final product with improved quality as in banana, strawberry, pineapple, etc. It also reduces the processing time by increasing the rate of dehydration, hence avoiding negative impacts in FAV due to longer immersion time in osmotic solution.

Cell disruption along with enhancement in permeabilization and change in cell wall structure occurs in a high-pressure treatment at 100 to 800 MPa, which has proven be beneficial in increasing the rate of mass transfer during the osmotic OD process.[31] The high-pressure application at 100, 300 and 500 MPa for 10 min causes a reduction in both the diffusion coefficients for solute diffusing out of the tissue and in the rehydration ratio, at different temperatures of 5, 25, and 35°C.[40]

Using HPP in combination with osmotic dehydration results in the enhancement of mass transfer rate in banana, strawberries, pineapple, and potato.[38,48,51] Up to 400 MPa of HPP is acceptable in biological materials, after which hindrance to water diffusion along with other substances is evident due to starch gelation. HHP above 500 MPa results in the enhancement of nonenzymatic browning, quality impairment, and alteration in nutritional composition in FAV.[31] HHP also enhances the volatile profile of Candied Wumei fruit (*Prunus mume*).[27] The significant retention of antioxidant compounds was observed under HHP treatment at 5 MPa for 1 min.

7.5.4 ULTRASOUND (US)-ASSISTED OD

Ultrasound-assisted osmotic dehydration (USOD) is used to enhance safety in food materials besides helping in food processing and food analysis. Using lower temperature during USOD is appreciable if higher power ultrasonic waves at low frequencies are applied for the moisture removal.[20] Small channels are formed by US waves in the intercellular spaces that direct the outflow of osmotic solution forming partially dried food products. The pressure and frequency are prominent factors creating internal turbulence by intense cavitation in the food materials. This leads to the enhancement in the water transfer rate in food products due to the application of high-frequency US waves for longer period and applied after dipping the food material (fruits/vegetables) in a hypertonic solution. Combining other treatments with US waves have some beneficial effects on food materials, for example, high sugar concentration minimizes the time of OD process by increasing the diffusion rate of water out of the foods.

This method also helps in the complete and rapid removal of gases thus minimizing food degradation.[43] The US waves result in the reduction of time required for the OD of FAV, for example, a 30 min time supported the complete OD of broccoli in addition to retention of L-ascorbic acid up to 80% by using US waves compared with OD alone. Moreover, exposure of kiwi fruits for about 30 min resulted in 50% of mass gain using US waves at 35 kHz.The application of US waves at low frequency helps in the color retention of FAV.[42] However, US waves at high frequency may cause damage to tissues of FAV due to the heat released during sonication.

7.5.5 HIGH ELECTRIC FIELD PULSE (HEP)-ASSISTED OD

In this technique, high voltage amplitude electric waves are used to cause breakage and to form minute pores in cell membranes. This method involves several mechanisms, such as membrane compression, critical transmembrane potential, viscoelastic properties, and cell membrane defects based on structural aspects. These lead to perforation in the cell membrane to boost the process of water and solute diffusivity.[247,53]

This method is energy efficient as the time of exposure is reduced to microseconds. An improvement in tenderness and reduction in shear force was observed in beef muscles processed with this technique.

Moreover, biocompounds are retained in dried samples that were dehydrated using pulsed electric waves. High electric pulse (HEP) reduces the loss of nutrients coupled with color and flavor, thereby preserves the nutritional, sensory, and freshness of the minimally processed foods. For best performance, pulse shape, field strength, pulse count, and duration of the pulse must be optimized. For example, enhanced solid gain and water loss were recorded in strawberry halves when exposed to high-intensity electric field pulse for a short period as a pretreatment for their preservation in the OD process.[48]

An increase in the number of electric field strength during the food material exposure to HEP coupled with the pulse rate increment also increases the water loss along with solid gain in apples.[4] HEP also helps in the retention of ascorbic acid by using HEP as a pretreatment before OD process in red peppers. It is suggested to use a pulse rate of 10 along with 5 kV/cm of electric field intensity.

7.6 SUMMARY

The OD is a potential process for food preservation through the removal of contained water partially that needs further treatments. Moreover, nutritional, organoleptic, textural, and functional properties are improved to an appreciable profile due to the food constituents leaching along with the solid uptake during the process of OD. It helps in maintaining the freshness of minimally processed FAV. The concentration, temperature, type of osmotic solution, and different pretreatments can affect the rate of mass transfer during OD. Based on the fact of the slow nature of OD, some promising pretreatments are carried out to improve the rate of water transfer. These include HHP, HEP, pulsed vacuum, microwave, and ultrasound which are used in combination with osmotic dehydration.

ACKNOWLEDGMENTS

The authors gratefully acknowledge the support from the Department of Postharvest Engineering and Technology, Faculty of Agricultural Sciences, Aligarh Muslim University, India. The first author (Arshied Manzoor) also acknowledges the award of ICMR-Senior Research Fellowship from the Indian Council of Medical Research, New Delhi, India.

KEYWORDS

- **food preservation**
- **food processing**
- **fruits and vegetables**
- **osmotic dehydration**
- **shelf life**
- **water activity**

REFERENCES

1. Ade-Omowaye, B. I. O.; Angersbach, A.; Taiwo, K. A.; Knorr, D. Use of Pulsed Electric Field Pre-Treatment to Improve Dehydration Characteristics of Plant-Based Foods. *Trends Food Sci. Technol.* **2001,** *12* (8), 285–295.
2. Akbarian, M.; Ghasemkhani, N.; Moayedi, F. Osmotic dehydration of Fruits in Food Industrial: A Review. *Int. J. Biosci.* **2014,** *4* (1), 42–57.
3. Ali, H. S.; Moharram, H. A.; Ramadan, M. T.; Ragab, G. H. Osmotic Dehydration of Banana Rings and Tomato Halves. *J. Am. Sci.* **2010,** *6* (9), 384–390.

4. Amami, E.; Vorobiev, E. Effect of pulsed electric field on the osmotic dehydration and mass transfer kinetics of apple tissue. *Dry. Technol.* **2005,** *23* (3), 581–595.
5. An, K.; Li, H.; Zhao, D.; Ding, S.; Tao, H.; Wang, Z. Effect of Osmotic Dehydration With Pulsed Vacuum on Hot-Air Drying Kinetics and Quality Attributes of Cherry Tomatoes. *Dry. Technol.* **2013,** *31* (6), 698–706.
6. Azoubel, P. M.; Da Silva, F. O. Optimization of Osmotic Dehydration of 'Tommy Atkins' Mango Fruit. *Int. J. Food Sci. Technol.* **2008,** *43* (7), 1276–1280.
7. Bekele, Y.; Ramaswamy, H. Going Beyond Conventional Osmotic Dehydration for Quality Advantage and Energy Savings. *Ethiopian J. Appl. Sci. Technol.* **2010,** *1* (1), 1–15.
8. Chabbouh, M.; Hajji, W.; Hadj Ahmed, S. B.; Farhat, A.; Bellagha, S.; Sahli, A. Combined Effects of Osmotic Dehydration and Convective Air Drying on Meats: Kinetics and Quality. *Dry. Technol.* **2011,** *29* (13), 1571–1579.
9. Chandra, S.; Kumari, D. Recent Development in Osmotic Dehydration of Fruit and Vegetables: A Review. *Crit. Rev. Food Sci. Nutr.* **2015,** *55* (4), 552–561.
10. Cocci, E.; Sacchetti, G.; Vallicelli, M.; Angioloni, A.; Dalla, R. M. Spaghetti Cooking by Microwave Oven: Cooking Kinetics and Product Quality. *J. Food Eng.* **2008,** *85* (4), 537–546.
11. Corrêa, J. L.; Pereira, L. M.; Vieira, G. S.; Hubinger, M. D. Mass Transfer Kinetics of Pulsed Vacuum Osmotic Dehydration of Guavas. *J. Food Eng.* **2010,** *96* (4), 498–504.
12. Drouzas, A. E.; Schubert, H. Microwave Application in Vacuum Drying of Fruits. *J. Food Eng.* **1996,** *28*, 203–209.
13. Falade, K. O.; Igbeka, J. C.; Ayanwuyi, F. A. Kinetics of Mass Transfer and Colour Changes During Osmotic Dehydration of Watermelon. *J. Food Eng.* **2007,** *80*, 979–985.
14. Giraldo, G.; Talens, P.; Fito, P.; Chiralt, A. Influence of Sucrose Solution Concentration on Kinetics and Yield During Osmotic Dehydration of Mango. *J. Food Eng.* **2003,** *58* (1), 33–43.
15. Gupta, R.; Singh, B.; Shivhare, U. S. Optimization of Osmo-Convective Dehydration Process for the Development of Honey-Ginger Candy Using Response Surface Methodology. *Dry. Technol.* **2012,** *30* (7), 750–759.
16. Islam, M. N.; Flink, J. N. Dehydration of Potato: II. Osmotic Concentration and Its Effect on Air Drying Behavior. *Int. J. Food Sci. Technol.* **1982,** *17* (3), 387–403.
17. Ispir, A.; Toğrul, Đ. T. Osmotic dehydration of apricot: Kinetics and the Effect of Process Parameters. *Chem. Eng. Res. Design* **2009,** *87*, 166–180.
18. Jackson, T. H.; Mohammed, B. B. The Shambat Process: New Development Arising from the Osmotic Dehydration of Fruits and Vegetables. *Sudan J. Food Sci. Technol.* **1971,** *3*, 18–22.
19. Jan, B.; Shams, R.; Rizvi, Q. E. H.; Manzoor, A. Ohmic Heating Technology for Food Processing: A Review of Recent Developments. *J. Postharvest Technol.* 2021, *9* (1), 20–34.
20. Kapturowska, A.; Stolarzewicz, I.; Chmielewska, I.; Białecka-Florjańczyk, E. Ultrasounds a Tool to Inactivate Yeast and to Extract Intracellular Protein. *Żywnosc Nauka Technology Jakość* 2011, *4* (77), 160–171.
21. Kaymak-Ertekin, F.; Sultanoğlu, M. Modelling of Mass Transfer During Osmotic Dehydration of Apples. *J. Food Eng.* **2000,** *46* (4), 243–250.
22. Khan, M. R. Osmotic Dehydration Technique for Fruits Preservation—A Review. *Pak. J. Food Sci.* **2012,** *22* (2), 71–85.

23. Lazarides, H. N. Osmotic Pre-Concentration: Developments and Prospects. In *Minimal Processing of Foods and Process Optimization: An Interface*; Singh, R. P., Oliveira, F. A. R., Eds.; CRC Press: Boca Raton, 1994; pp 17–33.
24. Lenart, A. Mathematical Modelling of Osmotic Dehydration of Apple and Carrot. *Polish J. Food Nutr. Sci.* **1992,** *1*, 1–33.
25. Lenart, A.; Flink, J. M. Osmotic Concentration of Potato. Criteria for the End Point of the Osmotic Process. *J. Food Sci. Technol.* **1984,** *19*, 45–63.
26. Li, H.; Ramaswamy, H. S. Osmotic Dehydration of Apple Cylinders: III. Continuous Medium Flow Microwave Heating Conditions. *Dry. Technol.* **2006,** *24* (5), 643–651.
27. Luo, W.; Tappi, S.; Wang, C.; Yu, Y.; Zhu, S.; Dalla, R. M.; Rocculi, P. Effect of High Hydrostatic Pressure (HHP) on the Antioxidant and Volatile Properties of Candied Wumei Fruit (*Prunus mume*) During Osmotic Dehydration. *Food Bioprocess Technol.* **2019,** *12* (1), 98–109.
28. Manzoor, A.; Dar, I. H.; Bhat, S. A.; Ahmad, S. Flavonoids: Health Benefits and Their Potential Use in Food Systems. In *Functional Food Products and Sustainable Health*; Springer: Singapore, 2020; pp 235–256.
29. Mohanty, S.; Cock, I. E. Bioactivity of *Syzygium jambos* methanolic Extracts: Antibacterial Activity and Toxicity. *Pharmacognosy Res.* **2010,** *2* (1), 4–9.
30. Mundada, M.; Hathan, B. S.; Maske, S. Mass Transfer Kinetics During Osmotic Dehydration of Pomegranate Arils. *J. Food Sci.* **2011,** *75*, 31–39.
31. Núñez-Mancilla, Y.; Vega-Gálvez, A.; Pérez-Won, M.; Zura, L.; García-Segovia, P.; Di Scala, K. Effect of Osmotic Dehydration Under High Hydrostatic Pressure on Microstructure, Functional Properties and Bioactive Compounds of Strawberry (FragariaVesca). *Food Bioprocess Technol.* 2014, *7* (2), 516–524.
32. Osorio, C.; Franco, M. S.; Castaño, M. P.; González-Miret, M. L.; Heredia, F. J.; Morales, A. L. Colour and Flavour Changes During Osmotic Dehydration of Fruits. *Innov. Food Sci. Emerg. Technol.* **2007,** *3* (8), 353–359.
33. Peiró-Mena, R.; Camacho, M. M.; Martínez-Navarrete, N. Compositional and Physicochemical Changes Associated to Successive Osmo-Dehydration Cycles of Pineapple (*Ananas comosus*). *J. Food Eng.* **2007,** *79* (3), 842–849.
34. Phisut, N. Factors Affecting Mass Transfer During Osmotic Dehydration of Fruits. *Int. Food Res. J.* **2012,** *19* (1), 7–18.
35. Piasecka, E.; Uczciwek, M.; Klewicki, R.; Konopacka, D., Mieszczakowska-Frac, M.; Szulc, M.; Bonazzi, C. Effect of Long-Time Storage on the Content of Polyphenols and Ascorbic Acid in Osmo-Convectively Dried and Osmo-Freeze-Dried Fruits. *J. Food Process. Preserv.* **2013,** *37* (3), 198–209.
36. Ponting, J. D. Osmotic Dehydration of Fruits. *Food Technol.* **1966,** *20*, 125–128.
37. Ramaswamy, H. S. Osmotic Drying. Unpublished Paper Presentation at Workshop on Drying of Food and Pharmaceuticals. Kolkata, India: Fourth Asia Pacific Drying Conference, 12 December 2005; p 8.
38. Rastogi, N. K.; Niranjan, K. Enhanced Mass Transfer During Osmotic Dehydration of High Pressure Treated Pineapple. *J. Food Sci.* 1998, *63* (3), 508–511.
39. Rastogi, N. K.; Raghavarao, K. S. M. S. Water and Solute Diffusion Coefficients of Carrot as a Function of Temperature and Concentration During Osmotic Dehydration. *J. Food Eng.* **1997,** *34* (4), 429–440.
40. Rastogi, N. K.; Angersbach, A.; Niranjan, K.; Knorr, D. Rehydration Kinetics of High Pressure Pretreated and Osmotically Dehydrated Pineapple. *J. Food Sci.* 2000, *65* (5), 838–841.

41. Rastogi, N. K.; Raghavarao, K. S. M. S.; Niranjan, K.; Knorr, D. Recent Developments in Osmotic Dehydration: Methods to Enhance Mass Transfer. *Trends Food Sci. Technol.* **2002,** *13* (2), 48–59.
42. Shamaei, S.; Emam-Djomeh, Z.; Moini, S. Ultrasound Assisted Osmotic Dehydration of Cranberries: Effect of Finish Drying Methods and Ultrasonic Frequency on Textural Properties. *J. Textural Stud.* **2011,** *43*, 133–141.
43. Silva, J. L.; Stojanovic, J. Influence of Osmotic Concentration, Continuous High Frequency Ultrasound and Dehydration on Antioxidants, Colour and Chemical Properties of Rabbiteye Blueberries. *Food Chem.* 2007, *101*, 898–906.
44. Stephen, N. M.; Shakila, J. R.; Jeyasekaran, G.; Sukumar, D. Effect of Different Types of Heat Processing on Chemical Changes of Tuna. *J. Food Sci. Technol.* **2010,** *47*, 174–181.
45. Sunjka, P. S.; Raghavan, G. S. V. Assessment of Pretreatment Methods and Osmotic Dehydration for Cranberries. *Can. Biosyst. Eng.* **2004,** *46* (1), 45–48.
46. Tadesse, T. F.; Abera, S.; Worku, S. Nutritional and Sensory Properties of Solar Dried Carrot Slices as Affected by Blanching and Osmotic Pre-Treatments. *Int. J. Food Sci. Nutr. Eng.* **2015,** *5* (1), 24–32.
47. Taiwo, K. A.; Angersbach, A.; Knorr, D. Effects of Pulsed Electric Field on Quality Factors and Mass Transfer During Osmotic Dehydration of Apples. *J. Food Process Eng.* **2003,** *26* (1), 31–48.
48. Taiwo, K. A.; Eshtiaghi, M. N.; Ade-Omowaye, B. I.; Knorr, D. Osmotic Dehydration of Strawberry Halves: Influence of Osmotic Agents and Pretreatment Methods on Mass Transfer and Product Characteristics. *Int. J. Food Sci. Technol.* 2003, *38* (6), 693–707.
49. Tiwari, R. B. Application of Osmo-Air-Dehydration for Processing of Tropical Fruits in Rural Areas. *Indian Food Ind.* **2005,** *24* (6), 62–69.
50. Tortoe, C. A Review of Osmo-Dehydration for Food Industry. *Afr. J. Food Sci.* **2010,** *4* (6), 303–324.
51. Verma, D.; Kaushik, N.; Rao, P. S. Application of High Hydrostatic Pressure as a Pretreatment for Osmotic Dehydration of Banana Slices (*Musa cavendishii*) Finish Dried by Dehumidified Air Drying. *Food Bioprocess Technol.* 2014, *7*, 1281–1297.
52. Wang, J.; Law, C. L.; Nema, P. K.; Zhao, J. H.; Liu, Z. L.; Deng, L. Z.; Xiao, H. W. Pulsed Vacuum Drying Enhances Drying Kinetics and Quality of Lemon Slices. *J. Food Eng.* **2018,** *224*, 129–138.
53. Wiktor, A.; Witrowa-Rajchert, D. Applying Pulsed Electric Field to Enhance Plant Tissue Dehydration Process. *Żywnosc Nauka Technology Jakość* **2012,** *19* (2), 22–32.
54. Yadav, A. K.; Singh, S. V. Osmotic Dehydration of Fruits and Vegetables: A Review. *J. Food Sci. Technol.* **2014,** *51* (9), 1654–1673.

CHAPTER 8

SCOPE OF ULTRASOUND TECHNOLOGY FOR PRESERVATION OF FRUITS AND VEGETABLES

GAZIA NASIR, SADAF ZAIDI, and FAIZAN AHMAD

ABSTRACT

The ultrasound technology has enumerated for minimizing processing, ensuring safety and enhancing the quality of food products. This chapter sheds some light on the mechanisms of ultrasound and its applications in decontamination, drying and freezing of fruits and vegetables (FAV). The chapter also discusses the influence of ultrasound pretreatment and ultrasound-assisted drying on the drying of FAV. The applications of ultrasound on initiation of ice nucleation, improving rate of freezing and controlling the size and shape of crystals has also been discussed in detail. Apart from this, the impact of ultrasound-assisted freezing on the food quality has been highlighted. In addition, the negative impacts of power ultrasound on FAV preservation has been presented.

8.1 INTRODUCTION

Currently, interest has been growing in the food sector concerning novel methods to preserve food depriving of additives, maintenance of sensory and nutritional attributes with decreased consumption of energy, at a reasonable price and making use of eco-friendly products with its safety.[78]

The ultrasound (US) methods have been introduced to reduce the time of processing, ensure safety, and improve the food quality. Ultrasonic technology

Quality Control in Fruit and Vegetable Processing: Methods and Strategies. Megh R. Goyal, & Faizan Ahmad (Eds.)

is an important field of research and development and it is based on mechanical waves operating at a frequency greater than 16 kHz and can be grouped into two frequency ranges: high and low.[52]

Ultrasonic waves with operating frequency >16 kHz are not perceived by the human ear. High-power ultrasonic waves are used by marine species to strike their prey while dolphins and bats make use of low-power ultrasonic waves. High- and low-power US have had certain implementations in the food preservation and processing sectors. High-power US may cause physical destruction of tissues, form emulsions, decontaminate instruments, and augment chemical reactions. Low-power US is a nondestructive technique to analyze the structure and composition of food substances. The various properties of sound waves (interference, scattering, reflection, and absorption) are observed in US. Also, US can propagate in all three media, such as solid, liquid, and gas.[61]

It has been confirmed by the U.S. Food and Drug Administration (FDA) that US has been efficient against microorganisms in fruit juices (for instance, *Escherichia coli*) and can fulfill the prerequisite of a 5-log reduction of certain pathogens.[110] Fruits and vegetables (FAV) have high level of moisture, which provides a favorable environment for the growth of microorganisms, causing immutable damage to the produce. Therefore, it is vital to minimize the moisture levels and loss of nutrients by applying efficient methods of preservation like drying. Thus, moisture removed in the drying process not only reduces the growth of microbes but also minimizes the packaging space requirement along with a reduction in total weight of the food material and shortening the transportation and packaging cost.[28]

The quality of the dried products can be improved by employing the process of pretreatment prior to drying.[25,53,64,71] Numerous methods have been established for the application of pretreatment before drying, as it is a widely researched area. To improve the effectiveness of drying technology, various pretreatment techniques have been evolved to draw the attention of the researchers.[55]

Wiktor et al.[92] investigated the synergistic influence of US pretreatment and pulsed electric field (PEF) on the drying properties of carrots. They found that the US pretreatments reduced the time of drying upto 40% with increased retention of carotenoids. Tao et al.[86] determined the influence of US pretreatment on the drying of mulberry leaves. A significant reduction was found in the time of drying, which in turn decreased the energy consumption.[28]

Freezing is a widely accepted technique for food product preservation to retain the nutritional and sensory properties.[100] During the storage of frozen foods, the physicochemical and biochemical reactions are slowed down that lead to nutritional loss, unappealing changes of texture, color, and formation of off-flavor.[11] The efficiency of freezing and the quality of frozen food products (FFP) are determined by ice crystallization. The original characteristics of frozen foods can be preserved by applying quick-freezing technique to form uniformly distributed fine ice crystals both inside and outside the cells and causing no damage to the food tissues.[5]

The freezing can be accelerated by intensifying ice crystals nucleation and enhancing the degree of heat and mass transfer in the case of US-assisted freezing method.[19] Additionally, the development of fine crystals in FFP in US-assisted freezing can enable efficient retention of food quality as compared with the traditional methods of freezing.[19]

This chapter highlights the applications of US technology in decontamination, drying, and freezing of FAV. The chapter also discusses the impact of US pretreatment (PRT) on drying and US-assisted drying. Furthermore, the impact of US-assisted freezing on the quality characteristics of foods have been discussed. In addition, the negative impacts of power US on the preservation of FAV have also been included in brief.

8.2 ULTRASOUND MECHANISM

The two phenomena that are associated with the US ability for the inactivation of microbes and in turn causing decontamination of food include acoustic cavitation and acoustic streaming.

In acoustic cavitation, areas of fluctuating compression and rarefaction are caused as a result of the generation of longitudinal waves. During the rarefaction phase, small bubbles in the liquid get enlarged as a result of reduced pressure in the local region, below the vapor pressure of the liquid, thus causing a change in intensities. In the compression cycle, the surface area of the bubble gets reduced. While there is an increase in the size of the bubble and the surface area becomes large, the liquid medium fails to absorb all the gases present in the bubbles thus resulting in further enlargement of the bubble. Thus, when the energy provided by US gets in line with the oscillation of the bubble wall, there is loss of stability and implosion, initiating the areas of high pressure (up to 1000 atmospheres) and high temperature (up to 5000°C).[56,62]

Acoustic streaming produces a dispersion in the acoustic energy causing an inclination in momentum, thus the fluid currents.[50] The velocity attained by the fluid enables a greater convection heat transfer coefficient near the boundaries of solid, occasionally resulting in turbulence and stimulating rate of heat transfer. This effect permits improved medium heating due to the dispersion in the mechanical energy.[47] All these cases lead to bacterial disturbances.[77] Enhancing pressure and temperature can lead to diverse chemical changes in both the surrounding fluid as well as in the vapor phase interior of the cavitation bubble.[37] Also, turbulence-located microcirculation is produced as a result of this phenomenon. The implosions generated are asymmetric nearer to a surface of the solid, producing a microjet that strikes the solid,[8] and these alterations give rise to the mechanism necessary in cleaning surfaces. The amount of energy liberated by the cavitation process depends upon the collapse and kinetics of expansion of the bubbles. Certainly, these microjets can remove the clinged bacteria on the surface of FAV.[78]

US produces and disseminates implosions of cavitation in a liquid medium. The energy liberated in the cavitation process enables the cleaning of those surfaces that are usually harder to reach by other methods.[24] Sometimes, chemical sanitizers are inefficient against microbes that are either tightly clinged on surfaces or found on biofilms. Under these situations, physical methods are preferred. US is a stable and steady method for the effective removal of contaminants. The application of ultrasonic technology in the food industrial sector is reassuring due to the presence of microbial biofilm on the exterior of FAV. US has several benefits, such as cost-effectiveness, improved processing time, minimized chemical damage and physical risks, selectivity and productivity, and enhanced quality. Nowadays, applications of US technology have gained popularity for their key role in causing a damage-free sustainable environment and therefore apply to the notion of green technology.[9]

8.3 APPLICATION OF POWER ULTRASOUND METHOD IN DECONTAMINATION OF FRUITS AND VEGETABLES

Ultrasound (US) technology is highly efficient in reducing the spoilage and pathogenic microbes present in FAV. In general, microbes are directly killed or removed by cavitation due to the combination of the following effects[35]: (1) Mechanical: involves the formation of turbulence, shear stresses, and liquid circulation currents. (2) Chemical: the formation of free radicals (H· and OH·) during cavitation held in an aqueous medium strike the chemical composition of the microbial cell wall and erode the cell wall till it disintegrates. (3) Physical: generation of acute local pressure and temperature.

Therefore, treating fresh FAV before storage using ultrasonic waves has been studied by several researchers. The research studies have been conducted on various FAV, such as green asparagus, plum fruit, spinach, strawberry, cherry tomatoes, apples, broccoli seeds, peach, lettuce, carrots, and organic lettuces [7,10,29,41,75,76,79,91,102,107] as given in Table 8.1. These findings confirm that the US treatment was very effective in reducing the amount of spoilage causing microbes, such as, bacteria, molds, and yeasts in fresh produce, and as a result, minimized the incidences of decay and prolonged the time of storage of FAV.[101]

Although treatment with US causes a reduction in the microbial count to a certain extent, yet it is not much effective for industrial application due to its low ability to sterilize.[76] To attain a satisfactory inactivation effect, an increased level of acoustic energy and prolonged time of treatment is needed. But this in turn leads to the deterioration of the tissues of fresh product and results in inflation in the susceptibility of the infestation of spoilage causing microbes.[76,81]

Additionally, the microbes that detach from FAV can be emancipated into the water used for washing them and thus have the possibility to cause cross-contamination.[81] It has been investigated that the ultrasonic technique along with other techniques might be advantageous in causing surface decontamination. US can be combined with ultraviolet radiation, high pressure, heat, or chemical methods for cleaning and sanitization. Heat and pressure are not endorsed in fruits/vegetables, as they can drastically alter the structure of plant tissues.[78] To overcome these limitations and to enhance sterilization efficiency, US is frequently combined with other techniques. The synergistic effect of US with sanitizers finds more applications as compared with other sterilization techniques.

Apart from the direct killing of microbial cells, US also enhances antimicrobial response by degrading the cell wall, which in turn causes the creeping of sanitizing agents. Moreover, apart from causing inactivation, US is an efficient technique for removing microbes on the surface of fruits/vegetables due to the nonsymmetric collapse of cavitation bubbles nearer to solid and forms a powerful microjet that dislodges the microbial cells. In particular, US cleaning is beneficial as it can easily reach crevices that are not often accessible by using other conventional techniques of cleaning.[54]

Thus, ultrasonic waves can loosen microbes from fresh commodities into sanitizer fluids, which evokes the exposure of microbes that are located in crevices. Hence, the combined effect of US and sanitizer proves effective synergy. In the year 2002, Seymour et al. [81]demonstrated the synergistic influence of US (25–70 kHz) and chlorinated water for their efficiency to decontaminate microbes. Consequently, many additional research studies were carried out involving the effect of various sanitizers on FAV.[1,29] These sanitizers may include artificial chemicals, such as chloric disinfectants (NaClO, aqueous ClO_2, and chlorinated water, etc.), ethanol, H_2O_2, and natural antimicrobials (few organic acids: acetic acid, citric acid, peracetic acid, lactic acid, and malic acid) have been frequently used (Table 8.1).

Recently, several research studies were conducted on the use of possibility and applicability of this postharvest technology and it has gained immense popularity.[10,69,76,81,107] These studies have confirmed that the synergistic treatment effectively removed the different microbial species and US method enhanced the antimicrobial efficiency of sanitizers. Moreover, this synergistic method was frequently used to decontaminate pathogenic bacteria, namely, *E. coli* O157:H7, *Listeria monocytogenes*, and *Salmonella typhimurium* for the sake of increased sterilization efficiency.[29,76,81,107]

8.4 POTENTIAL APPLICATION OF ULTRASOUND METHOD IN DRYING

8.4.1 INFLUENCE OF US PRETREATMENT ON DRYING

The hurdles in drying namely poorer quality of the product and prolonged time of drying can be overcome by involving pretreatments (PRTs) before drying of food produce. Numerous studies have documented the influence of different PRTs on the drying rate of fruits and their quality.[42,64] Manifold effects have been exhibited by various PRTs on final commodities. PRTs are frequently used to minimize the initial moisture content and to hasten the process of drying, leading to enhanced quality of the finished dried commodities with reduced cost. This can be attained by altering the structure of the substance to be dried so that there is an increase in the water diffusivity as well as hindering enzymatic activity or killing microbes and enzymes.[55] US-PRT before the drying process has gained a lot of attention recently and has exhibited the possibility of decreasing the time of drying significantly.

Wiktor et al.[92] investigated the collaborative effect of US-PRTs and PEF on the drying properties of carrots. They confirmed that the drying time was reduced up to 40% with greater retention of carotenoids. Tao et al.[86] also examined the influence of US-PRTs on drying of mulberry leaves. It

TABLE 8.1 Applications of US in Fruit and Vegetable Decontamination

Fruits and vegetables	Microorganism species	Conditions	Log reduction	Reference
Apple	*E. coli* O157: H7	US (170 kHz, 10 min) + 20 mg L^{-1} ClO_2	3.54 unit	[29]
	Salmonella		4 unit	
Broccoli	*E. coli* O157:H7	US (40 kHz, 30 min, 23°C)	1.04 unit	[41]
		US (40 kHz, 5 min, 55°C)	1.44 unit	
Cherry tomatoes	*Salmonella Typhimurium*	US (40 kHz, 10 min, 24°C) + 40 mg L^{-1} Peracetic acid	4 log unit	[79]
Green asparagus	Bacteria, mold	US 40 KHz, 350W, 250 g, 10 min, RT	-	[91]
Lettuce	*Bacillus cereus spores*	US (40 kHz, 5 min, 20 °C) + 0.1% Tween 20	2.49 unit	[75]
Carrots			2.22 unit	
Organic lettuce	*E. coli* O157: H7	US (40 kHz, 5 min, 20°C) + lactic acid (2%)	2.75 unit	[76]
	Salmonella typhimurium		2.71 unit	
	L. monocytogenes		2.50 unit	
Peach fruit	*Penicillium expansum*	US 40 kHz, 10 min, 20°C + salicylic acid (0.05 mM)	Reduce blue mold in peach fruit	[102]
Plum fruit	Total aerobic mesophilic bacteria; total aerobic psychrotrophic bacteria; yeasts and mold	US 40 kHz, 100W, 1 kg/5 L, 10 min + 40 mg/L [b]ClO_2	2.0 to 3.0	[10]
Spinach	*E. coli* O157:H7	US 21.1 kHz, 200 W/L, 25 g/500 mL, 2 min, RT[a]+ ASC (10% NaCl solution/5% citric acid); AEWc (80 mg/L); POAAd (80 mg/L); Chlorine solution (200 mg/L)	0.7–1.1 addition (compared with sanitizer alone)	[107]
Strawberry	Mold and yeast	US (40 kHz, 10 min, 20°C)	1.73 unit	[7]
	Mesophilic aerobic		1.49 unit	

[a]RT: Room temperature; [b]ClO_2: Chlorine dioxide; unit = log CFU per g

was found that PRTs resulted in reduced drying time and thus decreased the energy consumption significantly.

8.4.1.1 PARAMETRIC EFFECTS OF US-PRT

The drying properties and the quality of food substances during US-PRTs are significantly affected by the amplitude, frequency, time of sonication, and ultrasonic power of the sonicator probe. For instance, Nowacka et al.[60] examined the influence of US-PRT on convective drying (CD) of apples for a varying duration (10, 20, 30 min) in an ultrasonic bath. It was found that the time of drying was increased with sonication time at first, but then it was decreased, that is, amidst three pretreated samples, the apple treated with 20 min US-PRT showed the longest time of drying. The time of sonication presents diversified effects on food quality, water loss, and drying rate. Substantially, the employment of osmotic solution resulted in a loss of water in food products on pretreating with US.

As confirmed by Rodrigues et al.,[70] loss of water in the osmotic fluid in papaya was enhanced with an increase in the time of sonication. Similar findings were observed in guava, pineapple, melon, strawberry, and pomegranate.[2,3,16,20,21,36] A positive loss of water was shown in all these fruits and it was also observed that loss in water was intensified with the extension of the sonication time.

The time of sonication affects the quality of food and the rate of drying. Nowacka et al.[60] found that the rate of drying was slowest at 20 min in the case of pretreatment of apple cubes with US (10, 20 and 30 min). On increasing the time of sonication, a significant increase in the color change was observed for food materials.[3,36,66] And there was a decrease in antioxidant activity.[66,71] The effect of frequency, ultrasonic power, and amplitude was low in comparison to the influence of time of sonication.

Kek et al.[36] examined the impact of US-PRT on the guava slices in osmotic fluid and observed the loss of water. As the amplitude was increased, the loss of water was also increased. However, there was a significant reduction in the retention of antioxidants and the time of drying whereas the color change was increased with amplitude. Also, with the decrease in the level of frequency and an increase in the level of ultrasonic power, an increase in the weight loss and color change were observed. Therefore, food substances may either gain or lose moisture during US-PRT. However, these results are due to the synergistic influence of US-PRT and difference in the concentration between food substances and the liquefied media (osmotic fluid or distilled water).

8.4.1.2 INFLUENCE OF US-PRT ON DRYING KINETICS

There is always a rise in the drying kinetics of food materials on the application of US before drying. Fernandes et al.[21] applied ultrasonic-PRT on the CD of pineapple. They observed that there was an increase in the effective water diffusivity by 64%. Romero et al.[71] studied the influence of ultrasonic-PRT on the CD of Andean blackberry. They observed the enhanced rate of drying by five times when US-PRT was given to blackberries for 20 min and then further drying at 60°C. Even at low temperatures, the effective diffusivity was increased by 210%.

Ricce et al.[68] evaluated ultrasonic-PRT before the hot-air drying. It was determined that the rate of drying with US-PRT for 30 min and further drying at 40°C was low as compared with the control sample. The increment or uniform depletion of the drying phenomena with the ultrasonic-PRT may be the inclusive influence of the conditions of drying, the structure of food, and increase or decrease in water during the ultrasonic-PRT. As discussed before, the ultrasonic-PRT application led to cell damage and the occurrence of microscopic channels that make the movement of water easier. As there is a decrease in the resistance in the migration of water, the drying phenomena hastens.[28]

8.4.1.3 INFLUENCE OF US-PRT ON FOOD QUALITY

The quality of food materials gets influenced by ultrasonic-PRT. A major concern is involved in physical characteristics such as water activity. Kowalski et al.[46] observed that ultrasonic-PRT resulted in a remarkable reduction (66% for UDODIT) in the level of water activity in carrot. These results were analogous with the results of Kowalski and Szadzińska[45] in the drying of cherry using convective-intermittent method. They observed a major reduction of 18% in water activity of cherry. Similar findings were reported in the CD of mushroom (notable reduction of 19%) and cherry (slight reduction of 4–7%), [6,83] and microwave-assisted hot-air drying of cranberry.[82]

Varying results were obtained on the impact of ultrasonic-PRT on the water activity for drying of apples by convection with US at 35 kHz frequency and a major increase of 10–13% was observed.[22] The color change is also a vital parameter in the case of the dried food products that must be considered for evaluation of food quality. Dehsheikh et al.[17] demonstrated a major reduction of 33.6% in color change of the convective-dried banana slices in conjunction with ultrasonic-PRT. Nowacka et al.[60] determined that the rehydration ability of apple under US treatment was low in level as compared with the untreated sample.

8.4.2 US-ASSISTED DRYING (USAD) OF FOODS

8.4.2.1 INFLUENCE OF VARIOUS PARAMETERS ON US-ASSISTED DRYING

The quality of food products and the drying properties are affected mainly by the ultrasonic power, the type of application, and the distance of ultrasonic radiation. For instance, Schossler et al.[80] determined the intermittent and continuous USAD of bell pepper and apple. They found that the rate of drying was significantly enhanced even though the sonication time was reduced to half. Although, on reduction of net sonication time by 10%, there was no augmentation of the drying kinetics. The quality of food products was also affected by the US power.

Tao et al.[87] dried garlic slices by convection using contact US technology. They observed that color change was enhanced with an increase in the US power intensity as compared with the pure convective drying. It confirmed that the modern power by US is very effective in preserving the native white color of the garlic.

8.4.2.2 INFLUENCE OF US-ASSISTED DRYING ON DRYING KINETICS

On application of ultrasonic technology during drying, an increase in the drying kinetics of food materials has been observed. For instance, Baslar et al.[4] performed a comparative study of VD, OD, and USV drying of fish fillets. They observed the reduced time of drying in Salmon fillets by performing USV drying. The time of drying was reduced by 44 to 67% in comparison with OD drying, while 7.4–25.7% in comparison with VD. Also for trout fillets, the drying time was decreased by 52.5–75.8% in comparison with OD drying, while 21.9–27.4% in comparison with VD. The ultrasonic technology also finds application in reducing the consumption of energy.

In a study by Kowalski et al.,[43] there was 10% reduction in energy consumption. Sabarez et al.[72] studied the CD of apple by US-assisted technology. A decrease in the level of drying time by 46–57% and 42–54% decrease in energy consumption was observed while using the US technology.

8.4.2.3 *EFFECT OF US-ASSISTED DRYING ON FOOD QUALITY*

The chemical and physical properties of food materials is also influenced by US-assisted drying. One of the crucial physical characteristics is water activity. Szadzinska et al.[85] observed that CD of strawberries was improved with US and microwave combined and separated. The water activity was reduced (34% for CVMWUS) as compared with the CV drying. The color change is also a vital parameter while taking in to account the quality of dried food products.

8.5 CHARACTERISTICS OF US POWER-ASSISTED FREEZING

Three facets of US can be described as the mechanism of freezing stimulation. At first, the acoustic cavitation leads to the formation of microstreaming that results in a rigorous collision among microparticles of the medium, producing very thin solid–liquid linkage.[27,94] In the second place, the formation of cavitation bubbles in the medium of transport along with the disintegration of the bubbles gives rise to the generation of ultrahigh pressure (100 MPa). The rapidly increased pressure generates an increased degree of supercooling that contributes in the rapid nucleation.[26] In the third place, primary nucleation is persuaded by cavitation bubbles of definite size while secondary nucleation is perceived by the tiny sections from massive dendritic ice crystals.[14] There have been many investigations regarding the quality criterion of food materials, namely, color, texture, and nutritional profile which are nearly linked with the rate of heat transfer during the freezing.[11,105]

Ultrasonication is a highly efficient treatment in forming small ice crystals by causing acceleration in the heat transfer, thereby preserving the original quality characteristics of the FFP.[33,51] There are innumerable studies on the effect of US-assisted freezing on the changes in chemical, physical, and microstructure, depicting the quality attributes of food items.

Diverse facets of US applications in food freezing have been suggested that involve the control of size and shape of ice crystals,[40,73] initiation of ice nucleation,[97,38] enhancement in quality of FFP,[31,95] and rapid increase in freezing rate.[27,49]

8.5.1 ICE NUCLEATION

Nucleation is determined as the development of new crystals and it is a crucial element in optimization of the process of freezing and transmittance of enhanced quality of foods.[58] Though nucleation is unconstrained and probabilistic, yet the temperature of the emergence of nucleation cannot be anticipated with assurance. US treatment can instigate nucleation in both moist solutions and in solid food products.[15,30,39,103] Inada et al.[30] determined the control of phase change from supercooled water to the ice under US treatment. The findings confirmed that US treatment is effective in increasing the chances of nucleation in supercooled water tremendously. Although to obtain efficient results, the selection of intensity of US plays a vital role.

However, diverse facets of occurrence of nucleation incited by US have been widely explored, yet the mechanism of nucleation incited by US is not understood satisfactorily.[14,15,30,38,39,103] Several research studies have suggested to define the mechanism of nucleation incited by US treatment, amidst which the theory proposed by Hickling[26] is widely cited. This theory states that disintegration of cavitation bubbles violently, generated by ultrasonic treatment, produces huge localized pressure within small fractions of time. Therefore, this disintegration produces a high level of supercooling and supports spontaneous ice nucleation. Moreover, it has also been stated that there is an immediate occurrence of nucleation as soon as there is disintegration of cavitation bubbles.[14,38,98]

8.5.2 SHAPE AND SIZE OF ICE CRYSTALS

The size and distribution of size in ice crystals are a crucial criterion in FFP (FFP) according to several research studies on these facets.[40,58,73] The quality of FFP is enhanced generally in the case of small-sized ice crystals and their uniform distribution.[18] On the other side in the process of freeze-drying, comparatively larger and vertical crystals are favored because larger ice crystals contribute to rapid sublimation.[67] US treatment can induce nucleation at high temperatures during the process of freezing. Therefore, the size and distribution of crystals in frozen food produce can be controlled by utilizing this response of US. The process of freezing comprises of nucleation along with the growth of ice crystals, and both are dependent on the level of supercooling. Although, both nucleation and rate of crystal growth differ in their behavior on increasing the level of supercooling.[106]

8.5.3 FREEZING RATE

The freezing rate influences the quality of frozen food produce because it governs the level of dehydration in cells, the size of ice crystals and destruction in the structure of tissues. In general, quick-freezing can effectively control the quality of FFP as compared with slow freezing. However, its mechanism remains unexplored. US exhibits its ability to stimulate the process of freezing of various food substances.[18,27,94]

Hu et al.[27] examined the influence of US on the rate of freezing during freezing under dough cylinders. The findings demonstrated a significant improved rate of freezing and shorter time of freezing by the application of US treatment. It was deduced that freezing time was decreased by 11% when the level of power of US was 288W and 360W. In the process of freezing, the rate of heat transfer is closely linked to the efficacy of freezing and the end quality of FFP. It has been confirmed that an increased rate of heat transfer can preferably sustain the quality of FFP because of the development of small ice crystals.[84] Thus, US treatment has been validated to be a vital tool in enhancing the efficiency of heat transfer in the process of freezing.[49]

8.5.4 EFFECT OF US-ASSISTED FREEZING ON THE QUALITY OF FOODS

8.5.4.1 MICROSTRUCTURE

The microstructure firmly depicts the nutritional and sensory characteristics of food materials. Besides, the process of crystallization is an important process influencing the microstructure. US aids in both by improving the nucleation rate and the crystal growth.[38] The initial quality characteristics and cell structure can be better retained by US for the development of small and uniformly dispersed ice crystals. Numerous studies have been performed to study the variation in microstructure because of distinction in freezing of food materials. For example, Islam et al.[31] determined the effect of freezing assisted by US on the microstructure of both raw and freeze-thaw samples of mushrooms. The production of small-sized ice crystals in the tissues with improved cell structure was the key benefit of applying US irradiation during immersion freezing of mushrooms. Similar studies were performed by Islam et al.[32] in US-assisted contact freezing of mushrooms.

Xin et al.[96] studied the microstructural changes in broccoli (*Brassica oleracea*) as influenced by varying power levels in immersion freezing. The findings showed that the broccoli frozen by US-assisted immersion freezing has minimum shrinkage and disintegration in the tissues because of the formation of large amount of small-sized ice crystals in both the inner and outer portion of the cells. The quality attributes of frozen tissues are the key functions of the size and evenness of the ice crystals.

8.5.4.2 DRIP LOSS PARAMETER

Drip loss is a major parameter for determining the quality characteristics of thawed food items. The quality of FFP can be significantly enhanced by the reduction in the level of drip loss.[93] Tu et al.[89] determined the drip loss in lotus (*Nelumbo nucifera*) root by immersion freezing assisted by US. The findings showed that US treatment during immersion freezing significantly enhanced the rate of freezing, forming very small ice crystals, resulting in the declined value of drip loss of food products subsequent to thawing.

8.5.4.3 TEXTURE AND COLOR PROPERTIES

Texture and color are two important indices that significantly influence the acceptability of food products by consumers.[12] Color is a notable parameter for the estimation of the visual quality of food products which directly affects their appearance, acceptability, and presentation.[13] Xu et al.[97] determined the US-assisted osmotic dehydration freezing of radish and they observed that ultrasonic treatment was efficient in retaining the firmness of radish. Texture also plays an important role in influencing the consumer acceptability of food products.

Xin et al.[95] also examined the US-assisted osmotic dehydration of frozen broccoli (*Brassica oleracea*) and observed that there was better retention of firmness as compared with that without US-assisted osmotic dehydration. It has been confirmed that the textural characteristics are associated closely with the pectic composition and structure of the cell[88] due to the decrease in the amount of freezable water by osmotic dehydration assisted by US.[65]

8.5.4.4 CHEMICAL CHARACTERISTICS

Numerous research studies have focused on measuring vitamins, nutrients, and antioxidants along with their retention postthawing of food materials. Zhu et al.[109] studied the effect of US on the phenolic contents of potatoes, apples, and radishes postfreezing. With the aid of US, the decrease in phenolic contents in potatoes was reduced by 31.69% than the frozen potatoes without ultrasonic treatment (43.81%). This can be ascribed to the low value of drip loss as a result of US-assisted freezing. Although there was no remarkable reduction in the loss of phenolic contents in samples of radishes (54.41%) and apples (66.92%), which concludes that the retention of nutrients by US treatment depends on the type of food materials.

Xu et al.[99] demonstrated that the retention of vitamin C and total anthocyanin content of red radish was improved notably ($p < 0.05$) under US treatment during freezing as compared with slow immersion freezing without US. The radish frozen at a power level of 0.26 W/cm^2 recorded the increased retention of phenolic content (83.06 mg GAE/100 g FW). This may be due to the reduced value of drip loss in the case of red radish, frozen with the aid of US. The influence of drip loss in the reduction in anthocyanin content was higher than the degradation of anthocyanin caused by ultrasonic cavitation.

The quality of FFP can be firmly linked to the total amount and the size of ice crystals created in the interior of food items. The increase in the number of total ice crystals and decrease in the size of crystals contribute to the enhancement in the quality of frozen food items.

The selection of power and duration of US and the degree of supercooling and the ice bubble concentration in the mixture should be made strictly for improving the ice nucleation process in freezing of distinct food items. Initially, the higher concentration of gas bubbles is appreciable for ice crystals formation. Although the higher concentration of gas bubbles may lead to a delay in time amidst the inception of the ice nucleation process and treatment by US. Furthermore, an increased level of supercooling is preferable for the initiation of the ice nucleation process.

Moreover, higher power intensity of US at an adequate range is desirable for the process of ice nucleation, whereas at low-power/intensity of US, there is a delay in the initiation of ice nucleation. In the case of duration of US, increased duration is not preferred for the nucleation process due to the generation of heat, whereas the inception of the process of nucleation cannot happen by a shorter duration of US.[65]

8.6 NEGATIVE INFLUENCE OF US ON PRESERVATION OF FRUITS AND VEGETABLES

The fresh FAV contain high amount of moisture and are crispier in texture due to which ultrasonic waves may often lead to deterioration of cells and tissues of FAV. Zhu et al.[108] confirmed the eradication of pesticide residues from pakchoi leaves under US, but studies also confirmed the decrease in vitamin C levels and physical deterioration of the treated sample. Moreover, the proportion of spoiled leaves can be minimized by modifying the ultrasonic frequency, and no spoilage in the leaf was observed with an ultrasonic frequency of 135 kHz. However, a reduction in the levels of vitamin C was still observed with these treatments, signifying that spoilage in the leaf was not the sole cause for reduced vitamin C levels.

Neto et al.[59] observed the negative impact of US on the sensory properties of lettuce. Muzaffar et al.[57] determined that on increasing the ultrasonic treatment by 20 min, the firmness of the cherry was decreased significantly. Pieczywek et al.[63] demonstrated that ultrasonic waves increased the solubilization of total pectin in the cell walls of the apple, and led to the formation of intercellular spaces in its tissues. Large-sized cavitation bubbles are formed at a low frequency of US which therefore lead to increased pressure and higher energy, thereby increasing the temperature remarkably in some cases,[48] which is nonconducive during storage.

Neto et al.[59] observed that on increasing the temperature, there is a rapid increase in the number of microbes in the samples, and there is a decrease in the bactericidal effect of ultrasonic treatment. Zhang et al.[104] investigated that the structure of the cell wall in cherry tomatoes was loosened on increasing the temperature of US. Researchers confirmed that due to higher resistance by most enzymes to ultrasonic treatment, prolonged time is required for higher intensity treatment, which has a negative influence on the quality characteristics of processed foods and it decreased the sensory characteristics of the food products. The short-pulse US breaks large molecules, thereby creating ease for the enzymes to be in reaction with substrates, probably improving their activity.[61,90] Moreover, in few cases, the US processes necessitate the need for the high level of sound wave frequencies that can result in the operator's hearing loss if safety measures are neglected. Seldom, in industries, the effective generation of acoustic fields of adequate intensity may pose a problem.[34,55]

8.7 SUMMARY

US treatment is a profitable application in the preservation of FAV, such as decontamination, drying, and freezing of FAV. This chapter summarizes the explanatory literature on different techniques of preservation of various FAV. The competence of ultrasonic cavitation in the preservation of horticultural commodities is fortuitous. In ultrasonic treatment, the acoustic cavitation and streaming results in the implosion of the bubbles. The turbulence induced by streaming causes an increase in temperature and pressure, thus generating a massive amount of heat. This leads to the formation of microjets, which assists in decontaminating the surface of FAV. The US-PRT prior to drying and US-assisted drying have resulted in enhanced synergy with other food processing techniques. US contribution in adding value to the freezing of FAV is well acknowledged.

KEYWORDS

- **cavitation**
- **decontamination**
- **drying**
- **freezing**
- **preservation**
- **ultrasound**

REFERENCES

1. Ajlouni, S.; Sibrani, H.; Premier, R.; Tomkins, B. Ultrasonication and Fresh Produce (Lettuce) Preservation. *J. Food Sci.* **2006**, *71* (2), 62–68.
2. Allahdad, Z.; Nasiri, M.; Varidi, M.; Varidi, M. J. Effect of Sonication on Osmotic De-Hydration and Subsequent Air-Drying of Pomegranate Arils. *J. Food Eng.* **2019**, *244*, 202–211.
3. Amami, E.; Khezami, W.; Mezrigui, S.; Badwaik, L. S.; Bejar, A. K.; Perez, C. T.; Kechaou, N. Effect of Ultrasound-Assisted Osmotic Dehydration Pretreatment on the Convective Drying of strawberry. *Ultrasonic Sonochem.* **2017**, *36*, 286–300.
4. Baslar, M.; Kilicli, M.; Yalinkilic, B. Dehydration Kinetics of Salmon and Trout Fillets Using Ultrasonic Vacuum Drying as a Novel Technique. *Ultrasonic Sonochem.* **2015**, *27*, 495–502.

5. Cai, L.; Cao, M.; Regenstein, J.; Cao, A. Recent Advances in Food Thawing Technologies. *Comprehens. Rev. Food Sci. Food Safet.* **2019**, *18*, 953–970.
6. Çakmak, R. S.; Tekeoğlu, O.; Bozkır, H.; Ergün, A. R.; Baysal, T. Effects of Electrical and Sonication Pretreatments on the Drying Rate and Quality of Mushrooms. *LWT—Food Sci. Technol.* **2016**, *69*, 197–202.
7. Cao, S. F.; Hu, Z. C.; Pang, B.; Wang, H. O.; Xie, H. X.; Wu, F. Effect of Ultrasound Treatment on Fruit Decay and Quality Maintenance in Strawberry After Harvest. *Food Control* **2010**, *21* (4), 529–532.
8. Cárcel, J. A.; García-Pérez, J. V.; Benedito, J.; Mulet, A. Food Process Inno- Vation Through New Technologies: Use of Ultrasound. *J. Food Eng.* **2012**, *110* (2), 200–207.
9. Chemat, F.; Zill-e-huma; Khan, M. K. Applications of Ultrasound in Food Technology: Processing, Preservation and Extraction. *Ultrasonic Sonochem.* **2011**, *18* (4), 813–835.
10. Chen, Z.; Zhu, C. H. Combined Effects of Aqueous Chlorine Dioxide and Ultrasonic Treatments on Postharvest Storage Quality of Plum Fruit (*Prunus salicina L.*). *Postharv. Biol. Technol.* **2011**, *61* (2–3), 117–123.
11. Cheng, X.;Zhang, M.; Xu, B.; Adhikari, B.; Sun, J. The Principles of Ultrasound and its Application in Freezing Related Processes of Food Materials: a Review. *Ultrasonic Sonochem.* **2015**, *27*, 576–585.
12. Chiang, P. Y.; Luo, Y. Y. Effects of Pressurized Cooking on the Relationship Between the Chemical Compositions and Texture Changes of Lotus Root (*Nelumbo nucifera Gaertn.*). *Food Chem.* **2007**, *105*, 480–484.
13. Choi, E. J.; Park, H. W.; Chung, Y. B.; Park, S. H.; Kim, J. S.; Chun, H. H. Effect of Tem- Pering Methods on Quality Changes of Pork Loin Frozen by Cryogenic Immersion. *Meat Sci.* **2017**, *124*, 69–76.
14. Chow, R.; Blindt, R.; Chivers, R.; Povey, M. A Study on the Primary and Secondary Nucleation of Ice by Power Ultrasound. *Ultrasonic Sonochem.* **2005**, *43*, 227–230.
15. Chow, R.; Blindt, R.; Chivers, R.; Povey, M. The Sonocrystallisation of Ice in Sucrose Solutions: Primary and Secondary Nucleation. *Ultrasonics* **2003**, *41*, 595–604.
16. Corrêa, J. L. G.; Rasia, M. C.; Mulet, A.; Cárcel, J. A. Influence of Ultrasound Application on both the Osmotic Pretreatment and Subsequent Convective Drying of Pineapple (*Ananas comosus*). *Innov. Food Sci. Emerg. Technol.* **2017**, *41*, 284–291.
17. Dehsheikh, F. N.; Dinani, S. T. Coating Pretreatment of Banana Slices Using Carboxymethyl Cellulose in an Ultrasonic System Before Convective Drying. *Ultrasonics Sonochem.* **2019**, *52*, 401–413.
18. Delgado, A. E.; Zheng, L. Y.; Sun, D. W. Influence of Ultrasound on Freezing Rate of Immersion-Frozen Apples. *Food Bioprocess Technol.* **2009**, *2*, 263–270.
19. Deora, N. S.; Misra, N. N.; Deswal, A.; Mishra, H. N.; Cullen, P. J.; Tiwari, B. K. Ultrasound for Improved Crystallization In Food Processing. *Food Eng. Rev.* **2013**, *5*, 36–44.
20. Dias da Silva, G.; Barros, Z. M. P.; de Medeiros, R. A. B.; de Carvalho, C. B. O.; Brandão, S. C. R.; Azoubel, P. M. Pretreatments for Melon Drying Implementing Ultrasound And Vacuum. *LWT—Food Sci. Technol.* **2016**, *74*, 114–119.
21. Fernandes, F. A. N.; Gallão, M. I.; Rodrigues, S. Effect of Osmosis and Ultrasound on Pineapple Cell Tissue Structure During Dehydration. *J. Food Eng.* **2009**, *90*, 186–190.
22. Fijalkowska, A.; Nowacka, M.; Wiktor, A.; Sledz, M.; Witrowa-Rajchert, D. Ultrasound as A Pretreatment Method to Improve Drying Kinetics and Sensory Properties Of Dried Apple. *J. Food Process Eng.* **2016**, *39*, 256–265.

23. Galanakis, C. M. Recovery of High Added-Value Components from Food Wastes: Conventional, Emerging Technologies and Commercialized Applications. *Trends Food Sci. Technol.* **2012**, *26* (2), 68–87.
24. Gao, S.; Lewis, G. D.; Ashokkumar, M.; Hemar, Y. Inactivation of Microorganisms by Low-Frequency High-Power Ultrasound, Part 1: Effect of Growth Phase and Capsule Properties of the Bacteria. *Ultrasonics Sonochem.* **2014**, *21*, 446–453.
25. Harris, G. A.; Torgovnikov, G.; Vinden, P.; Brodie, G. I.; Shaginov, A. Microwave Pretreatment of Backsawn Messmate Boards to Improve Drying Quality. *Dry. Technol.* **2008**, *26* (5), 579–584.
26. Hickling, R. Nucleation of Freezing by Cavity Collapse and its Relation to Cavitation Damage. *Nature*, **1965**, *206*, 915–917.
27. Hu, S. Q.; Liu, G.; Li, L.; Li, Z. X.; Hou, Y. An Improvement in the Immersion Freezing Process for Frozen Dough via Ultrasound Irradiation. *J. Food Eng.* **2013**, *114*, 22–28.
28. Huang, D.; Men, K.; Li, D.; Wen, T.; Gong, Z.; Sunden, B.; Wu, Z. Application of Ultrasound Technology in the Drying of Food Products. *Ultrasonics Sonochem.* **2020**, *63*, E-article 104950; p 8.
29. Huang, T. S.; Xu, C. L.; Walker, K.; West, P.; Zhang, S. P.; Weese, J. Decontamination Efficacy of Combined Chlorine Dioxide with Ultrasonication on Apples and Lettuce. *J. Food Sci.* **2006**, *71* (4), M134–M139.
30. Inada, T.; Zhang, X.; Yabe, A.; Kozawa, Y. Active Control of Phase Change from Supercooled Water to ice by Ultrasonic Vibration, Part 1: Control of Freezing Temperature. *Int. J. Heat Mass Transf.* **2001**, *44*, 4523–4531.
31. Islam, M. N.; Zhang, M.; Adhikari, B.; Xinfeng, C.; Xu, B.-G. The Effect of Ultrasound-Assisted Immersion Freezing on Selected Physicochemical Properties of Mushrooms. *Int. J. Refrigeration* **2014**, *42*, 121–133.
32. Islam, M. N.; Zhang, M.; Fang, Z.; Sun, J. Direct Contact Ultrasound Assisted Freezing of Mushroom (Agaricus bisporus): Growth and Size Distribution of Ice Crystals. *Int. J. Refrigeration* **2015**, *57*, 46–53.
33. James,C.; Purnell, G.; James, S. J. A Review of Novel and Innovative Food Freezing Technologies. *Food Bioprocess Technol.* **2015**, *8*, 1616–1634.
34. Jiang, Q.; Zhang, M.; Xu, B. (2020). Application of Ultrasonic Technology in Postharvested Fruits and Vegetables Storage: A Review. *Ultrasonics Sonochem.* **2020**, *69*, E-article 105261, p 9.
35. Joyce, E.; Phull, S. S.; Lorimer, J. P.; Mason, T. J. The Development and Evaluation of Ultrasound for the Treatment of Bacterial Suspensions: A Study of Frequency, Power and Sonication Time on Cultured Bacillus Species. *Ultrasonics Sonochem.* **2003**, *10* (6), 315–318.
36. Kek, S. P.; Chin, N. L.; Yusof, Y. A. Direct and Indirect Power Ultrasound Assisted Pre-Osmotic Treatments in Convective Drying of Guava Slices. *Food Bioproducts Process.* **2013**, *91*, 495–506.
37. Kentish, S.; Ashokkumar, M. The Physical and Chemical Effects of Ultrasound. In *Ultrasound Technologies for Food and Bioprocessing*; Feng, H., Barbosa-Cánovas, G. V., Weiss, J., Eds.; Springer: New York, 2011; pp 1–12
38. Kiani, H.; Sun, D. W.; Delgado, A.; Zhang, Z. Investigation of the Effect of Power Ultrasound on the Nucleation of Water During Freezing of Agar Gel Samples in Tubing Vials. *Ultrasonics Sonochem.* **2012**, *19*, 576–581.

39. Kiani, H.; Zhang, Z.; Delgado, A.; Sun, D. W. Ultrasound Assisted Nucleation of Some Liquid and Solid Model Foods During Freezing. *Food Res. Int.* **2011**, *44*, 2915–2921.
40. Kiani, H.; Zhang, Z.; Sun, D. W. Effect of Ultrasound Irradiation on Ice Crystal Size Distribution in Frozen Agar Gel Samples. *Innov. Food Sci. Emerg. Technol.* **2013**, *18*, 126–131.
41. Kim, H. J.; Feng, H.; Kushad, M. M.; Fan, X. Effects of Ultrasound, Irradiation and Acidic Electrolyzed Water on Germination of Alfalfa and Broccoli Seeds and *Escherichia coli* O157:H7. *J. Food Sci.* **2006**, *71* (6), 168–173.
42. Kingsly, R. P.; Goyal, R. K.; Manikantan, M. R.; Ilyas, S. M. Effects of Pretreatments and Drying Air Temperature on Drying Behavior of Peach Slice. *Int. J. Food Sci. Technol.* **2007**, *42*, 65–69.
43. Kowalski, S. J.; Mierzwa, D. US-Assisted Convective Drying of Biological Materials. *Dry. Technol.* **2015**, *33*, 1601–1613.
44. Kowalski, S. J.; Pawłowski, A.; Szadzińska, J.; Łechtańska, J.; Stasiak, M. High Power Airborne Ultrasound Assist in Combined Drying of Raspberries. *Innov. Food Sci. Emerg. Technol.* **2016**, *34*, 225–233.
45. Kowalski, S. J.; Szadzińska, J. Convective-Intermittent Drying of Cherries Preceded by Ultrasonic Assisted Osmotic Dehydration. *Chem. Eng. Process.* **2014**, *82*, 65–70.
46. Kowalski, S. J.; Szadzińska, J.; Pawłowski, A. Ultrasonic-Assisted Osmotic Dehydration of Carrot Followed by Convective Drying with Continuous and Intermittent Heating. *Dry. Technol.* **2015**, *33*, 1570–1580.
47. Legay, M.; Gondrexon, N.; Le Person, S.; Boldo, P.; Bontemps, A. Enhancement of Heat Transfer by Ultrasound: Review and Recent Advances. *Int. J. Chem. Eng.* **2011**, *2011*, 1–17.
48. Leong, T.; Juliano, P.; Knoerzer, K. Advances in Ultrasonic and Megasonic Processing of Foods. *Food Eng. Rev.* **2017**, *9*, 237–256.
49. Li, B.; Sun, D. W. Effect of Power Ultrasound on Freezing Rate During Immersion Freezing of Potatoes. *J. Food Eng.* **2002**, *55*, 277–282.
50. Lighthill, S. J. Acoustic Streaming. *J. Sound Vibration* **1978**, *61*, 391–418.
51. Mahato, S.; Zhu, Z.; Sun, D. W. Glass Transitions as Affected by Food Compositions and by Conventional and Novel Freezing Technologies: A Review. *Trends Food Sci. Technol.* **2019**, *94*, 1–11.
52. Majid, I.; Nayik, G. A.; Nanda, V. Ultrasonication and Food Technology: A Review. *Cogent Food Agric.* **2015**, *1* (1), 1071022.
53. Martınez-Soto, G.; Ocanna-Camacho, R.; Paredes-Lopez, O. Effect of Pretreatment and Drying on the Quality of Oyster Mushrooms (*Pleurotus ostreatus*). *Dry. Technol.* **2001**, *19* (3–4), 661–672.
54. Mason, T. J.; Lorimer, J. P. Introduction to Applied Ultrasonics. In *Applied Sonochemistry: Uses of Power Ultrasound in Chemistry and Processing*; Weinheim: Wiley-VCH Verlag, 2002; pp 1–24.
55. Mothibe, K. J.; Zhang, M.; Nsor-atindana, J.; Wang, Y. Use of Ultrasound Pretreatment in Drying of Fruits: Drying Rates, Quality Attributes, and Shelf-Life Extension. *Dry. Technol.* **2011**, *29* (14), 1611–1621.
56. Mukhopadhyay, S.; Ramaswamy, R. Application of Emerging Technologies to Control Salmonella in Foods: A Review. *Food Res. Int.* **2012**, *45*, 666–677.
57. Muzaffar, S.; Ahmad, M.; Wani, S. M.; Gani, A.; Baba, W. N.; Shah, U.; Khan, A. A.; Masoodi, F. A.; Gani, A.; Wani, T. A. Ultrasound Treatment: Effect on Physicochemical,

Microbial and Antioxidant Properties of Cherry (*Prunus avium*). *J. Food Sci. Technol.* **2016**, *53*, 2752–2759.
58. Nakagawa, K.; Hottot, A.; Vessot, S.; Andrieu, J. Influence of Controlled Nucleation by Ultrasounds on Ice Morphology of Frozen Formulations for Pharmaceutical Proteins Freeze Drying. *Chem. Eng. Process.* **2006**, *45* (9), 783–791.
59. Neto,L.; Millan-Sango, D.; Brincat, J.; Cunha, L. M.; Valdramidis, V. P. Impact of Ultrasound Decontamination on the Microbial and Sensory Quality of Fresh Produce. *Food Control* **2019**, *104*, 262–268.
60. Nowacka, M.;Wiktor, A.; Śledź, M.; Jurek, N.; Witrowa-Rajchert, D. Drying of Ultrasound Pretreated Apple and Its Selected Physical Properties. *J. Food Eng.* **2012**, *113*, 427–433.
61. Ortega-Rivas, E. Ultrasound in Food Preservation. In *Non-thermal Food Engineering Operations*; Springer Science + Business Media: New York, 2012; pp 251–262.
62. Patist, A.; Bates, D. Ultrasonic Innovations in the Food Industry: From the Laboratory to Commercial Production. *Innov. Food Sci Emerg. Technol.* **2008**, *9*, 147–154.
63. Pieczywek, P. M.; Kozio, A.; Konopacka, D.; Cybulska, J.; Zdunek, A. Changes of cell wall stiffness and microstructure in ultrasonically treated apple. *J. Food Eng.* **2017**, *197*, 1–8.
64. Prajapati, V. K.; Nema, P. K.; Rathore, S. S. Effect of Pretreatment and Drying Methods on Quality of Value-Added Dried Anola (*Emblica officinalis*) Shreds. *J. Food Sci. Technol.* **2010**, *48* (1), 45–52.
65. Qiu, L.; Zhang, M.; Chitrakar, B.; Bhandari, B. Application of Power Ultrasound in Freezing and Thawing Processes: Effect on Process Efficiency and Product Quality. *Ultrasonics Sonochem.* **2020**, *68*, E-article 105230, p 7.
66. Ren, F.; Perussello, C. A.; Zhang, Z.; Kerry, J. P.; Tiwari, B. K. Impact of Ultrasound and Blanching on Functional Properties of Hot-Air Dried and Freeze Dried Onions. *LWT—Food Sci. Technol.* **2017**, *87*, 102–111.
67. Rhim, J. W.; Koh, S.; Kim, J. M. Effect of Freezing Temperature on Rehydration and Water Vapor Adsorption Characteristics of Freeze-Dried Rice Porridge. *J. Food Eng.* **2011**, *104*, 484–491.
68. Ricce, C.; Rojas, M. L.; Miano, A. C.; Siche, R.; Augusto, P. E. D. Ultrasound Pre-Treatment Enhances the Carrot Drying and Rehydration. *Food Res. Int.* **2016**, *89*, 701–708.
69. Rivera, C. S.; Venturini, M. E.; Oria, R.; Blanco, D. Selection of a Decontamination Treatment for Fresh *Tuber aestivum* and *Tuber melanosporum* Truffles Packaged in Modified Atmospheres. *Food Control* **2011**, *22* (3–4), 626–632.
70. Rodrigues, S.; Oliveira, F. I. P.; Gallão, M. I.; Fernandes, F. A. N. Effect of Immersion Time in Osmosis and Ultrasound on Papaya Cell Structure During Dehydration. *Dry. Technol.* **2009**, *27*, 220–225.
71. Romero, J. C.; Yepez, V. B. Ultrasound as Pretreatment to Convective Drying of Andean Black Berry (*Rubus glaucus*). *Ultrasonics Sonochem.* **2015**, *22*, 205–210.
72. Sabarez, H. T.; Gallego-Juarez, J. A.; Riera, E. Ultrasonic-Assisted Convective Drying of Apple Slices. *Dry. Technol.* **2012**, *30*, 989 997.
73. Saclier, M.; Peczalski, R.; Andrieu, J. Effect of Ultrasonically Induced Nucleation on Ice Crystals' Size and Shape During Freezing in Vials. *Chem. Eng. Sci.* **2010**, *65*, 3064–3071.
74. Sagar, V. R.; Suresh, K. P. Recent Advances in Drying and Dehydration of Fruits and Vegetables: A Review. *J. Food Sci. Technol.* **2010**, *47* (1), 15–20.

75. Sagong, H. G.; Cheon, H. L.; Kim, S. O.; Lee, S. Y.; Park, K. H.; Chung, M. S.; Choi, Y. J.; Kang, D. H. Combined Effects of Ultrasound and Surfactants to Reduce Bacillus Cereus Spores on Lettuce and Carrots. *Int. J. Food Microbiol.* **2013**, *160* (3), 367–372.
76. Sagong, H. G.; Lee, S. Y.; Chang, P. S.; Heu, S.; Ryu, S.; Choi, Y. J. Combined Effect of Ultrasound and Organic Acids to Reduce *Escherichia coli* O157:H7, *Salmonella typhimurium,* and *Listeria monocytogenes* on Organic Fresh Lettuce. *Int. J. Food Microbiol.* **2011**, *145* (1), 287–292.
77. Sango, D. M.; Abela, D.; Mcelhatton, A.; Valdramidis, V. P. Assisted Ultrasound Applications for the Production of Safe Foods. *J. Appl. Microbiol.* **2014**, *116* (5), 1067–1083.
78. São José, J. F. B. de; Andrade, N. J. de; Ramos, A. M.; Vanetti, M. C. D.; Stringheta, P. C.; Chaves, J. B. P. Decontamination by Ultrasound Application in Fresh Fruits and Vegetables. *Food Control* **2014**, *45*, 36–50.
79. São José, J. F. B.; Vanetti, M. C. D. Effect of Ultrasound and Commercial Sanitizers in Removing Natural Contaminants and *Salmonella enterica typhimurium* on Cherry Tomatoes. *Food Control* **2012**, *24* (1–2), 95–99.
80. Schössler, K.; Jäger, H.; Knorr, D. Effect of Continuous and Intermittent Ultrasound on Drying Time and Effective Diffusivity During Convective Drying of Apple and Red Bell Pepper. *J. Food Eng.* **2012**, *108*, 103–110.
81. Seymour, I. J.; Burfoot, D.; Smith, R. L.; Cox, L. A.; Lockwood, A. Ultrasound Decontamination of Minimally Processed Fruits and Vegetables. *Int. J. Food Sci. Technol.* **2002**, *37* (5), 547–557.
82. Shamaei, S.; Emam-Djomeh, Z.; Moini, S. Ultrasound-Assisted Osmotic Dehydration of Cranberries: Effect of Finish Drying Methods and Ultrasonic Frequency on Textural Properties. *J. Texture Stud.* **2012**, *43*, 133–141.
83. Siucińska, K.; Konopacka, D.; Mieszczakowska-Frąc, M.; Połubok, A. The Effects of Ultrasound on Quality and Nutritional Aspects of Dried Sour Cherries During Shelf-Life. *LWT—Food Sci. Technol.* **2016**, *68*, 168–173.
84. Sun, D. W.; Li, B. Microstructural Change of Potato Tissues Frozen by Ultrasound-Assisted Immersion Freezing. *J. Food Eng.* **2003**, *57*, 337–345.
85. Szadzińska, J.; Kowalski, S. J.; Stasiak, M. Microwave and Ultrasound Enhancement of Convective Drying of Strawberries: Experimental and Modeling Efficiency. *Int. J. Heat Mass Transf.* **2016**, *103*, 1065–1074.
86. Tao, Y.; Wang, P.; Wang, Y.; Kadam, S. U.; Han, Y.; Wang, J.; Zhou, J. Power Ultrasound as a Pretreatment to Convective Drying of Mulberry (*Morus alba L.*) Leaves: Impact on Drying Kinetics and Selected Quality Properties. *Ultrasonics Sonochem.* **2016**, *31*, 310–318.
87. Tao, Y.; Zhang, J.; Sirui, J.; Xu, Y.; Show, P. L.; Han, Y.; Ye, X.; Ye, M. Contacting Ultrasound Enhanced Hot-Air Convective Drying of Garlic Slices: Mass Transfer Modeling and Quality Evaluation. *J. Food Eng.* **2018**, *235*, 79–88.
88. Torreggiani, D.; Bertolo, G. Osmotic Pre-Treatments in Fruit Processing: Chemical, Physical and Structural Effects. *J. Food Eng.* **2001**, *49*, 247–253.
89. Tu, J.; Zhang, M.; Xu, B.; Liu, H. Effects of Different Freezing Methods on the Quality and Microstructure of Lotus (*Nelumbo nucifera*) root. *Int. J. Refrigeration* **2015**, *52*, 59–65.
90. Wang, D.; Yan, L.; Ma, X.; Wang, W.; Zou, M.; Zhong, J.; Ding, T.; Ye, X.; Liu, D. Ultrasound Promotes Enzymatic Reactions by Acting on Different Targets: Enzymes, Substrates and Enzymatic Reaction Systems. *Int. J. Biol. Macromol.* **2018**, *119*, 453–461.

91. Wei, Y. X. Effect of Postharvest Handling on Quality, Antioxidant Activity and Polyamines of Green Asparagus. PhD thesis; Zhejiang University – China, 2010; p 231.
92. Wiktor, A.; Dadan, M.; Nowacka, M.; Rybak, K.; Witrowa-Rajchert, D. The Impact of Combination of Pulsed Electric Field and Ultrasound Treatment on Air Drying Kinetics and Quality of Carrot Tissue. *LWT—Food Sci. Technol.* **2019**, *110*, 71–79.
93. Xia, X.; Kong, B.; Liu, Q.; Liu, J. Physicochemical Change and Protein Oxidation in Porcine Longissimus Dorsi as Influenced by Different Freeze–Thaw Cycles. *Meat Sci.* **2009**, *83*, 239–245.
94. Xin, Y.; Zhang, M.; Adhikari, B. The Effects of Ultrasound-Assisted Freezing on the Freezing Time and Quality of Broccoli (*Brassica oleracea*) During Immersion Freezing. *Int. J. Refrigeration* **2014**, *41*, 82–91.
95. Xin, Y.; Zhang, M.; Adhikari, B. Freezing Characteristics and Storage Stability of Broccoli (*Brassica oleracea*) Under Osmodehydrofreezing and Ultrasound-Assisted Osmodehydrofreezing Treatments. *Food Bioprocess Technol.* **2014**, *7*, 1736–1744.
96. Xin, Y.; Zhang, M.; Adhikari, B. Ultrasound Assisted Immersion Freezing of Broccoli (*Brassica oleracea*). *Ultrasonics Sonochem.* **2014**, *21*, 1728–1735.
97. Xu, B. G.; Zhang, M.; Bhandari, B.; Cheng, X. Influence of Ultrasound-Assisted Osmotic Dehydration and Freezing on the Water State, Cell Structure, and Quality of (*Raphanus sativus L.*) Cylinders. *Dry. Technol.* **2014**, *32*, 1803–1811.
98. Xu, B. G.; Zhang, M.; Bhandari, B.; Cheng, X. F. Influence of Power Ultrasound on Ice Nucleation of Radish Cylinders During Ultrasound-Assisted Immersion Freezing. *Int. J. Refrigeration* **2014**, *46*, 1–8.
99. Xu, B. G.; Zhang, M.; Bhandari, B.; Cheng, X. F.; Islam, M. N. Effect of Ultrasound-Assisted Freezing on the Physico-Chemical Properties and Volatile Compounds of Red Radish. *Ultrasonics Sonochem.* **2015**, *27*, 316–324.
100. Xu, Y.; Sismour, E.; Pao, S.; Rutto, L.; Grizzard, C.; Ren, S. X. Textural and Microbiological Qualities of Vegetable Soybean (Edamame) Affected by Blanching and Storage Conditions. *J. Food Process Technol.* **2012**, *3*, 2–6.
101. Xu, Y. T.; Zhang, L. F.; Zhong, J. J.; Shu, J.; Ye, X. Q.; Liu, D. H. Power Ultrasound for the Preservation of Postharvest Fruits and Vegetables. *Int. J. Agric. Biol. Eng.* **2013**, *6* (2), 116–125.
102. Yang, Z.; Cao, S.; Cai, Y.; Zheng, Y. Combination of Salicylic Acid and Ultrasound to Control Postharvest Blue Mould Caused by *Penicillium expansum* in Peach Fruit. *Innov. Food Sci. Emerg. Technol.* **2011**, *12* (3), 310–314.
103. Yu, D.; Liu, B.; Wang, B. The Effect of Ultrasonic Waves on the Nucleation of Pure Water and Degassed Water. *Ultrasonics Sonochem.* **2012**, *19*, 459–463.
104. Zhang, L.; Wang, P.; Sun, X.; Chen, F.; Lai, S.; Yang, H. Calcium Permeation Property and Firmness Change of Cherry Tomatoes Under Ultrasound Combined with Calcium Lactate Treatment. *Ultrasonics Sonochem.* **2020**, *60*, E-article 104784, p 9.
105. Zhang, P.; Zhu, Z.; Sun, D.-W. Using Power Ultrasound to Accelerate Food Freezing Processes: Effects on Freezing Efficiency and Food Microstructure. *Crit. Rev. Food Sci. Nutr.* **2018**, *58*, 2842–2853.
106. Zheng, L.; Sun, D. W. Ultrasonic Assistance of Food Freezing. In *Emerging Technologies for Food Processing*; Sun, D. W., Ed.; Elsevier Ltd.: London, 2005; pp 603–626.
107. Zhou, B.; Feng, H.; Luo, Y. G. Ultrasound Enhanced Sanitizer Efficiency in Reduction of *Escherichia coli* O157:H7 Population on Spinach Leaves. *J. Food Sci.* **2009**, *74* (6), 308–313.

108. Zhu, Y.; Zhang, T.; Xu, D.; Wang, S.; Yuan, Y.; He, S.; Cao, Y. The Removal of Pesticide Residues from Pakchoi (*Brassica rape*) by Ultrasonic Treatment. *Food Control* **2019**, *95*, 176–180.
109. Zhu, Z.; Chen, Z.; Zhou, Q.; Sun, D.-W.; Chen, H.; Zhao, Y.; Zhou, W.; Li, X.; Pan, H. Freezing Efficiency and Quality Attributes as Affected by Voids in Plant Tissues During Ultrasound-Assisted Immersion Freezing. *Food Bioprocess Technol.* **2018**, *11*, 1615–1626.
110. Zinoviadou, K. G.; Galanakis, C. M.; Brnčić, M.; Grimi, N.; Boussetta, N.; Mota, M. J.; Barba, F. J. Fruit Juice Sonication: Implications on Food Safety and Physicochemical and Nutritional Properties. *Food Res. Int.* **2015**, *77*, 743–752.

CHAPTER 9

SCOPE OF NANOEMULSIONS FOR QUALITY AND SAFETY OF FRUITS AND VEGETABLES: NANOTECHNOLOGY AND ESSENTIAL OILS

FARAH NAQASH, OMEERA AYOB, SADAF NAZIR, and B. N. DAR

ABSTRACT

Fruits and vegetables act as abundant reserves of bioactive compounds with positive health outcomes. However, a vast majority of fruits and vegetables are lost to the waste bins in the absence of adequate protection against factors causing quality defects in these commodities. Postharvest and during storage, fruits and vegetables undergo certain physical, biochemical, and biological processes that cause a loss in texture, color, sensory appeal, and nutritional value along with harboring a plethora of micro-organisms that can cause spoilage and be pathogenic as well, requiring novel preservation technologies. Among the emerging technologies, the application of nanoemulsions to extend shelf-life and preserve the quality of fruits and vegetables has taken up successfully in recent decades. It involves the application of antimicrobials and antioxidants, where the nanoemulsions act as their carriers. Various bioactive compounds encapsulated in nanoemulsions include antimicrobial and antioxidant essential oils and plant extracts. The nanoreduction of these active ingredients is known to enhance their stability, solubility, and activity attributed to their reach over an increased surface area. Nanoemulsions are applied as an edible coating on horticultural produce creating a

Quality Control in Fruit and Vegetable Processing: Methods and Strategies. Megh R. Goyal, & Faizan Ahmad (Eds.)

microenvironment around them and consequently arresting the stimuli antagonistic to their quality.

9.1 INTRODUCTION

Fruits and vegetables (FAVs) are consumed principally due to their positive health benefits. These are rich sources of antioxidants, phenols, flavonoids, pigments, and so on, providing antioxidant, anti-inflammatory, antidiabetic, and anticarcinogenic properties.[86] The quality of fresh produce should, however, be acceptable at the time of purchase to derive maximum benefits. On the contrary, it is known that horticultural produce after harvesting and minimal processing operations undergoes severe alterations in terms of integrity causing adverse changes in texture, color, nutritional and flavor loss, weight loss, and ultimately deterioration.[55] Microbial proliferation becomes another challenge as the fresh produce is rich in nutrients.[6] All these factors can reduce the shelf-life, reducing consumer acceptance, and causing wastage. It is estimated that about one-third of the agricultural production is wasted annually, in which FAV comprises the largest portion.[3]

According to estimates of Food and Agricultural Organization, about half of all the FAV produced is wasted [<www.fao.org>]. These losses can be due to physical or biological factors during postharvest handling, storage, and transportation, and therefore there is a need to put proper preservation technologies. In addition to these losses, FAV subjected to minimal processing also undergo severe losses due to cutting-induced changes and an exposed surface. For these commodities, effective measures must be adopted to ensure the availability of these convenience foods to consumers without compromising in quality. Methods, such as applying chemical fungicides, controlled atmospheric storage, modified atmospheric packaging is being used; however, owing to drawbacks like residual hazards, tolerant pathogens, anaerobiosis, and off-flavor, alternatives to overcome these issues are needed.[58]

Among various technologies available, application of edible coatings (ECs) has been found to be successful in enhancing shelf-life of fresh produce along with preserving their quality.[71] ECs are mostly elaborated from biopolymers including proteins, lipids, carbohydrates, and/or their mixtures, applied onto the product surface by dipping or spraying. These provide a barrier to the produce from gases and water vapor, act as texture enhancers by providing structural protection, and prevent chemical and microbial changes to avoid quality deterioration. ECs also function to improve the characteristics of product by incorporation of bioactive compounds of functional importance, such as antioxidants, antimicrobials, flavors, texture enhancers, and those providing mechanical and barrier properties.[19]

ECs are consumed with the food product itself and do not require disposal and thus mitigate the environmental burden, which is generally encountered by synthetic packaging. Along with the same concerns, mostly renewable and biodegradable materials are used to formulate ECs, which are not only abundantly available but also biocompatible.[73] Most commonly, alginates, starch, pectin, chitosan, gelatin, whey protein, collagen, lipids, and aliphatic polyesters are generally employed.[21]

To enhance the functionality of ECs, certain substances are added in their formulation, such as antimicrobials, antioxidants, nanoparticles, etc. Preferably, these components are of natural origin considering their safety, consumer acceptance, and better efficiency than the synthetic counterparts.[17] Among such functional components, essential oils (EOs) are added to coatings to confer antimicrobial properties to these, along with regulating respiration and metabolism.[73]

ECs carrying EOs are processed as emulsions that can encapsulate these oils, and aid in the distribution of such partially or fully hydrophobic compounds in hydrophilic matrices.[2] However, emulsions show pronounced instability against creaming, flocculation, and coalescence. To counter this, emulsion droplets are reduced to nanometer (nm) range producing a nano-emulsion (NE). NEs are metastable thermodynamically but show high kinetic stability because in these no gravitational separation and droplet aggregation are encountered.[44] Also, owing to their subcellular size, NEs confer stability to active compounds, enhanced bioactivity due to diffusion, and reduced effect on the organoleptic properties of a food product.[2] These are formulated by homogenizing two immiscible liquids, namely, oil and water, which are stabilized by an interfacial film supported by adding surfactant molecules. NEs can be water-in-oil (w/o) and oil-in-water (o/w), suitable for delivery of hydrophilic and hydrophobic agents, respectively.[31,73]

An increased application of NEs in food industries is being observed for various food processing and packaging systems. In food systems, NEs aid in encapsulating functional bioactive compounds, enable their controlled release, increase their solubility, and absorption in gastrointestinal tract.[1] NE-based ECs are widely used for maintaining the quality of various food products, such as meats, dairy, fresh, and fresh-cut products including FAV. These are found instrumental in regulating the metabolism during postharvest phase and, therefore, ensure desirable quality.[16,18,71]

This chapter focuses on the application of NEs for enhancing the safety and quality of FAVs, along with their formulation, stability, and mechanism of action. Their safety and sensory aspects are also discussed.

9.2 POSTHARVEST LOSSES OF FRUITS AND VEGETABLES

9.2.1 PHYSICOCHEMICAL LOSSES

FAVs are living systems that undergo deterioration in quality. The respiration is elevated in fresh-cut products as a result of cutting and wounding, reducing shelf-life. Ethylene accumulation in the product surroundings as a result of respiration is also detrimental to product quality. Fresh produce also encounters water loss through transpiration, reducing the product quality. This problem is more pronounced in the case of fresh-cut fruits, as the surface is readily exposed to the external environment. As a consequence, significant weight loss has been observed, which also affects the overall quality.[27] Loss in texture is also one of the problems in horticultural produce and has serious consequences for acceptability. Similar textural loss is seen in fresh-cut fruits more significantly, in addition to the browning of exposed flesh. Browning is mainly caused by the action of polyphenol oxidase and peroxidase on the plant phenolics, causing their oxidation and production of brown or black-colored compounds. In addition to these, severe nutritional losses, such as, that of sugars, ascorbic acid, and other bioactive compounds have been seen.[93]

9.2.2 LOSSES DUE TO MICROBIAL GROWTH

FAV after harvesting are subjected to losses due to spoilage caused by microbial growth. The presence of microbes not only spoils the fresh produce but also renders the products unsafe for consumption.[49] Consequently, foodborne diseases associated with FAV have also increased. Growth of human pathogens is favored in FAV with increase in pH. Damaged parts of plants and lesions on surfaces also promote the growth of pathogens, due to supporting the nutrients in the secretions from damaged tissues. Most of the microflora on FAV surfaces is actually present in soil. Pathogenic contamination of FAV is encountered during postharvest processing and transportation.

Common microbes commonly in fresh FAV include fungi (such as from genus *Aspergillus*, *Penicillium*, *Rhizopus*, and *Wallemia*); yeasts of genus

Pichia, Candida, Zygo saccharomyces; and Gram-negative bacteria, such as *Erwinina.*[56] Vegetable pathogens associated with disease outbreaks include *Salmonella, E. coli, Shigella,* and *Norovirus.*[82] *E. coli* O157: H7, *Salmonella,* and *Listeria monocytogenes* have been documented to cause diseases associated with the consumption of lettuce, Fenugreek, fresh-cut salads, apples, celery, lettuce, oranges, and watermelon.[37,50] The reader is referred to the review by Mostafidi et al.[49] for a better understanding of sources of microbial contamination of FAV. All these threats point toward the need for addressing the safety issues associated with FAV, and their minimally processed counterparts.

9.3 NANOEMULSIONS

NEs are emulsions with droplet size <100 nm and characterized by kinetic stability, attributed to their resistance to deformation owing to smaller size.[92] Their droplet size is characterized by colloidal interactions that enable the existence of long-order forces between droplets. This explains the potential viscoelastic and solid-like behavior of their phases and existence of ordered structures.[42]

9.3.1 FORMULATION METHODS

The formation of droplets in nanoscale is subservient to the creation of a huge interface requiring enormous energy inputs.[28] A minimum energy requirement needed to bring droplet size reduction is always a function of interfacial tension and volume fraction of the dispersed phase; and regardless of the emulsification method, this requirement remains the same. The administration of energy is carried out either by working on the fluid, or by changing the equilibrium state. The former method is referred to as "high-energy" and the latter is "low-energy" method.[89] High-energy methods include energy dissipation aided by high-pressure homogenization, microfluidization, and ultrasonication.[45]

Droplet size in such processes is limited by the efficiency in which energy is provided and the stresses deform the interface of fluid–fluid emulsion into smaller droplets. In case of methods based on thermodynamics of the components, the homogenous system of fluid and surfactant is transformed into a metastable region of two phases in which droplets undergo nucleation, growth, and ultimately saturation. Such methods are referred to as phase inversion methods and can be brought about by phase inversion composition involving dilution with one or another fluid or phase inversion temperature with well-controlled temperature change. Work-based methods produce NEs with more poly-dispersibility compared to that thermodynamic methods.[68]

9.3.2 STABILITY AND EFFICACY

NEs are destabilized by increasing the droplet size thus causing phase separation. The coalescence is a mechanism of instability that involves fusion of droplets due to collision, requiring the droplets to approach and drain the film between them causing interfacial fusion. The drainage of interfacial film leads to lubrication on droplet surfaces that can deform them even before contacting, causing contacting surfaces to increase in area.[4] Depending on the magnitude of relative attractive and repulsive forces existing between droplets, an increased surface area can either cause or oppose surfaces to contact.[42] For nanodroplets, the pressure that resists deformation is higher than that for micron-scale droplets; and therefore, lesser coalescence, consequently higher stability of NEs is expected.[28]

Another prime mechanism causing droplet instability is Ostwald ripening (OR), wherein a molecular transfer between droplets ensues owing to less soluble nature of dispersed phase in the dispersion medium. Due to polydisperse nature of NEs, the energy between droplets varies due to varying interfacial energy; therefore, a driving force is created to cause the flux of dispersed phase to pass into larger droplets at the cost of smaller ones.[84] However, nanodroplets have OR growth rates well-below those of larger droplets.

The third mechanism of instability is contact ripening which involves fluid–fluid interface fluctuations to enhance the transfer of molecules between them; however, it requires more insights into the exact mechanism.[69]

9.4 FUNCTIONAL COMPONENTS OF NANOEMULSIONS

NEs are functionalized by the incorporation of various compounds with antimicrobial, antioxidant, and other bioactivities to extend the shelf-life of food products. Earlier, these compounds were mostly of synthetic nature; however nowadays, synthetic compounds are being largely replaced by natural compounds owing to their safety toward health.

9.4.1 ANTIOXIDANTS AND PLANT EXTRACTS

The NEs act as effective delivery vehicles protecting their antioxidant activity and enabling their usage in food systems.[88] NEs constituted by corn oil were formulated to encapsulate Brazilian propolis extract with antiseptic, antioxidant, antibacterial, and antioxidant properties. The NEs could be used to mask the off-flavor of propolis, prevent its degradation and are used as a food preservative.[76]

Soybean oil was used to encapsulate olive leaf extract in NEs constituted by whey-protein concentrate and whey-protein concentrate–pectin complexes.[47] The NEs were found to possess high antioxidant activity in olive leaf extract as identified by peroxide formation. Acai berry anthocyanins encapsulated in water-in-oil NEs were found to retain antioxidant activity after storage for 30 days and a half-life of 385 days, indicating that such NEs could have food and pharmaceutical applications.[63] Polyphenolic extracts from grape marc were encapsulated in sunflower oil NEs. The NEs possessed high cellular antioxidant activity than the free extracts.[77]

NEs carrying jaboticaba extract (*Plinia peruviana*) were found to possess strong antioxidant activity as identified by the high concentration of flavonoids, phenolics, and its marked ability to scavenge free radicals.[43] Similarly, extracts obtained from leaves and stems of *Vellozia squamata* Pohl possessed high antioxidant activity and were encapsulated in NEs.[63]

Extracts obtained from *Physalis peruviana* calyx phenols (e.g., rutin) were encapsulated in NEs, stored at room temperature for 120 days, and were characterized by high antioxidant and reducing capacity compared to free extracts.[5] *Achyrocline satureioides* extracts exhibited high antioxidant and bioactive potential due to phenolic compounds, such as quercetin. The application of plant extract through NE has enhanced its solubility in aqueous systems and increased antioxidant activity.[94] Extracts obtained from *Phyllanthus emblica* were encapsulated in NEs and were characterized by presence of epigallocatechin, vanillic acid, ellagic acid, and gallic acid. The NE was stable at 4°C for 90 days and could act as a potential carrier for antioxidants.[9] Vitamin E was encapsulated in corn oil NEs and inhibited lipase activity in the small intestine.[39] These research studies demonstrate the success of NEs to enable the application of antioxidants in food systems.

9.4.2 ANTIMICROBIALS

The antimicrobials are encapsulated in suitable carrier systems, such as NEs.[60] Also, nanoencapsulation of EOs enhances their solubility and dispersion in aqueous media, which overcomes the barriers to their interaction with bacteria in such environments.[40] EOs are hydrophobic liquids synthesized as part of a defense mechanism against various pathogens, insects, and UV light.

Chemically, EOs are composed of terpenes (such as limonene), terpenoids (such as linalool and thymol), phenylpropanoids (e.g., eugenol), and cinnamaldehyde.[29] EOs have proven to be potent antimicrobial agents in-vitro and in many food systems, replacing the chemical sanitizers.[48] When encapsulated in NEs, EOs show enhanced bioactivity and minimal impact on the organoleptic properties of food products.[23] This was exemplified in the antibacterial activity of *Thymus saenensis* EO that was enhanced 10-folds by delivering through NEs.[43] The antimicrobial feature of EOs is attributed to their potency to penetrate bacterial membranes toward cell-interior and inhibit its functional properties. Phenolic compounds present in EOs elicit antimicrobial response against pathogens by disrupting cell-membrane and leaking cellular-constituents. These phenolic compounds actually are capable of damaging cytoplasmic membranes, disrupt proton motive force, alter cell permeability, and modify energy generation in cells, thus causing cell death mediated by one or more of the causes synergistically.[8]

Interaction of EOs with cells cause inhibition of Gram-positive bacteria (such as *L. monocytogenes*, *Bacillus cereus,* and *Staphylococcus aureus*) due to the interaction of membrane constituents with EO components directly, more pronouncedly than the Gram-negative bacteria (such as *Salmonella Enteritidis* and *Escherichia coli*).[12] Strong antimicrobial activity was reported[20] for (1) jasmine, celery seeds, carrot seeds, and orange oils against *Campylobacter jejuni*; (2) clove bud, cinnamon, oregano, and thyme EO against *E. coli* O157:H7; (3) oregano, cinnamon, clove bud, thyme, allspice, and marjoram EO against *L. monocytogenes*; and (4) thyme, clove bud, cinnamon, and oregano EO against *S. enteric.* Similarly, EO obtained from *Laurus nobilis* and *Myrtus communis* exhibited antagonism toward foodborne pathogens, with the former having the strongest activity and later showing moderate activity.[11]

EOs derived from *Cinnamonum cassia* showed potential antimicrobial activity against most dominant non-O157 Shiga toxin-producing *E. coli* O26, O45, O103, O111, O121, O145, mostly contributed by cinnamaldehyde (59.96%).[80] Citrus EOs rich in limonene and citrullene possess antimicrobial properties against foodborne pathogens, such as *Salmonella*, *Campylobacter*, *Arcobacter*, and *Cinnamomum coli.*[51,52] Anise oil NE was investigated for potential antimicrobial effects against *L. monocytogenes* and *E. coli* O157:H7 by identifying the minimum inhibitory concentration (MIC). The NE more effectively reduced the counts of pathogenic bacteria over bulk anise oil.[85]

Clove oil NEs stabilized with purity gum ultra-exhibited antimicrobial activity toward Gram-positive bacteria (such as *S. aureus* and *L. monocytogenes*), attributed to the presence of alpha-humelene, eugenyl acetate, and beta-caryophyllene.[40] Nanoemulsified basil oil possessed high antimicrobial activity against *Penicillium chrysogenum* and *Aspergillus flavus,* which are implicated in causing food spoilage. The fungicidal activity of NE at 1000 ppm was about 64–67% compared to Carbendazim (a standard fungicide).[26]

9.5 APPLICATION OF NANOEMULSIONS AS EDIBLE COATINGS

NE-based ECs have documented promising results with fresh produce enhancing their shelf-life and improving quality.[2,10]

9.5.1 FOR WHOLE FRUITS AND VEGETABLES

In fresh-cut Fuji apple, NE carrying lemongrass EO inhibited the growth of *E. coli* O157: H7 and *L. monocytogenes* during storage for 5 months. It was also effective in reducing undesirable color changes in the cut fruit.[30] In plums, lemongrass oil NE caused a decrease of >2.8 log CFU/g of *E. coli* O157:H7 and *Salmonella typhimurium*. Glossiness and fracturability also were not affected by applying this coating.[35]

Strawberries coated with pullulan-cinnamon EO and NE exhibited firmness retention, reduced weight loss, and pronounced antagonism toward molds and bacteria, compared to the control group and that without NE.[13] In another study,[67] strawberries coated with thymol NE increased their shelf-life by 4 days, and weight loss was also prevented as compared to control, during storage at 5°C for 16 days. Cinnamon EO NE applied on strawberries considerably reduced the fungal cartilage.[53]

Mango fruits coated with hydroxypropyl methylcellulose EO NE also retained weight, firmness, and color deterioration was also prevented.[36] Chitosan coatings, incorporated with thyme EO when applied on avocado, were found to maintain their quality, prevent weight loss, retain firmness and color, along with inhibiting microbial growth.[14] In tomatoes treated with alginate-based coating carrying *Citrus sinensis* EO NE, populations of mesophilic bacteria, *Salmonella* and *Listeria* were considerably reduced, firmness was enhanced, and weight loss was decreased 3-fold lower than the control.[15] Melons coated with cinnamon oil NE exhibited 7.7 and 5.5 log reductions in the counts of *L. monocytogenes* and *Salmonella* spp. inoculated artificially, respectively.[57]

In case of rucola leaf, the shelf-life was extended by 1 week as compared to control, by the application of chitosan NE.[77] Okra coated with alginate carrying nanoemulsified basil oil maintained texture, color, and overall acceptability better than uncoated samples.[26] Coating *Agaricus bisporus* with chitosan-cumin EO NE reduced PPO activity and microbial proliferation during storage.[32] Green beans coated with chitosan-mandarin EO NE caused microbial load reduction, delayed firming, maintenance of color, texture, and extension of shelf-life.[79]

9.5.2 FRESH CUT FRUITS AND VEGETABLES

NEs carrying active compounds have been applied for enabling the marketability and safety of fresh-cut FAV.[27,72] The effects of NE-based ECs carrying lemongrass EO (0.1, 0.5, 1% v/v) were found to completely inhibit the growth of natural microflora in fresh-cut apple during 2-week storage, along with lowering the browning index.[71]

Fresh-cut apple coated with chitosan-cinnamon EO NE exhibited reduced microbial counts and preservation of quality during storage.[90] NE carrying 0.5% orange peel EO was found to confer microbial stability to coated orange slices over that of uncoated ones and prevent discoloration due to the oxygen barrier properties of coating.[64] Similarly, promising results were observed in grape berries coated with carnauba wax NE carrying lemongrass EO, which significantly reduced the counts of *S. typhimurium* and *E. coli* during storage at 4 and 25°C for 28 days, compared to the control with zero inhibition.[34] Lemongrass oil/chitosan NE was more effective than the emulsion of the same composition in ensuring microbial safety against *S. typhimurium* in grape berries.[54]

NE containing xanthan gum, nopal mucilage, and α-tocopherol was found to maintain quality of fresh-cut apple by reducing polyphenol oxidase and pectin methylesterase activity by modifying the degradation metabolism of cell-wall.[22] In fresh-cut pineapples, citral NE (0.1–1%) incorporation in sodium alginate coatings enabled the retention of color, reduced respiration, and microbial growth along with storage. Artificially inoculated *Salmonella enterica* and *L. monocytogenes* was reduced by 2.52 and 2.23 log CFU/g when coated with 0.5% NE.[60] The efficacy of alginate-based coating carrying thyme oil NE effectively inhibited the growth of *L. monocytogenes*, *S. typhimurium*, *E. coli* O157:H7, and *S. aureus*; and improved the quality of fresh-cut cantaloupe along with extended shelf-life by reducing weight loss and respiration rate.[74]

In fresh-cut cucumber and carrot slices, carvacrol-loaded nanocoatings were effective in arresting the microbial proliferation during storage.[41,83] Similarly, lettuce coated with 0.1% oregano oil NE exhibited improved quality and microbial safety by inhibiting *E. coli* O157:H7, *S. Typhimurium*, and *L. monocytogenes* than the control.[7] Likewise, spinach leaves washed with carvacrol or eugenol NE exhibited considerable reduction of 4.5 log CFU/g in the counts of inoculated *E. coli* and *S. enterica.*[70] In minimally processed alfalfa, broccoli, mung bean, and radish and soaking in 4000 ppm carvacrol NE inactivated *E. coli* and *S. Enteritidis.*[38]

9.6 SENSORY AND SAFETY ASPECTS OF NANOEMULSIONS

Incorporating EOs in NEs may alter the sensory profile of FAV as EOs have a characteristic strong flavor easily perceived by the consumers.[27] For instance, in fresh-cut melon coated with cinnamon oil-based coatings, the sensory preference was recorded lower due to strong flavor.[66]

Radi et al.[64] observed higher sensory rating in orange slices coated with orange peel EO NE. In another study,[35] lemongrass oil concentration in NE applied onto grape berries did not affect flavor of berries, possibly due to the nanosize of oil droplets. Tomatoes coated with NE carrying *C. sinensis* EO were found to be acceptable despite a distinct odor.[15] The sensory effects of NEs, therefore, depend on the type of EO and its concentration. Thus, optimization studies for ensuring sensory acceptability of NE coated FAV are of prime importance.

The safety of NEs is defined by the manner in which they interact with humans. Numerous studies document the cytotoxicity of NEs at particular concentrations; however, some studies do show their safety toward biological systems. NEs carrying olmesartan at subchronic toxicity doses revealed less toxic behavior as identified by biochemical and hematologic analysis.[24] Resveratrol NEs on the other hand did not exhibit any kind of antiangiogenic activity in chick embryos.[61] Similarly, curcumin loaded NEs did not show any toxicity during intranasal transit to brain.[81]

NEs based on α-tocopherol and lemon oil were found to be nontoxic to HepG2 cells.[33] The toxicity of garlic oil NEs at 18.63 mL/kg caused immediate death in female rats; however, at 0.46 mL/kg, no such effects were seen.[65] Curcumin NEs were cyto-compatible in that no detrimental effects on fibroblasts were seen.[87] Most of the NEs are safe to administer at low concentrations. At excessively high loading concentrations, potential toxicity is induced.[91] In the preservation of FAV, low concentrations of EOs are always preferred and could, therefore, be assumed not to pose health concerns. However, deeper insights into the safety of NEs carrying EOs to be applied on FAVs should be carried out.

9.7 SUMMARY

NEs effectively carry natural antimicrobials (such as EOs) and preserve the quality of FAV during storage. Owing to an increased surface area, EOs produce more promising outcomes at lower concentrations compared to large droplet-sized emulsions. Increased area also enhances their solubility, stability, and acceptability from organoleptic perspective. Since NEs produce desirable effects at lower concentration of EOs, these are generally safe for human consumption. Therefore, an amalgamation of nanotechnology and EOs is instrumental in up keeping the safety and quality of fresh FAVs.

ACKNOWLEDGMENTS

The authors thank NPIU-MHRD, Government of India, for providing opportunity to work under World Bank-MHRD funded TEQIP Project.

KEYWORDS

- **antimicrobials**
- **antioxidants**
- **fruits and vegetables (FAV)**
- **nanoemulsions**
- **pathogens**
- **spoilage**

REFERENCES

1. Acevedo-Fani, A.; Soliva-Fortuny, R.; Martín-Belloso, O. Nanostructured Emulsions and Nanolaminates for Delivery of Active Ingredients: Improving Food Safety and Functionality. *Trends Food Sci. Technol.* **2017,** *60*, 12–22. https://doi.org/10.1016/j.tifs.2016.10.027.
2. Al-Tayyar, N. A.; Youssef, A. M.; Al-Hindi, R. R. Edible Coatings and Antimicrobial Nanoemulsions for Enhancing Shelf-life and Reducing Foodborne Pathogens of Fruits and Vegetables: A Review. *Sustain. Mater. Technol.* **2020,** *26*, E-article 0**0215,** p 6. https://doi.org/10.1016/j.susmat.2020.e00215.
3. Augustin, M. A.; Sanguansri, L.; Fox, E. M.; Cobiac, L.; Cole, M. B. Recovery of Wasted Fruit and Vegetables for Improving Sustainable Diets. *Trends Food Sci. Technol.* **2020,** *95* (April), 75–85. https://doi.org/10.1016/j.tifs.2019.11.010.
4. Baldessari, F.; Leal, L. G. Effect of Overall Drop Deformation on Flow-Induced Coalescence at Low Capillary Numbers. *Phys. Fluids* **2006,** *18* (1), 1–21. https://doi.org/10.1063/1.2158427.
5. Bazana, M. T.; da Silva, S. S.; Codevilla, C. F.; de Deus, C.; Lucas, B. N.; Ugalde, G. A.; Mazutti, M. A.; Moraes Flores, E. M.; Barin, J. S.; de Bona da Silva, C.; et al. Development of Nanoemulsions Containing Physalis Peruviana Calyx Extract: A Study on Stability and Antioxidant Capacity. *Food Res. Int.* **2019,** *125* (August), E-article 108645; p 7. https://doi.org/10.1016/j.foodres.2019.108645.
6. Berger, C. N.; Sodha, S. V.; Shaw, R. K.; Griffin, P. M.; Pink, D.; Hand, P.; Frankel, G. Fresh Fruit and Vegetables as Vehicles for the Transmission of Human Pathogens. *Environ. Microbiol.* **2010,** *12* (9), 2385–2397. https://doi.org/10.1111/j.1462-2920.2010.02297.x.
7. Bhargava, K.; Conti, D. S.; da Rocha, S. R. P.; Zhang, Y. Application of an Oregano Oil Nanoemulsion to the Control of Foodborne Bacteria on Fresh Lettuce. *Food Microbiol.* **2015,** *47*, 69–73. https://doi.org/10.1016/j.fm.2014.11.007.
8. Calo, J. R.; Crandall, P. G.; O'Bryan, C. A.; Ricke, S. C. Essential Oils as Antimicrobials in Food Systems—A Review. *Food Control* **2015,** *54*, 111–119. https://doi.org/10.1016/j.foodcont.2014.12.040.
9. Chaiittianan, R.; Sripanidkulchai, B. Development of a Nanoemulsion of *Phyllanthus Emblica* L. Branch Extract. *Drug Dev. Ind. Pharm.* **2014,** *40* (12), 1597–1606. https://doi.org/10.3109/03639045.2013.838580.
10. Chaudhary, S.; Kumar, S.; Kumar, V.; Sharma, R. Chitosan Nanoemulsions as Advanced Edible Coatings for Fruits and Vegetables: Composition, Fabrication and Developments in Last Decade. *Int. J. Biol. Macromol.* **2020,** *152*, 154–170. https://doi.org/10.1016/j.ijbiomac.2020.02.276.
11. Cherrat, L.; Espina, L.; Bakkali, M.; García-Gonzalo, D.; Pagán, R.; Laglaoui, A. Chemical Composition and Antioxidant Properties of *Laurus Nobilis* L. and *Myrtus Communis* L. Essential Oils from Morocco and Evaluation of Antimicrobial Activity Acting Alone or in Combined Processes for Food Preservation. *J. Sci. Food Agric.* **2014,** *94* (6), 1197–1204. https://doi.org/10.1002/jsfa.6397.
12. Chorianopoulos, N.; Kalpoutzakis, E.; Aligiannis, N.; Mitaku, S.; Nychas, G. J.; Haroutounian, S. A. Essential Oils of Satureja, Origanum, and Thymus Species: Chemical Composition and Antibacterial Activities against Foodborne Pathogens. *J. Agric. Food Chem.* **2004,** *52* (26), 8261–8267. https://doi.org/10.1021/jf049113i.

13. Chu, Y.; Gao, C. C.; Liu, X.; Zhang, N.; Xu, T.; Feng, X.; Yang, Y.; Shen, X.; Tang, X. Improvement of Storage Quality of Strawberries by Pullulan Coatings Incorporated with Cinnamon Essential Oil Nanoemulsion. *LWT* **2020,** *122* (January), E-article 109054; p 5. https://doi.org/10.1016/j.LWT.2020.109054.
14. Correa-Pacheco, Z. N.; Bautista-Baños, S.; Valle-Marquina, M. Á.; Hernández-López, M. The Effect of Nanostructured Chitosan and Chitosan-Thyme Essential Oil Coatings on *Colletotrichum* Gloeosporioides Growth *in Vitro* and on Avocado and Fruit Quality. *J. Phytopathol.* **2017,** *165* (5), 297–305. https://doi.org/10.1111/jph.12562.
15. Das, S.; Vishakha, K.; Banerjee, S.; Mondal, S.; Ganguli, A. Sodium Alginate-Based Edible Coating Containing Nanoemulsion of *Citrus Sinensis* Essential Oil Eradicates Planktonic and Sessile Cells of Food-Borne Pathogens and Increased Quality Attributes of Tomatoes. *Int. J. Biol. Macromol.* **2020,** *162*, 1770–1779. https://doi.org/10.1016/j.ijbiomac.2020.08.086.
16. Davidson, P. M.; Critzer, F. J.; Matthew Taylor, T. Naturally Occurring Antimicrobials for Minimally Processed Foods. *Annu. Rev. Food Sci. Technol.* **2013,** *4* (1), 163–190. https://doi.org/10.1146/annurev-food-030212-182535.
17. Divya, K.; Smitha, V.; Jisha, M. S. Antifungal, Antioxidant and Cytotoxic Activities of Chitosan Nanoparticles and Its Use as an Edible Coating on Vegetables. *Int. J. Biol. Macromol.* **2018,** *114* (2017), 572–577. https://doi.org/10.1016/j.ijbiomac.2018.03.130.
18. Donsì, F.; Annunziata, M.; Vincensi, M.; Ferrari, G. Design of Nanoemulsion-Based Delivery Systems of Natural Antimicrobials: Effect of the Emulsifier. *J. Biotechnol.* **2012,** *159* (4), 342–350. https://doi.org/10.1016/j.jbiotec.2011.07.001.
19. Falguera, V.; Quintero, J. P.; Jiménez, A.; Muñoz, J. A.; Ibarz, A. Edible Films and Coatings: Structures, Active Functions and Trends in Their Use. *Trends Food Sci. Technol.* **2011,** *22* (6), 292–303. https://doi.org/10.1016/j.tifs.2011.02.004.
20. Friedman, M.; Henika, P. R.; Mandrell, R. E. Bactericidal Activities of Plant Essential Oils and Some of Their Isolated Constituents against *Campylobacter Jejuni*, *Escherichia Coli*, *Listeria Monocytogenes*, and *Salmonella Enterica*. *J. Food Prot.* **2002,** *65* (10), 1545–1560. https://doi.org/10.4315/0362-028X-65.10.1545.
21. Gago, M. L.; Faleiro, M. L.; Miguel, M. G. C.; Guerreiro, A. C.; Antunes, M. D. C. The Effect of Edible Coatings on the Nutritional Quality of Fresh-Cut Apple through Shelf-Life. *LWT—Food Sci. Technol.* **2017,** *75*, 210–219. https://doi.org/10.1016/j.LWT.2016.08.052.
22. García-Betanzos, C. I.; Hernández-Sánchez, H.; Bernal-Couoh, T. F.; Quintanar-Guerrero, D.; Zambrano-Zaragoza, M. de la L. Physicochemical, Total Phenols and Pectin Methylesterase Changes on Quality Maintenance on Guava Fruit (*Psidium Guajava* L.) Coated with Candeuba Wax Solid Lipid Nanoparticles-Xanthan Gum. *Food Res. Int.* **2017,** *101*, 218–227. https://doi.org/10.1016/j.foodres.2017.08.065.
23. Ghosh, V.; Mukherjee, A.; Chandrasekaran, N. Eugenol-Loaded Antimicrobial Nanoemulsion Preserves Fruit Juice Against, Microbial Spoilage. *Colloids Surf. B Biointerf.* **2014,** *114*, 392–397. https://doi.org/10.1016/j.colsurfb.2013.10.034.
24. Gorain, B.; Choudhury, H.; Tekade, R. K.; Karan, S.; Jaisankar, P.; Pal, T. K. Comparative Biodistribution and Safety Profiling of Olmesartan Medoxomil Oil-in-Water Oral Nanoemulsion. *Regul. Toxicol. Pharmacol.* **2016,** *82*, 20–31. https://doi.org/10.1016/j.yrtph.2016.10.020.
25. Gundewadi, G.; Rudra, S. G.; Sarkar, D. J.; Singh, D. Nanoemulsion Based Alginate Organic Coating for Shelf-life Extension of Okra. *Food Packag. Shelf Life* **2018,** *18* (August), 1–12. https://doi.org/10.1016/j.fpsl.2018.08.002.

26. Gundewadi, G.; Sarkar, D. J.; Rudra, S. G.; Singh, D. Preparation of Basil Oil Nanoemulsion Using *Sapindus Mukorossi* Pericarp Extract: Physicochemical Properties and Antifungal Activity against Food Spoilage Pathogens. *Ind. Crops Prod.* **2018,** *125* (June), 95–104. https://doi.org/10.1016/j.indcrop.2018.08.076.
27. Hasan, S. M. K.; Ferrentino, G.; Scampicchio, M. Nanoemulsion as Advanced Edible Coatings to Preserve the Quality of Fresh-Cut Fruits and Vegetables: A Review. *Int. J. Food Sci. Technol.* **2020,** *55* (1), 1–10. https://doi.org/10.1111/ijfs.14273.
28. Helgeson, M. E. Colloidal Behavior of Nanoemulsions: Interactions, Structure, and Rheology. *Curr. Opin. Colloid Interf. Sci.* **2016,** *25*, 39–50. https://doi.org/10.1016/j.cocis.2016.06.006.
29. Hyldgaard, M.; Mygind, T.; Meyer, R. L. Essential Oils in Food Preservation: Mode of Action, Synergies, and Interactions with Food Matrix Components. *Front. Microbiol.* **2012,** *3* (Jan), 1–24. https://doi.org/10.3389/fmicb.2012.00012.
30. Jo, W. S.; Song, H. Y.; Song, N. B.; Lee, J. H.; Min, S. C.; Song, K. Bin. Quality and Microbial Safety of Apples Coated with Carnauba-Shellac Wax Containing Lemongrass Oil. *LWT—Food Sci. Technol.* **2014,** *55* (2), 490–497. https://doi.org/10.1016/j.LWT.2013.10.034.
31. Ju, J.; Xie, Y.; Guo, Y.; Cheng, Y.; Qian, H.; Yao, W. Application of Edible Coating with Essential Oil in Food Preservation. *Crit. Rev. Food Sci. Nutr.* **2019,** *59* (15), 2467–2480. https://doi.org/10.1080/10408398.2018.1456402.
32. Karimirad, R.; Behnamian, M.; Dezhsetan, S. Application of Chitosan Nanoparticles Containing *Cuminum Cyminum* Oil as a Delivery System for Shelf-life Extension of Agaricus Bisporus. *LWT* **2019,** *106* (Feb), 218–228. https://doi.org/10.1016/j.LWT.2019.02.062.
33. Kaur, K.; Kumar, R.; Arpita; Goel, S.; Uppal, S.; Bhatia, A.; Mehta, S. K. *Physiochemical and Cytotoxicity Study of TPGS Stabilized Nanoemulsion Designed by Ultrasonication Method*; **2017,** *34*. https://doi.org/10.1016/j.ultsonch.2016.05.037.
34. Kim, I. H.; Lee, H.; Kim, J. E.; Song, K. Bin; Lee, Y. S.; Chung, D. S.; Min, S. C. Plum Coatings of Lemongrass Oil-Incorporating Carnauba Wax-Based Nanoemulsion. *J. Food Sci.* **2013,** *78* (10), E-article 12244; p 9. https://doi.org/10.1111/1750-3841.12244.
35. Kim, I. H.; Oh, Y. A.; Lee, H.; Song, K. Bin; Min, S. C. Grape Berry Coatings of Lemongrass Oil-Incorporating Nanoemulsion. *LWT—Food Sci. Technol.* **2014,** *58* (1), 1–10. https://doi.org/10.1016/j.LWT.2014.03.018.
36. Klangmuang, P.; Sothornvit, R. Active Coating from Hydroxypropyl Methylcellulose-Based Nanocomposite Incorporated with Thai Essential Oils on Mango. *Food Biosci.* **2018,** *23*, 9–15. https://doi.org/10.1016/j.fbio.2018.02.012.
37. Kwak, T. Y.; Kim, N. H.; Rhee, M. S. Response Surface Methodology-Based Optimization of Decontamination Conditions for Escherichia Coli O157:H7 and *Salmonella Typhimurium* on Fresh-Cut Celery Using Thermo ultrasound and Calcium Propionate. *Int. J. Food Microbiol.* **2011,** *150* (2–3), 128–135. https://doi.org/10.1016/j.ijfoodmicro.2011.07.025.
38. Landry, K. S.; Komaiko, J.; Wong, D. E.; Xu, T.; McClements, D. J.; McLandsborough, L. Inactivation of Salmonella on Sprouting Seeds Using a Spontaneous Carvacrol Nanoemulsion Acidified with Organic Acids. *J. Food Prot.* **2016,** *79* (7), 1115–1126. https://doi.org/10.4315/0362-028X.JFP-15-397.
39. Lv, S.; Gu, J.; Zhang, R.; Zhang, Y.; Tan, H.; McClements, D. J. Vitamin E Encapsulation in Plant-Based Nanoemulsions Fabricated Using Dual-Channel Microfluidization: Formation, Stability, and Bioaccessibility. *J. Agric. Food Chem.* **2018,** *66* (40), 10532–10542. https://doi.org/10.1021/acs.jafc.8b03077.

40. Majeed, H.; Liu, F.; Hategekimana, J.; Sharif, H. R.; Qi, J.; Ali, B.; Bian, Y. Y.; Ma, J.; Yokoyama, W.; Zhong, F. Bactericidal Action Mechanism of Negatively Charged Food Grade Clove Oil Nanoemulsions. *Food Chem.* **2016,** *197*, 75–83. https://doi.org/10.1016/j.foodchem.2015.10.015.
41. Martínez-Hernández, G. B.; Amodio, M. L.; Colelli, G. Carvacrol-Loaded Chitosan Nanoparticles Maintain Quality of Fresh-Cut Carrots. *Innov. Food Sci. Emerg. Technol.* **2017,** *41*, 56–63. https://doi.org/10.1016/j.ifset.2017.02.005.
42. Mason, T. G.; Wilking, J. N.; Meleson, K.; Chang, C. B.; Graves, S. M. Nanoemulsions: Formation, Structure, and Physical Properties. *J. Phys. Condens. Matter* **2006,** *18* (41), E-article 1088; p 7; https://doi.org/10.1088/0953-8984/18/41/R01.
43. Mazzarino, L.; da Silva Pitz, H.; Lorenzen Voytena, A. P.; Dias Trevisan, A. C.; Ribeiro-Do-Valle, R. M.; Maraschin, M. Jaboticaba (*Plinia Peruviana*) Extract Nanoemulsions: Development, Stability, and in Vitro Antioxidant Activity. *Drug Dev. Ind. Pharm.* **2018,** *44* (4), 643–651. https://doi.org/10.1080/03639045.2017.1405976.
44. McClements, D. J. Recent Developments in Encapsulation and Release of Functional Food Ingredients: Delivery by Design. *Curr. Opin. Food Sci.* **2018,** *23*, 80–84. https://doi.org/10.1016/j.cofs.2018.06.008.
45. Meleson, K.; Graves, S.; Mason, T. G. Formation of Concentrated Nanoemulsions by Extreme Shear. *Soft Matter.* **2004,** *2* (2–3), 109–123. https://doi.org/10.1081/SMTS-200056102.
46. Moghimi, R.; Ghaderi, L.; Rafati, H.; Aliahmadi, A.; Mcclements, D. J. Superior Antibacterial Activity of Nanoemulsion of *Thymus Daenensis* Essential Oil against E. Coli. *Food Chem.* **2016,** *194*, 410–415. https://doi.org/10.1016/j.foodchem.2015.07.139.
47. Mohammadi, A.; Jafari, S. M.; Assadpour, E.; FaridiEsfanjani, A. Nano Encapsulation of Olive Leaf Phenolic Compounds through WPC-Pectin Complexes and Evaluating Their Release Rate. *Int. J. Biol. Macromol.* **2016,** *82*, 816–822. https://doi.org/10.1016/j.ijbiomac.2015.10.025.
48. Moore-Neibel, K.; Gerber, C.; Patel, J.; Friedman, M.; Ravishankar, S. Antimicrobial Activity of Lemongrass Oil against *Salmonella Enterica* on Organic Leafy Greens. *J. Appl. Microbiol.* **2012,** *112* (3), 485–492. https://doi.org/10.1111/j.1365-2672.2011.05222.x.
49. Mostafidi, M.; Sanjabi, M. R.; Shirkhan, F.; Zahedi, M. T. A Review of Recent Trends in the Development of the Microbial Safety of Fruits and Vegetables. *Trends Food Sci. Technol.* **2020,** *103* (July), 321–332. https://doi.org/10.1016/j.tifs.2020.07.009.
50. Mritunjay, S. K.; Kumar, V. Fresh Farm Produce as a Source of Pathogens: A Review. *Research Journal of Environmental Toxicology* **2015,** *2015*, 59–70. https://doi.org/10.3923/rjet.2015.59.70.
51. Muthaiyan, A.; Martin, E. M.; Natesan, S.; Crandall, P. G.; Wilkinson, B. J.; Ricke, S. C. Antimicrobial Effect and Mode of Action of Terpeneless Cold-Pressed Valencia Orange Essential Oil on Methicillin-Resistant *Staphylococcus Aureus*. *J. Appl. Microbiol.* **2012,** *112* (5), 1020–1033. https://doi.org/10.1111/j.1365-2672.2012.05270.x.
52. Nannapaneni, R.; Chalova, V. I.; Crandall, P. G.; Ricke, S. C.; Johnson, M. G.; O'Bryan, C. A. Campylobacter and Arcobacter Species Sensitivity to Commercial Orange Oil Fractions. *Int. J. Food Microbiol.* **2009,** *129* (1), 43–49. https://doi.org/10.1016/j.ijfoodmicro.2008.11.008.
53. Naserzadeh, Y.; Mahmoudi, N.; Pakina, E. Antipathogenic Effects of Emulsion and Nanoemulsion of Cinnamon Essential Oil against Rhizopus Rot and Grey Mold on Strawberry Fruits. *Foods Raw Mater.* **2019,** *7* (1), 210–216. https://doi.org/10.21603/2308-4057-2019-1-210-216.

54. Oh, Y. A.; Oh, Y. J.; Song, A. Y.; Won, J. S.; Song, K. Bin; Min, S. C. Comparison of Effectiveness of Edible Coatings Using Emulsions Containing Lemongrass Oil of Different Size Droplets on Grape Berry Safety and Preservation. *LWT* **2017,** *75*, 742–750. https://doi.org/10.1016/j.LWT.2016.10.033.
55. Oms-Oliu, G.; Rojas-Graü, M. A.; González, L. A.; Varela, P.; Soliva-Fortuny, R.; Hernando, M. I. H.; Munuera, I. P.; Fiszman, S.; Martín-Belloso, O. Recent Approaches Using Chemical Treatments to Preserve Quality of Fresh-Cut Fruit: A Review. *Postharvest Biol. Technol.* **2010,** *57* (3), 139–148. https://doi.org/10.1016/j.postharvbio.2010.04.001.
56. Paramithiotis, S.; Drosinos, E. H.; Skandamis, P. N. Microbial Ecology of Fruits and Fruit-Based Products. *Model. Microb. Ecol. Foods Quant. Microbiol. Food Process.* **2016,** *2016*, 358–381. https://doi.org/10.1002/9781118823071.ch18.
57. Paudel, S. K.; Bhargava, K.; Kotturi, H. Antimicrobial Activity of Cinnamon Oil Nanoemulsion against *Listeria Monocytogenes* and *Salmonella Spp.* on Melons. *LWT* **2019,** *111* (October 2018), 682–687. https://doi.org/10.1016/j.LWT.2019.05.087.
58. Paul, S. K. (Ed.) *Edible Films and Coatings for Fruits and Vegetables*; Elsevier Ltd.: New York, 2020; Online; p 10. https://doi.org/10.1016/b978-0-12-803581-8.11509-7.
59. Prakash, A.; Baskaran, R.; Paramasivam, N.; Vadivel, V. Department of Fruit and Vegetable Technology, Central Food Technological Research Institute. *Food Res. Int.* **2018**; online. https://doi.org/10.1016/j.foodres.2018.05.066.
60. Prakash, A.; Baskaran, R.; Vadivel, V. Citral Nanoemulsion Incorporated Edible Coating to Extend the Shelf-life of Fresh Cut Pineapples. *LWT* **2020,** *118* (Nov), E-article 108851; p 8. https://doi.org/10.1016/j.LWT.2019.108851.
61. Pund, S.; Thakur, R.; More, U.; Joshi, A. Lipid Based Nanoemulsifying Resveratrol for Improved Physicochemical Characteristics, *in Vitro* Cytotoxicity and *in Vivo* Antiangiogenic Efficacy. *Colloids Surf. B Biointerf.* **2014,** *120*, 110–117. https://doi.org/10.1016/j.colsurfb.2014.05.016.
62. Quintão, F. J. O.; Tavares, R. S. N.; Vieira-Filho, S. A.; Souza, G. H. B.; Santos, O. D. H. Hydroalcoholic Extracts of *Vellozia Squamata*: Study of Its Nanoemulsions for Pharmaceutical or Cosmetic Applications. *Brazilian J. Pharmacogn.* **2013,** *23* (1), 101–107. https://doi.org/10.1590/S0102-695X2013005000001.
63. Rabelo, C. A. S.; Taarji, N.; Khalid, N.; Kobayashi, I.; Nakajima, M.; Neves, M. A. Formulation and Characterization of Water-in-Oil Nanoemulsions Loaded with Açaí Berry Anthocyanins: Insights of Degradation Kinetics and Stability Evaluation of Anthocyanins and Nanoemulsions. *Food Res. Int.* **2018,** *106* (2017), 542–548. https://doi.org/10.1016/j.foodres.2018.01.017.
64. Radi, M.; Akhavan-Darabi, S.; Akhavan, H. R.; Amiri, S. The Use of Orange Peel Essential Oil Microemulsion and Nanoemulsion in Pectin-Based Coating to Extend the Shelf-life of Fresh-Cut Orange. *J. Food Process. Preserv.* **2018,** *42* (2), 1–9. https://doi.org/10.1111/jfpp.13441.
65. Ragavan, G.; Muralidaran, Y.; Sridharan, B.; Nachiappa Ganesh, R.; Viswanathan, P. Evaluation of Garlic Oil in Nano-Emulsified Form: Optimization and Its Efficacy in High-Fat Diet Induced Dyslipidemia in Wistar Rats. *Food Chem. Toxicol.* **2017,** *105*, 203–213. https://doi.org/10.1016/j.fct.2017.04.019.
66. Raybaudi-Massilia, R. M.; Rojas-Graü, M. A.; Mosqueda-Melgar, J.; Martín-Belloso, O. Comparative Study on Essential Oils Incorporated into an Alginate-Based Edible Coating to Assure the Safety and Quality of Fresh-Cut Fuji Apples. *J. Food Prot.* **2008,** *71* (6), 1150–1161. https://doi.org/10.4315/0362-028X-71.6.1150.

67. Robledo, N.; López, L.; Bunger, A.; Tapia, C.; Abugoch, L. Effects of Antimicrobial Edible Coating of Thymol Nanoemulsion/Quinoa Protein/Chitosan on the Safety, Sensorial Properties, and Quality of Refrigerated Strawberries under Commercial Storage Environment. *Food Bioprocess Technol.* **2018,** *11* (8), 1566–1574. https://doi.org/10.1007/s11947-018-2124-3.
68. Roger, K.; Cabane, B.; Olsson, U. Formation of 10-100Nm Size-Controlled Emulsions through a Sub-PIT Cycle. *Langmuir* **2010,** *26* (6), 3860–3867. https://doi.org/10.1021/la903401g.
69. Roger, K.; Olsson, U.; Schweins, R.; Cabane, B. Emulsion Ripening through Molecular Exchange at Droplet Contacts. *Angew. Chemie* **2015,** *127* (5), 1472–1475. https://doi.org/10.1002/ange.201407858.
70. Ruengvisesh, S.; Loquercio, A.; Castell-Perez, E.; Taylor, T. M. Inhibition of Bacterial Pathogens in Medium and on Spinach Leaf Surfaces Using Plant-Derived Antimicrobials Loaded in Surfactant Micelles. *J. Food Sci.* **2015,** *80* (11), M2522–M2529. https://doi.org/10.1111/1750-3841.13085.
71. Salvia-trujillo, L.; Rojas-graü, M. A.; Soliva-fortuny, R.; Martín-belloso, O. Postharvest Biology and Technology Use of Antimicrobial Nanoemulsions as Edible Coatings: Impact on Safety and Quality Attributes of Fresh-Cut Fuji Apples. *Postharvest Biol. Technol.* **2015,** *105*, 8–16. https://doi.org/10.1016/j.postharvbio.2015.03.009.
72. Salvia-Trujillo, L.; Soliva-Fortuny, R.; Rojas-Graü, M. A.; McClements, D. J.; Martín-Belloso, O. Edible Nanoemulsions as Carriers of Active Ingredients: A Review. *Annu. Rev. Food Sci. Technol.* **2017,** *8* (December), 439–466; https://doi.org/10.1146/annurev-food-030216-025908.
73. Sanchita, M.; Mishra, H. N. Engineering evaluation of thickness and type of packaging materials based on the modified atmosphere packaging requirements of guava. *LWT—Food Sci. Technol.* **2017,** *78* (May), 273-280; online; doi:10.1016/j.lwt.2016.12.043
74. Sarengaowa; Hu, W.; Feng, K.; Xiu, Z.; Jiang, A.; Lao, Y. Thyme Oil Alginate-Based Edible Coatings Inhibit Growth of Pathogenic Microorganisms Spoiling Fresh-Cut Cantaloupe. *Food Biosci.* **2019,** *32* (Sept), E-article 100467; p 6. https://doi.org/10.1016/j.fbio.2019.100467.
75. Saxena, A.; Sharma, L.; Maity, T. Enrichment of Edible Coatings and Films with Plant Extracts or Essential Oils for the Preservation of Fruits and Vegetables. Chapter 34; In: *Biopolymer-Based Formulations: Biomedical & Food Applications*; Elsevier Inc.: San Diego, 2020; pp 859–880. https://doi.org/10.1016/B978-0-12-816897-4.00034-5.
76. Seibert, J. B.; Bautista-Silva, J. P.; Amparo, T. R.; Petit, A.; Pervier, P.; dos Santos Almeida, J. C.; Azevedo, M. C.; Silveira, B. M.; Brandão, G. C.; de Souza, G. H. B.; et al. Development of Propolis Nanoemulsion with Antioxidant and Antimicrobial Activity for Use as a Potential Natural Preservative. *Food Chem.* **2019,** *287* (February), 61–67. https://doi.org/10.1016/j.foodchem.2019.02.078.
77. Sessa, M.; Casazza, A. A.; Perego, P.; Tsao, R.; Ferrari, G.; Donsì, F. Exploitation of Polyphenolic Extracts from Grape Marc as Natural Antioxidants by Encapsulation in Lipid-Based Nano delivery Systems. *Food Bioprocess Technol.* **2013,** *6* (10), 2609–2620. https://doi.org/10.1007/s11947-012-0911-9.
78. Sessa, M.; Ferrari, G.; Donsì, F. Novel Edible Coating Containing Essential Oil Nanoemulsions to Prolong the Shelf-life of Vegetable Products. *Chem. Eng. Trans.* **2015,** *43*, 55–60. https://doi.org/10.3303/CET1543010.

79. Severino, R.; Vu, K. D.; Donsì, F.; Salmieri, S.; Ferrari, G.; Lacroix, M. Antibacterial and Physical Effects of Modified Chitosan Based-Coating Containing Nanoemulsion of Mandarin Essential Oil and Three Non-Thermal Treatments against *Listeria Innocua* in Green Beans. *Int. J. Food Microbiol.* **2014,** *191*, 82–88. https://doi.org/10.1016/j.ijfoodmicro.2014.09.007.
80. Sheng, L.; Zhu, M. J. Inhibitory Effect of Cinnamomum Cassia Oil on Non-O157 Shiga Toxin-Producing Escherichia Coli. *Food Control* **2014,** *46*, 374–381. https://doi.org/10.1016/j.foodcont.2014.05.050.
81. Sood, S.; Jain, K.; Gowthamarajan, K. Optimization of Curcumin Nanoemulsion for Intranasal Delivery Using Design of Experiment and Its Toxicity Assessment. *Colloids Surf. B Biointerf.* **2014,** *113*, 330–337. https://doi.org/10.1016/j.colsurfb.2013.09.030.
82. Ssemanda, J. N.; Reij, M.; Bagabe, M. C.; Muvunyi, C. M.; Joosten, H.; Zwietering, M. H. Indicator Microorganisms in Fresh Vegetables from Farm to Fork in Rwanda. *Food Control* **2017,** *75*, 126–133. https://doi.org/10.1016/j.foodcont.2016.12.031.
83. Taştan, Ö.; Pataro, G.; Donsì, F.; Ferrari, G.; Baysal, T. Decontamination of Fresh-Cut Cucumber Slices by a Combination of a Modified Chitosan Coating Containing Carvacrol Nanoemulsions and Pulsed Light. *Int. J. Food Microbiol.* **2017,** *260*, 75–80. https://doi.org/10.1016/j.ijfoodmicro.2017.08.011.
84. Taylor, P. Ostwald Ripening in Emulsions. *Adv. Colloid Interface Sci.* **1998,** *75* (2), 107–163. https://doi.org/10.1016/S0001-8686(98)00035-9.
85. Topuz, O. K.; Özvural, E. B.; Zhao, Q.; Huang, Q.; Chikindas, M.; Gölükçü, M. Physical and Antimicrobial Properties of Anise Oil Loaded Nanoemulsions on the Survival of Foodborne Pathogens. *Food Chem.* **2016,** *203*, 117–123. https://doi.org/10.1016/j.foodchem.2016.02.051.
86. Vasco, C.; Ruales, J.; Kamal-Eldin, A. Total Phenolic Compounds and Antioxidant Capacities of Major Fruits from Ecuador. *Food Chem.* **2008,** *111* (4), 816–823. https://doi.org/10.1016/j.foodchem.2008.04.054.
87. Vecchione, R.; Quagliariello, V.; Calabria, D.; Calcagno, V.; De Luca, E.; Iaffaioli, R. V.; Netti, P. A. Curcumin Bioavailability from Oil in Water Nano-Emulsions: *In Vitro* and *in Vivo* Study on the Dimensional, Compositional and Interactional Dependence. *J. Control. Release* **2016,** *233*, 88–100. https://doi.org/10.1016/j.jconrel.2016.05.004.
88. Villalva, M.; Jaime, L.; Arranz, E.; Zhao, Z.; Corredig, M.; Reglero, G.; Santoyo, S. Nanoemulsions and Acidified Milk Gels as a Strategy for Improving Stability and Antioxidant Activity of Yarrow Phenolic Compounds after Gastrointestinal Digestion. *Food Res. Int.* **2020,** *130* (Dec), E-article 108922; p 5. https://doi.org/10.1016/j.foodres.2019.108922.
89. Vitale, S. A.; Katz, J. L. Liquid Droplet Dispersions Formed by Homogeneous Liquid-Liquid Nucleation: The Ouzo Effect. *Langmuir* **2003,** *19* (10), 4105–4110; https://doi.org/10.1021/la026842o.
90. Wang, W.; Zhang, Y.; Yang, Z.; He, Q. Effects of Incorporation with Clove (*Eugenia Caryophyllata*) Essential Oil (CEO) on Overall Performance of Chitosan as Active Coating. *Int. J. Biol. Macromol.* **2020,** 166:578-586. https://doi.org/10.1016/j.ijbiomac.2020.10.215.
91. Wani, T. A.; Masoodi, F. A.; Jafari, S. M.; McClements, D. J. Safety of Nanoemulsions and Their Regulatory Status; Chapter 19; In: *Nanoemulsions: Formulations, Applications & Characterization*; Elsevier Inc.: San Diego, 2018; pp 613–626. https://doi.org/10.1016/B978-0-12-811838-2.00019-9.

92. Wooster, T. J.; Golding, M.; Sanguansri, P. Impact of Oil Type on Nanoemulsion Formation and Ostwald Ripening Stability. *Langmuir* **2008,** *24* (22), 12758–12765. https://doi.org/10.1021/la801685v.
93. Yousuf, B.; Qadri, O. S. Preservation of Fresh-Cut Fruits and Vegetables by Edible Coatings; Chapter 11; In: *Technologies & Mechanisms for Safety Control: Fresh-Cut Fruits & Vegetables*; Elsevier Inc.: San Diego, 2019; pp 225–242. https://doi.org/10.1016/B978-0-12-816184-5.00011-2.
94. Zorzi, G. K.; Caregnato, F.; Moreira, J. C. F.; Teixeira, H. F.; Carvalho, E. L. S. Antioxidant Effect of Nanoemulsions Containing Extract of *Achyrocline Satureioides* (L.). *AAPS PharmSciTech* **2016,** *17* (4), 844–850. https://doi.org/10.1208/s12249-015-0408-8.

CHAPTER 10

ENGINEERED NANOMATERIALS FOR FOOD PRESERVATION AND PACKAGING: FOCUS ON ANTIMICROBIAL AND ANTIBIOFILM ACTIVITIES

SHAIBI SALEEM and SHAMS TABREZ KHAN

ABSTRACT

Nanomaterials (NMs)-based technologies are being used for food packaging, storage, transportation, and food functionality. Organic and inorganic and simple NMs or nanocomposites with bioactive agents are in use in the food industry. The nano-zinc dioxide NPs are most widely used followed by silver nanoparticles. These NMs can alter the oxygen availability, humidity, and freshness of food. However, this chapter focuses only on the antimicrobial activity of NMs. As foods are not only prone to microbial spoilage but can also serve as the carrier for pathogens. The biofilm formation by micro-organisms is also an issue especially associated with food packaging. Packaging films containing NMs coatings can protect foods from microbial spoilage, growth, and biofilm formation. However, the use of these NMs should be carefully evaluated, monitored, and regulated through research-based regulatory policies. Health and safety concerns should also be carefully addressed for ensuring the sustainable use of NMs in the food industry.

Quality Control in Fruit and Vegetable Processing: Methods and Strategies. Megh R. Goyal, & Faizan Ahmad (Eds.)

10.1 INTRODUCTION

The food packaging must be done properly to protect it from physical, chemical, or biological damage.[164] In this regard, human health is the first priority; therefore, it becomes necessary to provide the food with great hygienic value and in safe conditions because the contamination/spoilage of food can happen at any level in the supply chain, from output to consumption.[28] Food consists of essential nutrients, such as carbohydrates, proteins, fats, vitamins, minerals, etc. The pathogenic micro-organisms (spore-forming bacteria, fungi, aerobic psychrotrophic Gram-negative bacteria, molds, etc.) are responsible for spoilage and poisoning of food (*Clostridium perfringens* and *Staphylococcus aureus*),[17,187] which further reduces the quality and texture of the food; hence, the food becomes undesirable for human consumption.

Certain bacteria enter in the food chain during manufacturing and their persistence in food (toxins and spores) further leads to poisoning.[187] These micro-organisms can tolerate varying degrees of temperature (i.e., they are active at low and high temperatures), these micro-organisms further divide and form biofilm on the surfaces of food that deteriorates the complete food.[20] The food poisoning results in serious illness and various diseases; 43% of the diseases occur due to contaminated food basically in youngsters under the age of 5.[127]

The use of antimicrobial agents in food packaging is an adequate method to reduce the growth of micro-organisms in food controlling both microbial spoilage and the growth of pathogenic bacteria. A number of antimicrobial agents can be used for the advancement of such antimicrobial foodstuff packaging, such as organic acids, enzymes, bacteriocins, and other natural edible extracts. These have been successfully used for packaging shredded cheese, beef, Ham, fish, and many other foods. These packaging were shown to minimize and protect the food against micro-organisms, such as *L. monocytogens, S. aureus, S. typhimurium*, etc.[8]

Commonly used engineered nanomaterials (NMs) in foodstuff packaging include nano-zinc oxide (nano-ZnO), nano-titanium dioxide (nano-TiO_2), silver NPs (Ag NPs), nanoclay, titanium nitride NPs (nano-TiN), etc.[38] These NMs impart mechanical barriers, gas barriers, thermal and antimicrobial properties. In spite of the fact that the mechanisms of nanoantimicrobials are unknown, some of them have been linked to structural damage caused by nanoparticles (NPs), the production of reactive oxygen species (ROS), or the discharge of toxic ions from NPs.[13,61] NPs also have antibiofilm properties as they avoid bacterial adhesion to avoid biofilm arrangement/formation, eradicating existing biofilm, or destroying bacteria without making antimicrobial resistance.[3] In addition to this, NMs also enhance nutrient availability and increase shelf-life of foods.[38,164]

Various antibiofilm nanocoatings are being developed for preventing infections on the basis of antimicrobial activities, which help to kill and disrupt the growth mechanisms of pathogens. The antibiofilm nanocoatings can respond to minute changes in the microenvironment and are known to hinder the biofilm of *S. mutans, Escherichia coli,* and *S. aureus*.[13,95] Nanocoatings can also work as antibiofilm as specialized coatings with quorum sensing inhibitory activity targeting the c-di-GMP-dependent signal transduction pathway.[13] As the food is always at the risk of microbial spoilage and poisoning, resulting in the food losses and health-related problems, the use of nanotechnology is providing new and better possibilities than conventional packaging for minimizing contamination as they enhance flavors, color, and quality and help in informing the consumers about the food quality and status.[200]

Incorporation of nanosensors along with antimicrobials in food packaging is being developed to keep the food safe. The nanosensors are also widely used in packaging to detect the changed conditions of food. Food gases can be detected using nanosensors in plastic packaging, alarming consumers about the food spoilage.[13] It is also utilized for toxin and deterioration detection. It is necessary to develop NP-based packaging with minimum cost for its extensive use.

This chapter describes the details of the food preservation and food safety strategies with the help of engineered NMs. The chapter also discusses the use of nanotechnology to revolutionize the food industry and its possible risks and safety assessments of exposure of NMs to human health.

10.2 FOOD SHORTAGE AND MALNUTRITION

The increased population is resulting in drastic alterations in the food production and consumption level including rise in grains, meat, and cereal production.[152,185] Proper nutrition is important for human health. The state of food security (SOFI) and nutrition in the world in 2020 indicated that India had the highest population of food-insecure persons. India needs more improvements in the nutritional and security level of food.

India is positioned 94th among 107 nations within the current Global Hunger Index (GHI).[145] GHI highlights the areas and countries, which require necessary actions and where there is an urgent need to give attention to global hunger and undernutrition. India has seen few advances in which the frequency of extreme undernutrition among children and the frequency of certain nutritional-deficiency infections (such as kwashiorkor, marasmus, and pellagra) have been significantly decreased.[14]

However, there is an issue of malnutrition as India is one of 88 nations on the verge of failing to meet global nutrition goals by 2025 as suggested by the World Health Organization (WHO).[145] Hence, for any country, it is not possible to achieve economic growth, and social development without eliminating the problem of malnutrition. Undernutrition (deficiencies) and overnutrition are both considered as malnutrition.[105] Food shortage, food insecurity, and nutritional-related diseases are threat to developing and underdeveloped countries.[218] These conditions usually occur due to low quality of food, high prices of good quality food and poverty.[110]

There are two types of undernutrition: protein-energy malnutrition and dietary deficiencies. Protein-energy malnutrition has two forms: kwashiorkor (deficiency of protein) and marasmus (deficiency of protein and calories).[11] Lack of zinc, iodine, iron, vitamin A, and vitamin B_2 are all common micro-nutrient deficiencies.[24,60,103,175,252,255] On the other hand, overnutrition can result in obesity, which raises the risk of heart disease, hypertension, cancer, and type-2 diabetes.[11]

It has been observed that India is the world's second-largest food producer, still home to 37.9% stunted children. India is included among world's three worst countries counting Nigeria and Indonesia.[145] Obesity is also on the rise, affecting one-fifth of the population (21.6% of women and 17.8% of men).[145] Malnutrition is a large and common issue that affects individuals of all ages, income levels, and genders. Malnutrition kills about 45% of youngsters under the age of 5 in low-income nations every year[25] (Table 10.1).

10.2.1 FOOD CRISES DUE TO COVID-19

In order to avoid the spread of COVID-19 [2019], many local governments imposed national lockdowns and international travel restrictions.[240] Various food-insecure countries suffered from food crises as there is no transport or manufacturing of food and other essential goods, expanding conveyance period, and decreasing accessibility of indeed the foremost essential food items during lockdown.[70,159] Nutrition assistance was most vulnerable because of the pandemic during 2019. People are facing problems to cope with the health requirements or economic consequences; as in the worst cases they are saving life from the corona virus to die from hunger.

TABLE 10.1 Some of the Prevalent Micronutrient Deficiencies and the Consequences on Human Health.

Micronutrient deficiency	Effects	No. of people effected (2020)	References
Iodine	Goitre, weight gain, problems during pregnancy, brain damage in childhood, reduced work productivity in adults	>1.9 billion people have insufficient iodine nourishment, of which 285 million are school-aged children	[54, 144, 224, 253, 254]
Iron	Anemia, cognitive impairment, Extreme fatigue in adults, repeat infections and loss of appetite in children, premature births	It affects 614 million women and 280 million children worldwide	[72, 144, 146, 235]
Vitamin A	Infertility, delayed growth and respiratory infections, liver disorders, Xerophthalmia, inflamed skin, night blindness	Vitamin A deficiency causes between 250,000 and 500,000 youngsters to go blind each year	[156, 206, 211, 223]
Zinc	Diarrhea, loss of appetite, hair loss, drowsiness, and a diminished sense of taste and smell are all symptoms	Every year, 800,000 people die from zinc deficiency, with 450,000 of them being children under the age of 5	[55, 101, 116, 176, 177]

Unemployment or underemployment is also one of the reasons, as these problems are reducing the purchasing power, regulations on social distancing, and government controls to limit transmission placed daily wage earners in the informal economy and service sector workers at risk of losing their income sources.[151,239] After comparing 50 countries (2019 and 2020 reports), the population in crisis rose from 112 to 123 million.[170] Approximately 183 million people in 47 countries were rated as depressed, putting them at risk of entering a crisis, or worsening their situation.[170] Around one in every two children under the age of five endured from "hidden" starvation due to micronutrient insufficiencies.[51,243,244] These youngsters lack adequate dietary energy, nutritionally diversified meals, clean drinking water, cleanliness, and health services, compromising their health and sustenance, with serious effects on long-term efficiency and progress. Africa had the greatest number of people, who were severely food insecure and in need of assistance. Furthermore, Africa, Central America, and Pakistan also faced the difficulty in nutritional supply due to climatic events.[81]

To build nutrition capability, governments and multilateral organizations must finance in and improve people health and financial resources, expand cooperation, and make consistent observation and evaluation. Governments must also establish a regulatory framework that prioritizes healthy eating.[34,217,236] Private companies should be ordered to grow more nutritious seeds or food to overcome nutrient deficiency and erase the hunger. A number of community nutritional programs and NGOs are in place in India to fight against malnutrition.

10.2.2 FOOD WASTAGE

According to the UN data, the world's population would rise to 9.3 billion by 2050, and the increased population will require additional food generation of around 70% along with land and resources. The growing food wastage at a different level of food production has put the pressure over urban areas to manage food wastage in different countries.[41,140] Agriculture, forestry, and fisheries are likely industries with the greatest cumulative effect on social well-being and the environment.[199] Producing food, fiber, fuel and a growing number of other bioproducts are critical to our survival. More than one-third of the world's labor force work in agribusiness, which produces almost 23.7 million tons of food per day.[50] Still nearly, 11% of the global population lives under food insecurity.

On the contrary, food wastage continues to be one of the contributing reasons for global food insecurity. Food waste includes the use of food meant for human consumption for nonhuman purposes, the diversion of food to feed animals, and the discarding of healthy food. According to Dhir et al.,[58] food waste can be grouped into three different types:

- avoidable waste, this is food that was once edible but has now become inedible by the time it meets its expiration date,
- unavoidable waste, which refers to food or food parts that are not edible, such as eggshells, and
- potentially avoidable food waste, which refers to specific wastes that are eaten sometimes, such as potato skins.

According to Thyberg et al.,[222] the wasted food is divided into two types:

- Food loss: Any food that goes uneaten at any point is considered food loss. Some food is lost at almost every stage of the food processing from production to consumption level.
- Food waste includes food discarded by retailers due to a change of color or appearance, as well as food wasted by customers. In the United States, food waste is discarded at a rate of over 0.6 pounds per person per day.

According to the United Nations, more than 40% of India's food production is wasted, discarded, destroyed, or lost each year, costing the country more than 10^5 crore rupees.[97] Every year, about 21 million tons of wheat is rotten in India which is equal to or sometimes larger than Australia's total annual production.[228] FAO estimated that around one-third of the world's food is lost or wasted each year. We must watch the food supply chain to avoid future food crises. Proper care must be taken to prevent the food wastage to save the resources.[89,222] There is a need for appropriate transportation as a part of resource management, with the correct environmental and hygiene conditions during this time. Processing waste, the requirement for cold storage facilities, process deterioration, inappropriate packing, transportation delays, and higher inventory owing to poor forecasting all contribute to food waste in the supply chain.[166]

10.3 RECENT ADVANCES IN FOOD PRESERVATION

The food deterioration progresses during packaging and storage resulting in public health concerns.[74,173] The various factors resulting in the deterioration of food are enzymatic changes (pH, temperature, chemicals),[19] chemical changes (lipid oxidation, nonenzymatic browning, color change, flavor change),[129] physical changes, biological changes (microbiological and macrobiological),[242] etc. Several food experts have been working to develop safe and hygienic food processing practices to prevent diseases caused by biological changes and to ensure that physical changes do not occur.[132,229]

The FoodNet identified 25,866 cases of infections, 6164 hospitalizations, and 122 deaths in the United States. Compared with 2016–2018 data, the incidence in 2019 was increased significantly for *Cyclospora* (1209%), *Yersinia* (153%), *Vibrio* (79%), Shiga Toxin producing *E. Coli* (34%), and *Campylobacter* (13%). According to these findings, the targets for food-borne illness reduction goals will not be fulfilled.[229] Microbial quality and other food processes (such as chemical reactions, enzymatic reactions) are frequently linked to food safety, shelf-life, and also have indirect impact on consumer perception of food quality.[160,237] Various methods have been used earlier to protect food naturally.[1,40] The bottling, canning, drying, curing, pickling, chilling, freezing, salting, brining, and jam-making are all used to keep food safe from micro-organisms.[76,102,179,203]

One way of protecting the food is through the use of proper packaging. Plastic packaging is one of the most popular methods today. However, due to its nonbiodegradable nature, the growing use of plastic packaging is causing increased environmental problems.[75] Several food degradation reactions (such as lipid oxidation and enzymatic browning) are triggered by ROS.[134] Various amended materials for food packaging are being developed for food protection. Antioxidant compounds incorporated into food packaging materials can control the negative oxidative effect of oxygen.[27] Films are formed freely and then applied to the surface of food.[163] The edible film made from vegetables and fruits is healthy, cost-effective, and is used as carriers of common bioactive compounds (such as antioxidants).[163]

Carica papaya L. fruit has a short shelf-life, and it helps to develop edible films using papaya and papaya processing waste with low water activity and longer shelf-life.[138] According to Rodriguez et al.,[190] ascorbic acid (vitamin C) has been incorporated into whey-protein films, alginate, and in *Aloe vera* coating gel for food preservation.[87] The use of high hydrostatic pressure (HPP) is also one of the methods in food processing.[195] HPP helps to preserve the various fruits and vegetables (such as dried strawberry, aloe vera gel, orange and pomegranate juice, fruit smoothies) at different pressure conditions.[78,119,231,234] HPP also helps to preserve sea foods,[178] and dairy products.[186,56]

The use of essential oil (EO) of lemongrass in food preservation is considered as a better alternative in comparison to synthetic preservatives.[64] The *C. citratus* (CC) has antibacterial,[155] antibiofilm,[221] and antifungal properties.[191] Ekpenyong et al.[64] reported that CC–EO exhibits antifungal activity and it was effective against many yeast species (e.g., *Candida oleophila, C. albicans, Saccharomyces uvarum, S. cerevisiae, Schizosaccharomyces pombe,* and *Metschnikowia fructicola*) and fungi (e.g., *Aspergillus spp., Fusarium oxysporum, Alternaria alternata, Penicillium roqueforti, Phytophthora infestans, Phaeoramularia angolensis, Phoma sorghina,* etc.) responsible for spoilage of food. The *C. citratus* extract and CC–EO components have free radical scavenging activity, implying that they can be used in both domestic and industrial food preservation.[64]

For the effective preservation of food, hurdle technology is also used in developing countries.[209] The hurdle technology includes various hurdles or factors, such as temperature, pH, water activity, acidity, redox potential, preservatives, etc., that are important for microbial stability.[209]

10.4 USE OF BIOFILMS IN FOOD CONTAMINATION

Various health-related diseases are associated with biofilms.[44,150,198] Micro-organisms on wet surfaces can aggregate, form microcolonies, and produce biofilms.[47,82,128] Biofilms are primarily made up of viable and nonviable micro-organisms embedded in extracellular polymeric substances that help in attachment with different abiotic and biotic surfaces.[131,235,241] Extracellular polymeric substances can be composed of phospholipids, polysaccharides, teichoic, proteins, and nucleic acids, and other polymeric substances hydrated with 85–95% water.[162,193] Various factors influence biofilm formation and growth, such as bacterial strain, material surface properties, and environmental parameters, such as nutrient levels, pH, and temperature.[59,213,230] Antimicrobials, environmental stress, and immune responses from hosts discourage biofilm formation.[57,85,189]

Polymeric substances and multicellular nature of the biofilm provide protection to its inhabitants by concentrating nutrients, preventing biocides from entering the biofilm, sequestering metals and contaminants, and preventing desiccation.[80,91] Biofilms can infect food, drinking water, and are responsible for various infections due to their antibiotic resistance.[83,135] In food processing conditions, the growth of pathogenic biofilms lead to microbial contamination and increase food safety risk,[42,87,143]] because it decreases shelf-life, quality, and facilitates disease transmission.[91,154]

The contaminated food collected from market results in the economic loss to the food industry.[149] The food processing unit must follow the proper sanitation of food contact equipment and surfaces because there is high risk of disease outbreaks from food processing environments. Food spoilage bacteria (such as *Listeria monocytogenes, Pseudomonas spp., Bacillus spp.,* and *Salmonella spp.*) is commonly found on food substrates and in food processing plants.[92,184,214] *L. monocytogenes* surviving in *Pseudomonas* biofilms is an example of pathogenic bacteria coexisting with environmental microflora inside a biofilm.[42]

A variety of micro-organisms is able to grow on food matrixes; hence, they give rise to biofilms, which contaminate the diverse food products and result in diseases. Because of these bacterial characteristics that contribute to antibiotic resistance, as well as the negative effects on humans, there is an urgent need for new strategies to overcome the bacterial formation of biofilms and to inhibit their activities.[46,48]

Disease caused by polluted food continues to be a significant challenge to public health and a major impediment to global socioeconomic growth. The use of antimicrobial EOs in the treatment of biofilms is one of the effective procedures.[119] The antibiofilm activity of CC–EO against biofilms of *L. monocytogenes, S. aureus,* and *Candida albicans* has been reported in the literature.[6,64,183,232] Different EOs (such as marjoram, lemon, and basil) can prevent biofilm formation and quorum sensing in a wide range of foodborne micro-organisms.[5]

10.5 GENERAL PROPERTIES OF NANOMATERIALS

NMs have nanoscale dimensions. Molecules interact (Vander Waals interactions) to form a molecular architecture that exist on a nm scale.[21,35,45,134] The NMs are unique because of their novel physicochemical properties. Nobel Laureate Richard P. Feynman first introduced the term "nanotechnology" in his lecture, "there's Plenty of Room at the Bottom".[79] These NMs are divided into 0-dimensional (e.g., NPs: quantum dots, nanoclusters, and fullerenes), 1-dimensional (e.g., nanotubes and nanorods), 2-dimensional (e.g., thin films), or 3-dimensional (e.g., nanocomposites and nanofibers).[158]

NPs have small size, which leads to an exponential increase in surface area. They are chemically reactive having low levels of agglomeration and high dispersibility and thermal conductivity.[249] In the case of noble metal–NPs due to their size and shape have interesting optical properties.[37] They possess strong UV–visible extinction band in their spectrum that is lacking in the spectrum of bulk metal.[26] Their mechanical parameters include elastic modulus, hardness, stress and strain, adhesion, surface coating, coagulation, lubrication, etc..[94] Owing to such properties, NMs are widely used in research and innovation. Industrial and consumer nanoproducts continue to grow, and the global market size of nanotechnology is projected to reach $90.5 billion in 2021.[7]

Paints, filters, insulation, lubricant additives, and healthcare are only a few of the items that use NMs. In healthcare, they are used as nanozymes (artificial enzyme).[90], which have wide range of applications in biosensing, bioimaging, tumor diagnosis, etc. NM-based therapeutics are widely used to lower the antibiotic resistance and target the biofilm-based bacterial infections that are hard to cure.[65,96,141] Biofilm-related infections are resulting in serious illness in humans, for example, persistent lung infections of cystic fibrosis patients, chronic rhinosinusitis, chronic osteomyelitis, chronic otitis media, chronic prostatitis, periodontitis, dental caries, etc.[43,53,165] NMs have been documented to improve food protection, food quality, fortification, shelf-life, and biosensors for tainted or spoiled food products due to their applications in several areas, such as medical drugs, food-preservatives, and packaging materials like nanosized fibers or plates.[194]

The NM databases are being generated to provide the information about the NMs through published data and are also used for modeling studies. The database, which can be accessed at http://www.pubvinas.com/, consists of 705 unique NMs covering 11 different material types.[248]

Nanotechnology in agriculture offers new agrochemical agents and delivery mechanisms to raise crop productivity while also reducing pesticide use.[201] Nanosensors are being developed for use in crop defense to detect diseases and agrochemical residues.[125] Nanodevices are also used for the genetic engineering of plants and plant disease diagnostics, animal health, animal breeding, poultry production, and postharvest management.[120] NP-mediated gene or DNA transfer in plants for the growth of insect-resistant varieties is one of nanotechnology's applications.[181]

10.6 ENGINEERING NANOMATERIALS FOR FOOD PRESERVATION AND PACKAGING

NMs have a variety of properties, such as mechanical, catalytic, optical, and antimicrobial properties, and recycling and barrier properties, which make them a viable option for food packaging.[61,121,122,250] As we know the food-borne diseases caused by bacteria, virus, and fungi affect human health.[130,136]

According to the WHO, 600 million people were infected with foodborne illnesses in 2015, with 420,000 cases of death reported every year.[66] The purpose of food packaging is to overcome all these issues.[204,207]

The main aim of food packaging includes food safety, longer shelf-life, better food quality, protection from spoilage by micro-organisms, and contamination by extended exposure of environmental sources like pollution, atmospheric gases, moisture, light, temperature, etc..[3,187]

The use of NMs in food packaging helps to fulfill all the necessary requirements including mechanical (carbon nanotubes), barrier (nanoclay), antimicrobial (silver NPs), biodegradable and nontoxic properties (nanoclay), active (iron NPs) and intelligent packaging (cellulosic NMs), and detection of foodborne pathogens (*E. coli, Salmonella typhimurium, Salmonella enterica, Cronobacter Sakazakiiis*).[12,32,33,107,139,226] NMs can be used as biosensors to trace and monitor the food condition during storage and transport, additives, antimicrobial agents, and nutrient carriers while their mechanical strength and durability can help to improve packaging quality.[3,151,207,210]

Food packaging includes NMs, such as silver nanoparticles (Ag NPs), nanoclay, zinc oxide (nano-ZnO), titanium dioxide (nano-TiO_2), titanium nitride (nano-TiN), silicon dioxide (nano- SiO_2), copper oxide (nano-CuO), and magnesium oxide (nano-MgO), nanosized fibers or plates, nanocomposites, assembled NMs or shell structures, etc.[32,38,188,207]

Ag NPs are widely used as antimicrobial agent in food packaging.[36,52,109,121,197,244,250] Ag NPs are widely used for sterilization and have antimicrobial, antiyeast, antifungal, and antiviral properties as they have a greater surface area per mass compared to microscale silver particles or bulk silver.[37] During ancient times, silver vessels were used to preserve beverages, such as milk or water. There are 16 bacterial strains against which silver has shown antimicrobial activity, such as *E. coli, S. aureus, Staphylococcus epidermis, Klebsiella pneumonia, Leuconostocmes enteroides, Klebsiella mobilis*, and *Bacillus subtilis.*[180]

In another study, AgNPs have been demonstrated to kill about 650 disease-causing pathogens.[208] Different mechanisms of this antimicrobial effect of silver NPs are already known.[63] Ag NPs may attach to surface of cells and degrade lipopolysaccharides, thus result in the disruption of the cell membrane; Ag NPs allow silver ions to enter inside the bacteria and cause DNA damage.[251] Ag NPs can be easily introduced in the plastic polymers for packaging using porous substrate, such as zeolite.[33]

Nanoclay is used to improve the physical properties of plastic packaging, and gas and moisture barriers.[10] In the 1990s, nanoclay was first used in food processing.[33] Montmorillonite is a nanoclay made up of two layers of tetrahedral silica sandwiched within an octahedral edge-shaped sheet of aluminum hydroxide.[71,95] It is readily accessible and made from volcanic ash and rocks. It has been widely used in polymer composites due to its high cation exchange performance, large surface area, and good swelling behavior.[228] Durethan® KU2-2601 (a nanoclay engineered polyamide film produced by Bayer Co.) is commercialized under the Lanxess Deutschland brand.[38]

Nano-zinc oxides (ZnO NPs) are used in medical devices, cosmetics, drug delivery, and atmospheric processed packaging because of its excellent antimicrobial and UV protection activities.[122] ZnO nanostructures have been more effective against *E. coli, Bacillus atrophaeus,* and *Salmonella aureus* than the other metal oxides and it inhibits the growth of bacteria rather than killing the bacteria. The effect of ZnO is concentration-dependent, if the concentration is increased it will act as bactericidal. Zinc NPs are used nowadays in food packaging in comparison to Ag NPs due to its less toxic nature. The bacterial cellulose film with polypyrrole-ZnO nanocomposite (BC–PPy–ZnO) as a smart and active food kit can control microbial growth and regulate the pH increment in chicken thighs increasing its shelf-life.[174] Emamifar et al.[69] observed the effect of LDPE-ZnO nanocomposite packaging on inhibition of *Lactobacillus plantarum* for orange juice. The growth rate of *L. plantarum* at 4°C was significantly decreased in a ZnO nanocomposite packaging.

Nano-titanium dioxide: The US FDA approved the use of nano-TiO_2 as a food additive and food contact medium in 1996.[115,122] TiO_2 NPs have a number of benefits, such as inexpensive, nontoxic, and photostable.[220] It is gaining attention as a superior photocatalyst material for applications in energy and atmosphere, including air and water purification.[124]

The TiO_2 NPs are used in food packaging as antimicrobial agents,[108,117,250] for example, it acts as protective covering and help to preserve chocolates for longer duration and provide protection from micro-organisms.[123] (Figure 10.1). Because of its strong UV-blocking strength and ability to absorb short-wavelength light with considerable photostability, TiO_2 is a widely studied semiconductor.[2] Incorporating TiO_2 NPs into polymeric films for food processing can protect food components.[216] TiO_2 integration into chitosan and ethylene vinyl alcohol yields nanocomposite materials.[88]

Nano-silicon dioxide (SiO_2) was used to create Durethan which is a light, durable, heat-resistant plastic with gas barrier properties that prevent food spoilage and extend its shelf-life.[15]

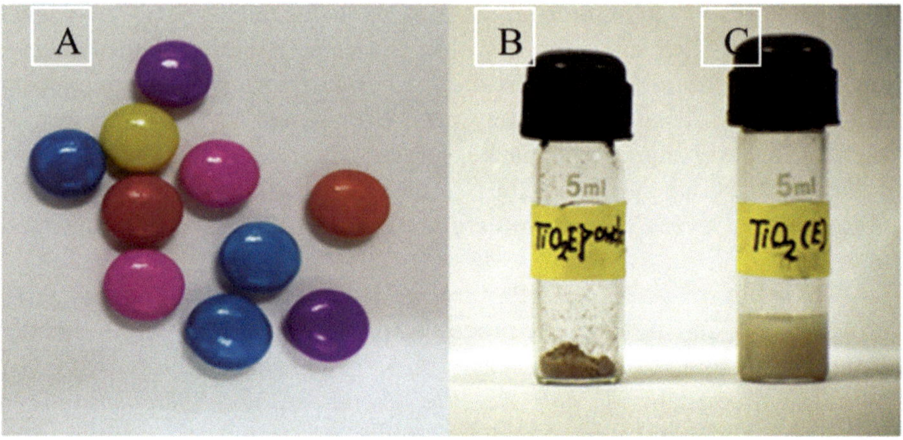

FIGURE 10.1 Titanium dioxide (TiO_2) nanoparticle obtained from chocolate (gems) (A); in powder (B); or in aqueous form (C).

The European Food Safety Authority has authorized three NMs for use in food packaging: titanium nitride NPs (TiN NPs) to deter CO_2 leakage from carbonated beverages in PET bottles, carbon black, and SiO_2 NPs.[203] ZnO and MgO NPs offer affordable, antimicrobial, and healthy food packaging solutions in the future than nanosilver NPs.[67] Starch-based nanocomposite films are also prepared for food packaging applications by using a single or combination of Ag, CuO, and ZnO NPs and have higher tensile strength and Young's modulus and antimicrobial activity.[169]

10.6.1 NANOSENSORS

Nanobiosensors are sensors that incorporate knowledge of biology, chemistry, and nanotechnology to analyze food.[39] The quality control of the food product is related to its production technology, microbiological protection, physical properties, chemical composition, and nutritional value.[73,225]

The inspection of developed food is very important at every step of its production. Berekaa[18] described how nanotechnology was used in the design of smart or intelligent packaging to improve the contact aspect of the package. Hence, food safety requires smart and active packaging for the successful transport of required food in market.[171,247]

In the food industry, nanotechnology plays an important role in food manufacturing, food storage, preservation, and pathogen identification.[62] The use of NM-based sensors to check the quality of food during transport, technology for the detection of common food pathogens like *Salmonella, E. coli,* and *listeria* will enable consumers to select fresh and healthy foods.[61] These devices exhibit high sensitivity, quick response, and recovery. Nanosensors have been designed for food analysis, flavor or color analysis,[149] drinking water analysis,[121] and clinical diagnostics.[9]

Nanosensors are being widely used for the detection of mycotoxins and pathogens in the food.[245] Guillermo et al.[84] estimated that the global demand for active and intelligent packaging will double in 2021, rising at an annual rate of 7% in 2021, will be reaching the US $24.65 billion.[84,215] In the next 10 years, worldwide demand for electronic smart packaging will exceed $1.45 billion.[215] Many countries have built intelligent systems to assist customers in selecting suitable food, storing it, and evaluating its true conditions using nanosensors, tags, and stickers.[159] Internal and external metrics provided by intelligent packaging technology can track interactions between food, packaging, and the atmosphere.[22]

Since various food packaging are opaque, customers are unaware of the quality and condition of the food package and must rely on expiration dates issued by manufacturers. If the quality of the food deteriorates at some stage, the customer would be unaware. Because of their unusual chemical and electro-optical properties, nanosensors can help to detect such changes.[204] Gases, aromas, organic toxins, bacteria, and even variations in environments can be detected by nanosensors.[114] Food protection can be improved by using nanosensors, to ensure that customers use fresh food products, and also help to reduce the incidence of food-borne illnesses.[161]

For example, an intelligent packaging containing nanobiosensor (which fluoresces in different colors when in contact with food pathogens) and AgroMicron developed the nanobioluminescence detection spray that contains a luminescent protein designed to bind to the surface of *salmonella* and *E. coli*.[84] Foodborne contaminants and toxins can also be detected using nanosensors.[205] For detection of *E. coli* O157:H7, *Salmonella* sp., and *L. monocytogenes* in foods, researchers developed a universal G-liposomal nanovesicles-based immune-magnetic bead sandwich assay.[18] The use of a magnetic nano-glo-dimmuno sensor could detect aflatoxins generated by *Aspergillus flavus* and *Aspergillus parasiticus* that contaminate food items.[147]

10.7 HEALTH RISKS AND OTHER CONCERNS

The use of NMs is increasing in the food industry.[49,106] Despite benefits of NMs, the toxicity and environmental issues affecting human life remain to be discussed.[123,192] The NPs used in food will increase human exposure through its potential migration into the food matrix and soft drinks.[212] NPs have a small size, enhanced toxicity, and high chemical reactivity due to which its exposure can cause various health-related problems.[113]

The study of total and precise migration to decide whether any unnecessary or even harmful compounds migrate to the packed product is important step during manufacturing of new contact materials.[212] Modern techniques have shown that NPs with increased reactivity and ability to cross membrane barriers and capillaries can result in a wide range of toxic-kinetic and toxic-dynamic properties.[30] Following NP administration, certain NPs can interfere with proteins and enzymes, inducing oxidative stress and the generation of ROS, resulting in mitochondrial degradation and apoptosis.[126,142]

For production of efficient NMs-based food packaging, it is essential to study the nature and mechanism of action of NMs in biological systems. Various studies have demonstrated the toxicity of NMs, such as TiO_2 in-vitro on cell lines.[29] NPs are also responsible for the oxidative damage in cells; particles smaller than 70 nm in size will reach the cell nuclei and can disrupt DNA replication and transcription.[137] Hence, it is necessary to regulate the use of NMs in food, and all countries must establish regulatory systems to regulate NPs exposure to humans and to protect consumers from the risks associated with the use NMs in food items.[111]

10.8 SUMMARY

NMs in the food industry provide barrier properties, antimicrobial activity, reduced water vapor permeability, UV blocking, gas scavenging, and biodegradability; and these advantages of NMs have improved spoilage control in food products. NMs have also increased the shelf-life of foods. The antimicrobial activity of NMs also helps to control the disease-causing pathogenic micro-organisms thus decreasing the microbial load. Nanotechnology offers a variety cost-effective, eco-friendly, degradable, and renewable packaging materials, which are gaining popularity to address the issues of food security. Since nanosensors and nanoelectronic technologies can be used to detect changes in food, the presence of pathogens or its composition, they can be considered as important devices for the food safety. In the United States and Asia, NMs-based food packaging has been adopted at the industrial level. However, NM migration from packaging into the food matrix, its consequent toxicity, and other environment issues are a matter of concern requiring detailed investigations and strict regulations for the use of NMs in food industry.

KEYWORDS

- **antibiofilm activity**
- **antimicrobial activity**
- **food loss**
- **food packaging**
- **food spoilage**
- **malnutrition**
- **nanomaterials**
- **nanosensor**

REFERENCES

1. Abdulmumeen, H. A.; Risikat, A. N.; Sururah, A. R. Food: Its Preservatives, Additives and Applications. *Int. J. Chem. Biochem. Sci.* **2012,** *1,* 36–47.
2. Alebeid, O. K.; Zhao, T. Review on: Developing UV Protection for Cotton Fabric. *J. Textile Inst.* **2017,** *108* (12), 2027–2039.
3. Alfadul, S. M.; Elneshwy, A. A. Use of Nanotechnology in Food Processing, Packaging and Safety: A Review. *Afr. J. Food Agric. Nutr. Dev.* **2010,** *10* (6), online. DOI:10.4314/ajfand.v10i6.58068
4. Ali, A.; Ovais, M.; Cui, X.; Rui, Y.; Chen, C. Safety Assessment of Nanomaterials for Antimicrobial Applications. *Chem. Res. Toxicol.* **2020,** *33* (5), 1082–1109.
5. Alibi, S.; Selma, W. B.; Ramos-Vivas, J.; Smach, M. A.; Touati, R.; Boukadida, J.; Mansour, H. B. Anti-Oxidant, Antibacterial, Anti-Biofilm, and Anti-Quorum Sensing Activities of Four Essential Oils Against Multidrug-Resistant Bacterial Clinical Isolates. *Curr. Res. Transl. Med.* **2020,** *68* (2), 59–66.
6. Altaf, M.; Manoharadas, S.; Zeyad, M. T. Green Synthesis of Cerium Oxide Nanoparticles Using Acorus calamus Extract and Their Antibiofilm Activity Against Bacterial Pathogens. *Microsc. Res. Techn.* **2021,** 84 (8), 1638–1648.
7. Al-Tayyar, N. A.; Youssef; A. M.; Al-Hindi, R. Antimicrobial Food Packaging Based on Sustainable Bio-Based Materials for Reducing Foodborne Pathogens: A Review. *Food Chem.* **2020,** *310*, E-article 125915; p 8.
8. Appendini, P.; Hotchkiss, J. H. Review of Antimicrobial Food Packaging. *Innov. Food Sci. Emerg. Technol.* **2002,** *3* (2), 113–126.
9. Arndt, N.; Tran, H. D.; Zhang, R.; Xu, Z. P.; Ta, H. T. Different Approaches to Develop Nanosensors for Diagnosis of Diseases. *Adv. Sci.* **2020,** *7* (24), E-article 2001476, p 8.

10. Arora, A.; Padua, G. W. Nanocomposites in Food Packaging. *J. Food Sci.* **2010,** *75* (1), R43–R49.
11. Awuchi, C. G.; Igwe, V. S.; Amagwula, I. O. Ready-to-Use Therapeutic Foods for Remedying Malnutrition and Preventable Nutritional Diseases. *Int. J. Adv. Acad. Res.* **2020,** *6* (1), 47–81.
12. Bajpai, V. K..; Kamle, M.; Shukla, S.; Mahato, D. K.; Chandra, P.; Hwang, S. K.; Han, Y. K. Prospects of Using Nanotechnology for Food Preservation, Safety, and Security. *J. Food Drug Analys.* **2018,** *26* (4), 1201–1214.
13. Balaure, P. C.; Grumezescu, A. M. Recent Advances in Surface Nanoengineering for Biofilm Prevention and Control, Part II: Active, Combined Active and Passive, and Smart Bacteria—Responsive Antibiofilm Nanocoatings. *Nanomater.* **2020,** *10* (8), 1527–1532.
14. Banik, D. *Starvation and India's Democracy*, Vol. 8; Routledge: London, 2007; E-book; p 244.
15. Basavegowda, N.; Mandal, T. K.; Baek, K. H. Bimetallic and Trimetallic Nanoparticles for Active Food Packaging Applications: A Review. *Food Bioprocess Technol.* **2020,** *13* (1), 30–44.
16. Belas, R. Biofilms, Flagella, and Mechano-Sensing of Surfaces by Bacteria. *Trends Microbiol.* **2014,** *22* (9), 517–527.
17. Bennett, S. D.; Walsh, K. A.; Gould, L. H. Foodborne Disease Outbreaks Caused by *Bacillus cereus*, *Clostridium perfringens*, and *Staphylococcus aureus*: United States, 1998–2008. *Clin. Infect. Dis.* **2013,** *57* (3), 425–433.
18. Berekaa, M. M. Nanotechnology in Food Industry; Advances in Food Processing, Packaging and Food Safety. *Int. J. Curr. Microbiol. App. Sci.* **2015,** *4* (5), 345–357.
19. Bharate, S. S., Bharate, S. B. Non-Enzymatic Browning in Citrus Juice: Chemical Markers, Their Detection and Ways to Improve Product Quality. *J. Food Sci. Technol.* **2014,** *51* (10), 2271–2288.
20. Bhunia, A. K. *Foodborne Microbial Pathogens: Mechanisms and Pathogenesis*, 2nd ed.; Springer: New York, 2018; p 365.
21. Bianco, A.; Cheng, H. M.; Enoki, T.; Gogotsi, Y.; Hurt, R. H.; Koratkar, N.; Zhang, J. All in the Graphene Family: A Recommended Nomenclature for Two-Dimensional Carbon Materials. *Carbon* **2013,** *65* (6), 1–6.
22. Biji, K. B.; Ravishankar, C. N.; Mohan, C. O.; Gopal, T. S. Smart Packaging Systems for Food Applications: A Review. *J. Food Sci. Technol.* **2015,** *52* (10), 6125–6135.
23. Bjarnsholt, T. The Role of Bacterial Biofilms in Chronic Infections. *APMIS* **2013,** *121*, 1–58.
24. Black, M. M. Micronutrient Deficiencies and Cognitive Functioning. *J. Nutr.* **2003,** *133* (11), 3927S–3931S.
25. Black, R. E.; Victora, C. G.; Walker, S. P.; Bhutta, Z. A.; Christian, P.; de-Onis, M. Maternal and Child Nutrition Study Group Maternal and Child Undernutrition and Overweight in Low-Income and Middle-Income Countries. *Lancet*, **2013,** *382* (9890), 427–451.
26. Blemker, M. A.; Gibbs, S. L.; Raulerson, E. K.; Milliron, D. J.; Roberts, S. T. Modulation of the Visible Absorption and Reflection Profiles of ITO Nanocrystal Thin Films by Plasmon Excitation. *ACS Photon.* **2020,** *7* (5), 1188–1196.
27. Bonilla, J.; Atarés, L.; Vargas, M.; Chiralt, A. Edible Films and Coatings to Prevent the Detrimental Effect of Oxygen on Food Quality: Possibilities and Limitations. *J. Food Eng.* **2012,** *110* (2), 208–213.

28. Bouwmeester, H.; Dekkers, S.; Noordam, M. Y.; Hagens, W. I.; Bulder, A. S.; De Heer, C.; Sips, A. J. Review Of Health Safety Aspects of Nanotechnologies in Food Production. *Regul. Toxicol. Pharmacol.* **2009,** *53* (1), 52–61.
29. Brandão, F.; Fernández-Bertólez, N.; Rosario, F.; Bessa, M. J.; Fraga, S.; Pásaro, E.; Costa, C. Genotoxicity of TiO_2 Nanoparticles in Four Different Human Cell Lines (A549, HEPG2, A172 and SH-SY5Y). *Nanomaterials* **2020,** *10* (3), 412–416.
30. Bratovcic, A. Nanomaterials in Food Processing and Packaging, Its Toxicity and Food Labeling. *Acta Sci. Nutr. Health* **2020,** *4*, 7–13.
31. Bridier, A.; Briandet, R.; Thomas, V.; Dubois-Brissonnet, F. Resistance of Bacterial Biofilms to Disinfectants: A Review. *Biofouling* **2011,** *27* (9), 1017–1032.
32. Bumbudsanpharoke, N.; Ko, S. Nano-Food Packaging: An Overview of Market, Migration Research, and Safety Regulations. *J. Food Sci.* **2015,** *80* (5), R910–R913.
33. Bumbudsanpharoke, N.; Choi, J.; Ko, S. Applications of Nanomaterials in Food Packaging. *J. Nanosci. Nanotechnol.* **2015,** *15* (9), 6357–6372.
34. Burchi, F.; Fanzo, J.; Frison, E. The Role of Food and Nutrition System Approaches in Tackling Hidden Hunger. *Int. J. Environ. Res Public Health* **2011,** *8* (2), 358–373.
35. Calandra, P.; Caschera, D.; Liveri, V. T.; Lombardo, D. How Self-Assembly of Amphiphilic Molecules Can Generate Complexity in the Nanoscale. *Colloids Surf. A: Physicochem. Eng. Aspects* **2015,** *484*, 164–183.
36. Carbone, M.; Donia, D. T.; Sabbatella, G.; Antiochia, R. Silver Nanoparticles in Polymeric Matrices for Fresh Food Packaging. *J. King Saud University-Sci.* **2016,** *28* (4), 273–279.
37. Chakraborty, I.; Parak, W. J. Protein-Induced Shape Control of Noble Metal Nanoparticles. *Adv. Mater. Interf.* **2019,** *6* (6), E-article 1801407; p 5.
38. Chaudhary, P.; Fatima, F.; Kumar, A. Relevance of nanomaterials in Food Packaging and Its Advanced Future Prospects. *J. Inorg. Organometallic Polym. Mater.* **2020,** *30* (12), 5180–5192.
39. Chaudhry, Q.; Scotter, M.; Blackburn, J.; Ross, B.; Boxall, A.; Castle, L.; Watkins, R. Applications and Implications of Nanotechnologies for the Food Sector. *Food Add. Contam.* **2008,** *25* (3), 241–258.
40. Chemat, F.; Khan, M. K. Applications of Ultrasound in Food Technology: Processing, Preservation and Extraction. *Ultrasonics Sonochem.* **2011,** *18* (4), 813–835.
41. Chen, P.; Xie, Q.; Addy, M.; Zhou, W.; Liu, Y.; Wang, Y.; Ruan, R. Utilization of Municipal Solid and Liquid Wastes for Bioenergy and Bioproducts Production. *Bioresour. Technol.* **2016,** *215*, 163–172.
42. Chmielewski, R. A. N.; Frank, J. F. Biofilm Formation and Control in Food Processing Facilities. *Compreh. Rev. Food Sci. Food Safet.* **2003,** *2*, 22–32.
43. Ciofu, O.; Tolker-Nielsen, T.; Jensen, P. O.; Wang, H.; Hoiby, N. Antimicrobial Resistance, Respiratory Tract Infections and Role of Biofilms in Lung Infections in Cystic Fibrosis Patients. *Adv. Drug Deliv. Rev.* **2015,** *85*, 7–23.
44. Colombo, A. P. V.; Magalhães, C. B.; Hartenbach, F. A. R. R.; do Souto, R. M.; da Silva-Boghossian, C. M. Periodontal Disease Associated Biofilm: A Reservoir for Pathogens of Medical Importance. *Microb. Pathogen.* **2016,** *94*, 27–34.
45. Comandella, D.; Gottardo, S.; Rio-Echevarria, I. M.; Rauscher, H. Quality of Physicochemical Data on Nanomaterials: An Assessment of Data Completeness and Variability. *Nanoscale* **2020,** *12* (7), 4695–4708.

46. Cortés, M. E.; Bonilla, J. C.; Sinisterra, R. D. Biofilm Formation, Control and Novel Strategies for Eradication. *Sci. Against Microbial. Pathog. Commun. Curr. Res. Technol. Adv.* **2011,** *2*, 896–905.
47. Costerton, W.; Veeh, R.; Shirtliff, M.; Pasmore, M.; Post, C.; Ehrlich, G. The Application of Biofilm Science to the Study and Control of Chronic Bacterial Infections. *J. Clin. Investig.* **2003,** *112* (10), 1466–1477.
48. Coughlan, L. M.; Cotter, P. D.; Hill, C.; Alvarez-Ordóñez, A. New Weapons to Fight Old Enemies: Novel Strategies for the Biocontrol of Bacterial Biofilms in the Food Industry. *Front. Microbiol.* **2016,** *7*, 1641–1647.
49. Cushen, M.; Kerry, J.; Morris, M.; Cruz-Romero, M.; Cummins, E. Nanotechnologies in the Food Industry: Recent Developments, Risks and Regulation. *Trends Food Sci. Technol.* **2012,** *24* (1), 30–46.
50. Dastagiri, M. B.; Chand, R.; Immanuelraj, T. K.; Hanumanthaiah, C. V.; Paramsivam, P.; Sidhu, R. S.; Kumar, B. G. Indian Vegetables: Production Trends, Marketing Efficiency and Export Competitiveness. *Am. J. Agric. Forest.* **2013,** *1* (1), 1–11.
51. De Moura, M. R.; Mattoso, L. H.; Zucolotto, V. Development of Cellulose-Based Bactericidal Nanocomposites Containing Silver Nanoparticles and Their Use as Active Food Packaging. *J. Food Eng.* **2012,** *109* (3), 520–524.
52. de Onis, M.; Borghi, E.; Blössner, M. *Trends in Child Malnutrition Malnutrition (UNICEF-WHO): The World Bank Joint Child Malnutrition Estimates*; Rome: UNICEF; **2021**; online. https://www.who.int/publications/i/item/9789240025257]
53. Del-Pozo, J. L. Biofilm-Related Disease. *Expert Rev.Anti-infect. Therap.* **2018,** *16* (1), 51–65.
54. Delange, F. Iodine Deficiency as a Cause of Brain Damage. *Postgraduate Med. J.* **2001,** *77* (906), 217–220.
55. Deshpande, J. D.; Joshi, M. M.; Giri, P. A. Zinc: The Trace Element of Major Importance in Human Nutrition and Health. *Int. J. Med. Sci. Public Health* **2013,** *2*, 1–6.
56. Devi, A. F.; Buckow, R.; Hemar, Y.; Kasapis, S. Structuring Dairy Systems Through High Pressure Processing. *J. Food Eng.* **2013,** *114* (1), 106–122.
57. D'Haeze, W.; Holsters, M. Surface Polysaccharides Enable Bacteria to Evade Plant Immunity. *Trends Microbiol.* **2004,** *12* (12), 555–561.
58. Dhir, A.; Talwar, S.; Kaur, P.; Malibari, A. Food Waste in Hospitality and Food Services: A Systematic Literature Review and Framework Development Approach. *J. Clean. Prod.* **2020,** *270*, E-article 122861; p 8.
59. Donlan, R. M. Biofilms: Microbial Life on Surfaces. *Emerg. Infect. Dis.* **2002,** *8* (9), 881–890.
60. Dowling, J. E.; Wald, G. Vitamin A Deficiency and Night Blindness. *Proc. Natl. Acad. Sci. USA* **1958,** *44* (7), 648–652.
61. Duncan, T. V. Applications of Nanotechnology in Food Packaging and Food Safety: Barrier Materials, Antimicrobials and Sensors. *J. Coll. Interf. Sci.* **2011,** *363* (1), 1–24.
62. Durán, N.; Marcato, P. D. Nanobiotechnology Perspectives. Role of Nanotechnology in the Food Industry: A Review. *Int. J. Food Sci. Technol.* **2013,** *48* (6), 1127–1134.
63. Durán, N.; Durán, M.; De Jesus, M. B.; Seabra, A. B.; Fávaro, W. J.; Nakazato, G. Silver Nanoparticles: A New View on Mechanistic Aspects on Antimicrobial Activity. *Nanomed. Nanotechnol. Biol. Med.* **2016,** *12* (3), 789–799.
64. Ekpenyong, C. E.; Akpan, E. E. Use of *Cymbopogon citratus* Essential Oil in Food Preservation: Recent Advances and Future Perspectives. *Cri. Rev. Food Sci. Nutr.* **2017,** *57* (12), 2541–2559.

65. Eleraky, N. E.; Allam, A.; Hassan, S. B.; Omar, M. M. Nanomedicine Fight Against Antibacterial Resistance: An Overview of the Recent Pharmaceutical Innovations. *Pharmaceutics* **2020,** *12* (2), 142–148.
66. Emamhadi, M. A.; Sarafraz, M.; Akbari, M.; Fakhri, Y.; Linh, N. T.; Khaneghah, A. M. Nanomaterials for Food Packaging Applications: A Systematic Review. *Food Chem. Toxicol.* **2020,** E-article 111825, p 7.
67. Emamifar, A. Applications of Antimicrobial Polymer Nanocomposites in Food Packaging. Chapter 13. In *Advances in Nanocomposite Technology*; Hashim, A., Ed.; IntechOpen Online; 2011; pp 299–308.
68. Emamifar, A.; Kadivar, M.; Shahedi, M.; Soleimanian-Zad, S. Evaluation of Nanocomposite Packaging Containing Ag and ZnO on Shelf-Life of Fresh Orange Juice. *Innov. Food Sci. Emerg. Technol.* **2010,** *11* (4), 742–748.
69. Emamifar, A.; Kadivar, M.; Shahedi, M.; Soleimanian-Zad, S. Effect of Nanocomposite Packaging Containing Ag and ZnO on Inactivation of Lactobacillus Plantarum in Orange Juice. *Food Control* **2011,** *22* (3), 408–413.
70. Erokhin, V.; Gao, T. Impacts of COVID-19 on Trade and Economic Aspects of Food Security: Evidence from 45 Developing Countries. *Int. J. Environ. Res Public Health* **2020,** *17* (16), E-article 5775, p 5.
71. Erol, M.; Hum, J.; Boccaccini, A. R. Bioactive Nanocomposites Containing Silicate Phases for Bone Replacement and Regeneration. *Biomimetic Approach. Biomater. Dev.* **2012,** *18*, 19–23.
72. Estrada, J. A.; Contreras, I.; Pliego-Rivero, F. B.; Otero, G. A. Molecular Mechanisms of Cognitive Impairment in Iron Deficiency: Alterations in Brain-Derived Neurotrophic Factor and Insulin-Like Growth Factor Expression and Function in the Central Nervous System. *Nutr. Neurosci.* **2014,** *17* (5), 193–206.
73. Falasconi, M.; Concina, I.; Gobbi, E.; Sberveglieri, V.; Pulvirenti, A.; Sberveglieri, G. Electronic Nose for Microbiological Quality Control of Food Products. *Int. J. Electrochem.* **2012,** 2012 (Feb), E-article 715763; https://doi.org/10.1155/2012/715763; accessed on August 2, 2021.
74. Farber, J. M. Microbiological Aspects of Modified-Atmosphere Packaging Technology-A Review. *J. Food Protect.* **1991,** *54* (1), 58–70.
75. Farhan, A.; Hani, N. M. Characterization of Edible Packaging Films Based on Semi-Refined Kappa-Carrageenan Plasticized with Glycerol and Sorbitol. *Food Hydrocoll.* **2017,** *64*, 48–58.
76. Farrer, K. T. H. The Thermal Destruction of Vitamin B1 in Foods. *Adv. Food Res. (Academic Press)*, **1955,** *6*, 257–311.
77. Fernández, A.; Picouet, P.; Lloret, E. Cellulose-Silver Nanoparticle Hybrid Materials to Control Spoilage-Related Microflora in Absorbent Pads Located in Trays of Fresh-Cut Melon. *Int. J. Food Microbiol.* **2010,** *142* (1), 222–228.
78. Fernandez, M. V.; Denoya, G. I.; Agüero, M. V.; Jagus, R. J.; Vaudagna, S. R. Optimization of HPP Parameters to Preserve Quality Attributes of a Mixed Fruit and Vegetable Smoothie. *Innov. Food Sci. Emerg. Technol.* **2018,** *47*, 170–179.
79. Feynman, R. P. There's Plenty of Room at the Bottom. *Calif. Inst. Technol.—Eng. Sci. Magaz.* **1960,** *23* (5), 22–36.
80. Flemming, H. C.; Wingender, J. The Biofilm Matrix. *Nat. Rev. Microbiol.* **2010,** *8* (9), 623–633.

81. Food Security Information Network. Global Report on Food Crises: *Technical Report*. Rome: World Food Programs—United Nations (FAO), 2020. https://www.wfp.org/publications/2020-global-report-food-crises (accessed August 02, 2021).
82. Frank, J. F. Microbial Attachment to Food and Food Contact Surfaces. *Adv. Food Nutr. Res*. **2001,** *43*, 319–370.
83. Frieri, M.; Kumar, K.; Boutin, A. Antibiotic resistance. *J. Infect. Public Health* **2017,** *10* (4), 369–378.
84. Fuertes, G.; Soto, I.; Carrasco, R.; Vargas, M.; Sabattin, J.; Lagos, C. Intelligent Packaging Systems: Sensors and Nanosensors to Monitor Food Quality and Safety. *J. Sens.* **2016,** *2016* (2), 1–8.
85. Fux, C. A.; Costerton, J. W.; Stewart, P. S.; Stoodley, P. Survival Strategies of Infectious Biofilms. *Trends Microbiol.* **2005,** *13* (1), 34–40.
86. Fysun, O.; Kern, H.; Wilke, B.; Langowski, H. C. Evaluation of Factors Influencing Dairy Biofilm Formation in Filling Hoses of Food-Processing Equipment. *Food Bioprod. Process.* **2019,** *113*, 39–48.
87. Gao, H.; Zhang, Z. K.; Chai, H. K.; Cheng, N.; Yang, Y.; Wang, D. N.; Cao, W. Melatonin Treatment Delays Postharvest Senescence and Regulates Reactive Oxygen Species Metabolism in Peach Fruit. *Postharvest Biol. Technol.* **2016,** *118*, 103–110.
88. Ghosal, K.; Agatemor, C.; Špitálsky, Z.; Thomas, S.; Kny, E. Electrospinning Tissue Engineering and Wound Dressing Scaffolds from Polymer-Titanium Dioxide Nanocomposites. *Chem. Eng. J.* **2019,** *358*, 1262–1278.
89. Girotto, F.; Alibardi, L.; Cossu, R. Food Waste Generation and industrial Uses: A Review. *Waste Manage.* **2015,** *45*, 32–41.
90. Golchin, J.; Golchin, K.; Alidadian, N.; Ghaderi, S.; Eslamkhah, S.; Eslamkhah, M.; Akbarzadeh, A. Nanozyme Applications in Biology and Medicine: An Overview. *Artif. Cells Nanomed. Biotechnol.* **2017,** *45* (6), 1069–1076.
91. González-Rivas, F.; Ripolles-Avila, C.; Fontecha-Umaña, F.; Ríos-Castillo, A. G.; Rodríguez-Jerez, J. J. Biofilms in the Spotlight: Detection, Quantification, and Removal Methods. *Compreh. Rev. Food Sci. Food Safet.* **2018,** *17* (5), 1261–1276.
92. Gram, L.; Ravn, L.; Rasch, M.; Bruhn, J. B.; Christensen, A. B.; Givskov, M. Food Spoilage: Interactions Between Food Spoilage Bacteria. *Int. J Food Microbiol.* **2002,** *78* (1–2), 79–97.
93. Grumezescu, V.; Socol, G.; Grumezescu, A. M.; Holban, A. M.; Ficai, A.; Truşcă, R.; Chifiriuc, M. C. Functionalized Antibiofilm Thin Coatings Based on PLA–PVA Microspheres Loaded with Usnic Acid Natural Compounds Fabricated by MAPLE. *Appl. Surf. Sci.* **2014,** *302*, 262–267.
94. Guo, D.; Xie, G.; Luo, J. Mechanical Properties of Nanoparticles: Basics and Applications. *J. Phys. D Appl. Phys.* **2013,** *47* (1), E-article 013001, p 7.
95. Guo, G.; Park, C. B.; Lee, Y. H.; Kim, Y. S.; Sain, M. Flame Retarding Effects of Nanoclay on Wood–Fiber Composites. *Polym. Eng. Sci.* **2007,** *47* (3), 330–336.
96. Gupta, A.; Mumtaz, S.; Li, C. H.; Hussain, I.;Rotello, V. M. Combatting Antibiotic-Resistant Bacteria Using Nanomaterials. *Chem. Soc. Rev.* **2019,** *48* (2), 415–427.
97. Gustavsson, J.; Cederberg, C.; Sonesson, U.; Van Otterdijk, R.; Meybeck, A. Global Food Losses and Food Waste: Extend, Causes and Prevention. Presented at InterPack 2011 Dusseldorf, Germany; Rome Food & Agriculture Organization of the United Nations, 2011; p 37.

98. Haes, A. J.; Duyne, R. P. V. Preliminary Studies and Potential Applications of localized Surface Plasmon Resonance Spectroscopy in Medical Diagnostics. *Expert Rev. Mol. Diagnost.* **2004,** *4* (4), 527–537.
99. Haes, A. J.; Van Duyne, R. P. A Nanoscale Optical Biosensor: Sensitivity and Selectivity of an Approach Based on the Localized Surface Plasmon Resonance Spectroscopy of Triangular Silver Nanoparticles. *J. Am. Chem. Soc.* **2002,** *124* (35), 10596–10604.
100. Hagens, W. I.; Oomen, A. G.; de Jong, W. H.; Cassee, F. R.; Sips, A. J. What Do We (Need to) Know About The Kinetic Properties of Nanoparticles in the Body? *Regul. Toxicol. Pharmacol.* **2007,** *49* (3), 217–229.
101. Hagmeyer, S.; Haderspeck, J. C.; Grabrucker, A. M. Behavioral Impairments in Animal Models for Zinc Deficiency. *Front. Behav. Neurosci.* **2015,** *8*, 443–448.
102. Hamad, S. H. Twenty Factors Affecting the Growth of Microorganisms in Food. *Progress Food Preserv.* **2012,** *2012*, 405–410.
103. Hambidge, M. Human zinc deficiency. *J. Nutr.* **2000,** *130* (5), 1344S-1349S.
104. Handford, C. E.; Dean, M.; Henchion, M.; Spence, M.; Elliott, C. T.; Campbell, K. Implications of Nanotechnology for the Agri-Food Industry: Opportunities, Benefits and Risks. *Trends Food Sci. Technol.* **2014,** *40* (2), 226–241.
105. Hanson, M. A.; Bardsley, A.; De-Regil, L. M.; Moore, S. E.; Oken, E.; Poston, L.; Morris, J. L. The International Federation of Gynecology and Obstetrics (FIGO) Recommendations on Adolescent, Preconception, and Maternal Nutrition: Think Nutrition First. *Int. J. Gynecol. Obstetr.* **2015,** *131*, S213–S253.
106. Higashisaka, K.; Yoshioka, Y.; Tsutsumi, Y. Applications and Safety of Nanomaterials Used in the Food Industry. *Food Safet.* **2015,** *3* (2), 39–47.
107. Huang, Y.; Mei, L.; Chen, X.; Wang, Q. Recent Developments in Food Packaging Based on Nanomaterials. *Nanomaterials* **2018,** *8* (10), 830–836.
108. Ibrahim, S.; El-Naggar, M. E.; Youssef, A. M.; Abdel-Aziz, M. S. Functionalization of Polystyrene Nanocomposite with Excellent Antimicrobial Efficiency for Food Packaging Application. *J. Cluster Sci.* **2019,** *2019*, 1–12.
109. Istiqola, A.; Syafiuddin, A. A Review of Silver Nanoparticles in Food Packaging Technologies: Regulation, Methods, Properties, Migration, and Future Challenges. *J. Chinese Chem. Soc.* **2020,** *67* (11), 1942–1956.
110. Ivanic, M.; Martin, W. Implications of Higher Global Food Prices for Poverty in Low-Income Countries. *Agric. Econ.* **2008,** *39* (1), 405–416.
111. Jain, A.; Ranjan, S.; Dasgupta, N.; Ramalingam, C. Nanomaterials in Food and Agriculture: An Overview on Their Safety Concerns and Regulatory Issues. *Crit. Rev. Food Sci. Nutr.* **2018,** *58* (2), 297–317.
112. Jamal, M.; Ahmad, W.; Andleeb, S.; Jalil, F.; Imran, M.; Nawaz, M. A.; Kamil, M. A. Bacterial Biofilm and Associated Infections. *J. Chinese Med. Assoc.* **2018,** *81* (1), 7–11.
113. Jeevanandam, J.; Barhoum, A.; Chan, Y. S.; Dufresne, A.; Danquah, M. K. Review on Nanoparticles and Nanostructured Materials: History, Sources, Toxicity and Regulations. *Beilstein J. Nanotechnol.* **2018,** *9* (1), 1050–1074.
114. Joyner, J. J.; Kumar, D. V. Nanosensors and Their Applications in Food Analysis: A Review. *Int. J Sci. Technol.* **2015,** *3* (4), 80–88.
115. Kamat, P. V. Graphene-Based Nano Assemblies for Energy Conversion. *J. Phys. Chem. Lett.* **2011,** *2* (3), 242–251.
116. Kambe, T.;Fukue, K., Ishida, R.; Miyazaki, S. Overview of Inherited Zinc Deficiency in Infants and Children. *J. Nutr. Sci. Vitaminol.* **2015,** *61* (Supplement), S44–S46.

117. Kanmani, P.; Rhim, J. W. Nano and Nanocomposite Antimicrobial Materials for Food Packaging Applications. *Future Science Book Series: Progress in Nanomaterials for Food Packaging*, 2014. https://www.futuremedicine.com/doi/abs/10.4155/ebo.13.303. Online. https://doi.org/10.4155/ebo.13.303. (accessed Aug 2, 2021).
118. Kavanaugh, N. L.; Ribbeck, K. Selected Antimicrobial Essential Oils Eradicate *Pseudomonas* spp. and *Staphylococcus aureus* Biofilms. *Appli. Environ. Microbiol.* **2012,** *78* (11), 4057–4061.
119. Keenan, D. F.; Rößle, C.; Gormley, R.; Butler, F.; Brunton, N. P. Effect of High Hydrostatic Pressure and Thermal Processing on the Nutritional Quality and Enzyme Activity of Fruit Smoothies. *LWT—Food Sci. Technol.* **2012,** *45* (1), 50–57.
120. Kerry, R. G.; Gouda, S.; Das, G.; Vishnuprasad, C. N.; Patra, J. K. Agricultural Nanotechnologies: Current Applications and Future Prospects. In *Microbial Biotechnology*; Springer: Singapore, 2017; pp 3–28.
121. Khan, S. T.; Malik, A. Engineered Nanomaterials for Water Decontamination and Purification: From Lab to Products. *J. Hazard. Mater.* **2019,** *363*, 295–308.
122. Khan, S. T.; Musarrat, J.; Al-Khedhairy, A. A. Countering Drug Resistance, Infectious Diseases, and Sepsis Using Metal and Metal Oxides Nanoparticles: Current Status. *Coll. Surf. B: Biointerf.* **2016,** *146*, 70–83.
123. Khan, S. T.; Saleem, S.; Ahamed, M.; Ahmad, J. Survival of Probiotic Bacteria in the Presence of Food Grade Nanoparticles from Chocolates: An In Vitro and In Vivo Study. *Appl. Microbiol. Biotechnol.* **2019,** *103* (16), 6689–6700.
124. Khataee, A.; Mansoori, G. A. *Nanostructured Titanium Dioxide Materials: Properties, Preparation and Applications*; World Scientific Publishing Co. Ltd.: Singapore, 2011 (Nov.); p 204. https://doi.org/10.1142/8325
125. Khot, L. R.; Sankaran, S.; Maja, J. M.; Ehsani, R.; Schuster, E. W. Applications of Nanomaterials in Agricultural Production and Crop Protection: A Review. *Crop Protect.* **2012,** *35*, 64–70.
126. Kim, S.; Ryu, D. Y. Silver Nanoparticle-Induced Oxidative Stress, Genotoxicity and Apoptosis in Cultured Cells and Animal Tissues. *J. Appl. Toxicol.* **2013,** *33* (2), 78–89.
127. Kirk, M. D.; Pires, S. M.; Black, R. E.; Caipo, M.; Crump, J. A.; Devleesschauwer, B.; Angulo, F. J. World Health Organization Estimates of the Global and Regional Disease Burden of 22 Foodborne Bacterial, Protozoal, and Viral Diseases, 2010 Data Synthesis. *PLoS Med.* **2015,** *12* (11), E-article 1001921, p 15.
128. Kokare, C. R., Chakraborty, S., Khopade, A. N., Mahadik, K. R. Biofilm: Importance and Applications. *Indian J. Biotechnol.* **2009,** *8* (April), 159–168.
129. Kong, F.; Singh, R. P. Chemical Deterioration and Physical Instability of Foods and Beverages. In *The Stability and Shelf-life of Food*; Woodhead Publishing: London, 2016; pp 43–76.
130. Kumar, P.; Mahato, D. K.; Kamle, M.; Mohanta, T. K.; Kang, S. G. Aflatoxins: A Global Concern for Food Safety, Human Health and Their Management. *Front. Microbiol.* **2017,** *7*, 2170–2177.
131. Kurzbaum, E.; Iliasafov, L.; Kolik, L.; Starosvetsky, J.; Bilanovic, D.; Butnariu, M.; Armon, R. From the Titanic and Other Shipwrecks to Biofilm Prevention: The Interesting Role of Polyphenol-Protein Complexes in Biofilm Inhibition. *Sci. Total Environ.* **2019,** *658*, 1098–1105.
132. Langiano, E.; Ferrara, M.; Lanni, L.; Viscardi, V.; Abbatecola, A. M.; De Vito, E. Food Safety at Home: Knowledge and Practices of Consumers. *J. Public Health,* **2012,** *20* (1), 47–57.

133. Lead, J. R.; Wilkinson, K. J. Aquatic Colloids and Nanoparticles: Current Knowledge and Future Trends. *Environ. Chem.* **2006,** *3* (3), 159–171.
134. Lee, J. S.; Chang, Y.; Lee, E. S.; Song, H. G.; Chang, P. S.; Han, J. Ascorbic Acid-Based Oxygen Scavenger in Active Food Packaging System for Raw Meat-Loaf. *J. Food Sci.* **2018,** *83* (3), 682–688.
135. Li, B.; Webster, T. J. Bacteria Antibiotic Resistance: New Challenges and Opportunities for Implant-Associated Orthopedic Infections. *J. Orthopedic Res.* **2018,** *36* (1), 22–32.
136. Liew, W. P. P.; Mohd-Redzwan, S. Mycotoxin: Its Impact on Gut Health and Microbiota. *Front. Cell. Infect. Microbiol.* **2018,** *8*, 60–67.
137. Magdolenova, Z.; Collins, A.; Kumar, A.; Dhawan, A.; Stone, V.; Dusinska, M. Mechanisms of Genotoxicity. A Review of In Vitro and In Vivo Studies with Engineered Nanoparticles. *Nanotoxicology* **2014,** *8* (3), 233–278.
138. Mahajan, B. C.; Tandon, R.; Kapoor, S.; Sidhu, M. K. Natural Coatings for Shelf-Life Enhancement and Quality Maintenance of Fresh Fruits and Vegetables: A Review. *J. Postharvest Technol*, **2018,** *6* (1), 12–26.
139. Mahalik, N. P.; Nambiar, A. N. Trends in Food Packaging and Manufacturing Systems and Technology. *Trends Food Sci. Technol.* **2010,** *21* (3), 117–128.
140. Maina, S.; Kachrimanidou, V.; Koutinas, A. A roadmap towards a Circular and Sustainable Bioeconomy Through Waste Valorization. *Curr. Opin. Green Sustain. Chem.* **2017,** *8*, 18–23.
141. Makabenta, J. M. V.; Nabawy, A., Li, C. H.; Schmidt-Malan, S.; Patel, R.; Rotello, V. M. Nanomaterial-Based Therapeutics for Antibiotic-Resistant Bacterial Infections. *Nat. Rev. Microbiol.* **2020,** *2020*, 1–14.
142. Manke, A.; Wang, L.; Rojanasakul, Y. Mechanisms of Nanoparticle-Induced Oxidative Stress and Toxicity. *BioMed Res. Int.* **2013,** *2013* (August), E-article 942916; Online. https://doi.org/10.1155/2013/942916. (accessed August 2, 2021).
143. Marchand, S.; De Block, J.; De Jonghe, V.; Coorevits, A.; Heyndrickx, M.; Herman, L. Biofilm Formation in Milk Production and Processing Environments; Influence on Milk Quality and Safety. *Compreh. Rev. Food Sci. Food Safet.* **2012,** *11* (2), 133–147.
144. Mathews, D. M.; Johnson, N. P.; Sim, R. G.; O'Sullivan, S.; Peart, J. M.; Hofman, P. L. Iodine and Fertility: Do We Know Enough. *Human Reprod.* **2021,** *36* (2), 265–274; online. DOI: 10.1093/humrep/deaa312 (accessed August 2, 2021).
145. Micha, R.; Mannar, V.; Afshin, A.; Allemandi, L.; Baker, P.; Battersby, J.; Grummer-Strawn, L. Global Nutrition Report: Action on Equity to End Malnutrition. Online, 2020. https://reliefweb.int/report/world/2020-global-nutrition-report-action-equity-end-malnutrition (accessed Aug 2, 2021).
146. Miller, J. L. Iron Deficiency Anemia: A Common and Curable Disease. *Cold Spring Harbor Perspect. Med.* **2013,** *3* (7), E-article 011866, p 7.
147. Mishra, V.; Dubey, N.; Kumar, S.; Singh, N. Nano-Enabled Packing of Food Products. *Innov. Packag. Fruits Veg: Strategies Safet. Qual. Mainten.* **2018,** *2018*, 315–320.
148. Mitchell, N. J.; Bowers, E.; Hurburgh, C.; Wu, F. Potential Economic Losses to the US Corn Industry from Aflatoxin Contamination. *Food Add. Contam. A*, **2016,** *33* (3), 540–550.
149. Momin, J. K.; Jayakumar, C.; Prajapati, J. B. Potential of Nanotechnology in Functional Foods. *Emirates J. Food Agric. (EJFA)*, **2013,** *25* (1), 10–19.
150. Motta, J. P.; Wallace, J. L.; Buret, A. G.; Deraison, C.; Vergnolle, N. Gastrointestinal Biofilms in Health and Disease. *Nat. Rev. Gastroenterol. Hepatol.* **2021,** *2021*, 1–21.

151. Mottaleb, K. A.; Mainuddin, M.; Sonobe, T. COVID-19 Induced Economic Loss and Ensuring Food Security for Vulnerable Groups: Policy Implications from Bangladesh. *PloS One* **2020,** *15* (10), E-article 0240709, p 8.
152. Müller, A.;Schmidhuber, J.; Hoogeveen, J.; Steduto, P. Some Insights in the Effect of Growing Bio-Energy Demand on Global Food Security and natural resources. *Water Policy*, **2008,** *10* (S1), 83–94.
153. Mustafa, F.; Andreescu, S. Nanotechnology-Based Approaches for Food Sensing and Packaging Applications. *RSC Adv.* **2020,** *10* (33), 19309–19336.
154. Myszka, K.; Czaczyk, K. Bacterial Biofilms on Food Contact Surfaces-A Review. *Polish J. Food Nutr. Sci.* **2011,** *61* (3), 173–181.
155. Naik, M. I.; Fomda, B. A.; Jaykumar, E.; Bhat, J. A. Antibacterial Activity of Lemongrass (*Cymbopogon citratus*) Oil Against Some Selected Pathogenic Bacteria. *Asian Pacific J. Trop. Med.* **2010,** *3* (7), 535–538.
156. Nazrana, N.; Jain, T.; Verma, S. Role of Nutrition in Maintaining Normal Eyesight: A Review. *Int. J. Med. Biomed. Stud.* **2020,** *4* (3); online. https://www.ijmbs.info/index.php/ijmbs/article/view/1068. (accessed Aug 2, 2021).
157. Nemes, S. A.; Szabo, K.; Vodnar, D. C. Applicability of Agro-Industrial By-Products in Intelligent Food Packaging. *Coatings* **2020,** *10* (6), 550–556.
158. Nersisyan, H. H.; Lee, J. H.; Ding, J. R.; Kim, K. S.; Manukyan, K. V.; Mukasyan, A. S. Combustion Synthesis of zero-, One-, Two-and Three-Dimensional Nanostructures: Current Trends and Future Perspectives. *Progress Energy Combustion Sci.* **2017,** *63*, 79–118.
159. Nicola, M.; Alsafi, Z.; Sohrabi, C.; Kerwan, A.; Al-Jabir, A.; Iosifidis, C.; Agha, R. The Socio-Economic Implications of the Coronavirus Pandemic (Covid-19): A Review. *Int. J. Surg. (London, England)* **2020,** *78*, 185.
160. Olatunde, O. O.; Benjakul, S. Nonthermal Processes for Shelf-Life Extension of Seafoods: A Revisit. *Compreh. Rev. Food Sci. Food Safet.* **2018,** *17* (4), 892–904.
161. Omanović-Mikličanina, E.; Maksimović, M. Nanosensors Applications in Agriculture and Food Industry. *Bull Chem Technol Bosnia Herzegovina* **2016,** *47*, 59–70.
162. Omoike, A.; Chorover, J. Spectroscopic Study of Extracellular Polymeric Substances from *Bacillus subtilis*: Aqueous Chemistry and Adsorption Effects. *Biomacromolecules* **2004,** *5* (4), 1219–1230.
163. Otoni, C. G.; Avena-Bustillos, R. J.; Azeredo, H. M.; Lorevice, M. V.; Moura, M. R.; Mattoso, L. H.; McHugh, T. H. Recent advances on Edible Films Based on Fruits and Vegetables: A Review. *Compreh. Rev. Food Sci. Food Safet.* **2017,** *16* (5), 1151–1169.
164. Pal, M.; Devrani, M.; Hadush, A. Recent Developments in Food Packaging Technologies. *Bever. Food World* **2019,** *46* (1), 21–25.
165. Palmer, J. N. Bacterial Biofilms: Do They Play a Role in Chronic Sinusitis? *Otolaryngologic Clin. North Am.* **2005,** *38* (6), 1193–1201.
166. Papargyropoulou, E.; Lozano, R.; Steinberger, J. K.; Wright, N.; Bin Ujang, Z. The Food Waste Hierarchy as a Framework for the Management of Food Surplus and Food Waste. *J. Clean. Prod.* **2014,** *76*, 106–115.
167. Parfitt, J.; Barthel, M.; Macnaughton, S. Food Waste Within Food Supply Chains: Quantification and Potential for Change to 2050. *Phil. Trans. R. Soc. B Biol. Sci.* **2010,** *365* (1554), 3065–3081.
168. Parsek, M. R.; Singh, P. K. Bacterial Biofilms: An Emerging Link to Disease Pathogenesis. *Annu.Reviews in Microbiology*, **2003,** *57* (1), 677–701.

169. Peighambardoust, S. J;Peighambardoust, S. H.; Pournasir, N.; Pakdel, P. M. Properties of Active Starch-Based Films Incorporating a Combination of Ag, ZnO and CuO Nanoparticles for Potential Use in Food Packaging Applications. *Food Packag. Shelf-Life* **2019,** *22*, E-article 100420, p 7.
170. Pellegrini, M. COVID-19 Pandemic: A Brief Review on How Computer Science and Information Technology Can Help Shape the Future of Food Safety. *COVID-19 Pandemic: Case Studies Opin. 01(05)*, 2020, *2020* (October), 94–98.
171. Pereira de Abreu, D. A.; Cruz, J. M.; Paseiro-Losada, P. Active and Intelligent Packaging for the Food Industry. *Food Rev. Int.* **2012,** *28* (2), 146–187.
172. Pfalzgraff, A.; Brandenburg, K.; Weindl, G. Antimicrobial Peptides and Their Therapeutic Potential for Bacterial Skin Infections and Wounds. *Front. Pharmacol.* **2018,** *9*, 281–288.
173. Phillips, C. A. Modified atmosphere packaging and its effects on the microbiological quality and safety of produce. *International Journal of Food Science Technology*, **1996,** *31* (6), 463–479.
174. Pirsa, S.; Shamusi, T. Intelligent and Active Packaging of Chicken Thigh Meat by Conducting Nano Structure Cellulose—Polypyrrole—ZnO Film. *Mater. Sci. Eng. C* **2019,** *102*, 798–809.
175. Powers, H. J. Riboflavin (Vitamin B-2) and Health. *Am. J. Clin. Nutr.* **2003,** *77* (6), 1352–1360.
176. Prasad, A. S. Impact of the Discovery of Human Zinc Deficiency on Health—Part I. *J. Am. College Nutr.* **2009,** *28* (3), 257–265.
177. Prasad, A. S. Impact of the Discovery of Human Zinc Deficiency on Health – Part II. *J. Trace Elements Med. Biol.* **2014,** *28* (4), 357–363.
178. Puértolas, E.; Lavilla, M. HPP in Seafood Products: Impact on Quality and Applications. In *Present and Future of High-Pressure Processing*; Elsevier: San Diego, 2020; pp 201–220.
179. Qiu, L.; Zhang, M.; Tang, J.; Adhikari, B.; Cao, P. Innovative Technologies for Producing and Preserving Intermediate Moisture Foods: A Review. *Food Res. Int.* **2019,** *116*, 90–102.
180. Rahimi, G.; Alizadeh, F.; Khodavandi, A. Mycosynthesis of Silver Nanoparticles from Candida Albicans and Its Antibacterial Activity Against *Escherichia coli* and *Staphylococcus aureus*. *Trop. J. Pharm. Res.* **2016,** *15* (2), 371–375.
181. Rai, M.; Ingle, A. Role of Nanotechnology in Agriculture with Special Reference to Management of Insect Pests. *Appl. Microbiol. Biotechnol.* **2012,** *94* (2), 287–293.
182. Rai, M.; Ingle, A. P.; Gupta, I.; Pandit, R.; Paralikar, P.; Gade, A.; dos-Santos, C. A. Smart Nanopackaging for the Enhancement of Food Shelf-Life. *Environ. Chem. Lett.* **2019,** *17* (1), 277–290.
183. Ramasamy, M.; Lee, J. Recent Nanotechnology Approaches for Prevention and Treatment of Biofilm-Associated Infections on Medical Devices. *BioMed Res. Int.* **2016,** *2016* (October), online. https://www.hindawi.com/journals/bmri/2016/1851242/ (accessed Aug 2, 2021).
184. Raposo, A.; Pérez, E.; de Faria, C. T.; Ferrús, M. A.; Carrascosa, C. Food Spoilage by *pseudomonas* spp.: An Overview. *Food Borne Pathogens Antibiotic Resistance* **2017,** *2017*, 41–58.
185. Rask, K. J.; Rask, N. Economic Development and Food Production–Consumption Balance: A Growing Global Challenge. *Food Policy* **2011,** *36* (2), 186–196.
186. Ravash, N.; Peighambardoust, S. H.; Soltanzadeh, M.; Pateiro, M.; Lorenzo, J. M. Impact of High-Pressure Treatment on Casein Micelles, Whey Proteins, Fat Globules

and Enzymes Activity in Dairy Products: A Review. *Crit. Rev. Food Sci. Nutr.* **2020,** ***2020,*** 1–21.
187. Rawat, S. Food Spoilage: Microorganisms and Their Prevention. *Asian J. Plant Sci. Res.* **2015,** *5* (4), 47–56.
188. Reig, C. S.; Lopez, A. D.; Ramos, M. H.; Ballester, V. A. C. Nanomaterials: A Map for Their Selection in Food Packaging Applications. *Packag. Technol. Sci.* **2014,** *27* (11), 839–866.
189. Ricciardi, B. F.; Muthukrishnan, G.; Masters, E.; Ninomiya, M.; Lee, C. C.; Schwarz, E. M. *Staphylococcus aureus* Evasion of Host Immunity in the Setting of Prosthetic Joint Infection: Biofilm and Beyond. *Curr. Rev. Musculoskeletal Med.* **2018,** *11* (3), 389–400.
190. Rodríguez, G. M.; Sibaja, J. C.; Espitia, P. J.; Otoni, C. G. Antioxidant Active Packaging Based on Papaya Edible Films Incorporated with *Moringa oleifera* and Ascorbic Acid for Food Preservation. *Food Hydrocoll.* **2020,** *103*, E-article 105630, p 7.
191. Sahal, G.; Woerdenbag, H. J.; Hinrichs, W. L.; Visser, A.; Tepper, P. G.; Quax, W. J.; Bilkay, I. S. Antifungal and Biofilm Inhibitory Effect of *Cymbopogon citratus* (Lemongrass) Essential Oil on Biofilm Forming by *Candida tropicalis* Isolates; an in Vitro Study. *J. Ethnopharmacol.* **2020,** *246*, E-article 1**12188,** p 8.
192. Sajid, M.; Ilyas, M.; Basheer, C.; Tariq, M.; Daud, M.; Baig, N.; Shehzad, F. Impact of Nanoparticles on Human and Environment: Review of Toxicity Factors, Exposures, Control Strategies, and Future Prospects. *Environ. Sci. Pollut. Res.*, **2015,** *22*(6), 4122–4143.
193. Salama, Y.; Chennaoui, M.; Sylla, A.; Mountadar, M.; Rihani, M.; Assobhei, O. Characterization, Structure, and Function of Extracellular Polymeric Substances (EPS) of Microbial Biofilm in Biological Wastewater Treatment Systems: A Review. *Desalination Water Treatment* **2016,** *57* (35), 16220–16237.
194. Samal, D. Use of Nanotechnology in Food Industry: A Review. *Int. J. Environ. Agric. Biotechnol.* **2017,** *2* (4), E-article 238902, p 9.
195. San Martin, M. F.; Barbosa-Canovas, G. V.; Swanson, B. G. Food Processing by High Hydrostatic Pressure. *Crit. Rev. Food Sci. Nutr.* **2002,** *42* (6), 627–645.
196. Sangkhae, V.; Fisher, A. L.; Wong, S.; Koenig, M. D.; Tussing-Humphreys, L.; Chu, A.; Nemeth, E. Effects of Maternal Iron Status on Placental and Fetal Iron Homeostasis. *J. Clin. Investig.* **2020,** *130* (2), 625–640.
197. Sarwar, M. S.; Niazi, M. B. K.; Jahan, Z.; Ahmad, T.; Hussain, A. Preparation and Characterization of PVA/Nanocellulose/Ag Nanocomposite Films for Antimicrobial Food Packaging. *Carbohydr. Polym.* **2018,** *184*, 453–464.
198. Sbordone, L.; Bortolaia, C. Oral Microbial Biofilms and Plaque-Related Diseases: Microbial Communities and Their Role in the Shift from Oral Health to Disease. *Clin. Oral Investig.* **2003,** *7* (4), 181–188.
199. Scialabba, N. (Ed.). *Integrated Coastal Area Management and Agriculture, Forestry and Fisheries*; Food & Agricultural Organization: Rome, 1998, online. http://www.fao.org/3/w8440e/w8440e00.htm. (accessed Aug 2, 2021).
200. Sekhon, B. S. Food Nanotechnology–An Overview. *Nanotechnol. Sci. Appl.* **2010,** *3*, 1–9.
201. Sekhon, B. S. Nanotechnology in Agri-Food Production: An Overview. *Nanotechnol. Sci. Appl.* **2014,** *7*, 31–38.
202. Sen, S.; Pathak, Y., Eds. *Nanotechnology in Nutraceuticals: Production to Consumption*; CRC Press: Boca Raton, 2016; p 487.
203. Shajil, S.; Mary, A.; Juneius, C. E. R. Recent Food Preservation Techniques Employed in the Food Industry. In *Microbial Biotechnology*; Springer: Singapore, 2018; pp 3–21.

204. Sharma, C.; Dhiman, R.; Rokana, N.; Panwar, H. Nanotechnology: An Untapped Resource for Food Packaging. *Front. Microbiol.* **2017,** *8*, 1735–1741.
205. Sharma, H.; Mutharasan, R. Review of Biosensors for Foodborne Pathogens and Toxins. *Sens. Actuat B: Chem.* **2013,** *183*, 535–549.
206. Sherwin, J. C.; Reacher, M. H.; Dean, W. H.; Ngondi, J. Epidemiology of Vitamin A Deficiency and Xerophthalmia in at-Risk Populations. *Trans. R. Soc. Trop. Med. Hygiene.* **2012,** *106* (4), 205–214.
207. Silvestre, C.; Duraccio, D.; Cimmino, S. Food Packaging Based on Polymer Nanomaterials. *Progress Polym Sci.* **2011,** *36* (12), 1766–1782.
208. Singh, M.; Singh, S.; Prasad, S.; Gambhir, I. S. Nanotechnology in Medicine and Antibacterial Effect of Silver Nanoparticles. *Digest J. Nanomater. Biostruct.* **2008,** *3* (3), 115–122.
209. Singh, S.; Shalini, R. Effect of Hurdle Technology in Food Preservation: A Review. *Crit. Rev. Food Sci. Nutr.* **2016,** *56* (4), 641–649.
210. Singh, T.; Shukla, S.; Kumar, P.; Wahla, V.; Bajpai, V. K.; Rather, I. A. Application of Nanotechnology in Food Science: Perception and Overview. *Front. Microbiol.* **2017,** *8*, 1501–1510.
211. Sommer, A. Vitamin A Deficiency and Clinical Disease: An Historical Overview. *J. Nutr.* **2008,** *138* (10), 1835–1839.
212. Souza, V. G. L.; Fernando, A. L. Nanoparticles in Food Packaging: Biodegradability and Potential Migration to Food—A Review. *Food Packag. Shelf-Life*, **2016,** *8*, 63–70.
213. Speranza, B.; Corbo, M. R.; Sinigaglia, M. Effects of Nutritional and Environmental Conditions on Salmonella sp. Biofilm Formation. *J. Food Sci.* **2011,** *76* (1), M12–M16.
214. Stepanović, S.; Ćirković, I.; Ranin, L.; S.vabić-Vlahović, M. Biofilm Formation by *Salmonella spp.* and *Listeria monocytogenes* on Plastic Surface. *Lett. Appl. Microbiol.* **2004,** *38* (5), 428–432.
215. Sundramoorthy, A. K.; Kumar, T. H. V.; Gunasekaran, S. Graphene-Based Nanosensors and Smart Food Packaging Systems for Food Safety and Quality Monitoring. In *Graphene Bioelectronics*; Elsevier: New York, 2018; pp 267–306.
216. Sung, S. Y.; Sin, L. T.; Tee, T. T.; Bee, S. T.; Rahmat, A. R.; Rahman, W. A.; Vikhraman, M. Antimicrobial Agents for Food Packaging Applications. *Trends Food Sci. Technol.* **2013,** *33* (2), 110–123.
217. Swinburn, B.; Kraak, V.; Rutter, H.; Vandevijvere, S.; Lobstein, T.; Sacks, G.; Magnusson, R. Strengthening of Accountability Systems to Create Healthy Food Environments and Reduce Global Obesity. *Lancet*, **2015,** *385* (9986), 2534–2545.
218. Szabo, S. Urbanisation and Food Insecurity Risks: Assessing the Role of Human Development. *Oxford Dev. Stud.* **2016,** *44* (1), 28–48.
219. Tack, D. M.; Ray, L.; Griffin, P. M.; Cieslak, P. R.; Dunn, J.; Rissman, T.; Payne, D. C. Preliminary Incidence and Trends of Infections with Pathogens Transmitted Commonly Through Food: Foodborne Diseases. *Active Surveillance Netw.* (10 US Sites, 2016–2019)—*Morbidity Mortality Weekly Rep.*, **2020,** *69* (17), 509–516.
220. Tahir, M.; Tahir, B.; Amin, N. A. S.; Zakaria, Z. Y. Photo-Induced Reduction of CO2 to CO with Hydrogen Over Plasmonic Ag-NPs/TiO2 NWs Core/Shell Hetero-Junction under UV and Visible Light. *J. CO_2 Utilization*, **2017,** *18*, 250–260.
221. Taweechaisupapong, S.; Ngaonee, P.; Patsuk, P.; Pitiphat, W.; Khunkitti, W. Antibiofilm Activity and Post Antifungal Effect of Lemongrass Oil on Clinical *Candida dubliniensis* Isolate. *SA J. Bot.* **2012,** *78*, 37–43.

222. Thyberg, K. L.; Tonjes, D. J. Drivers of Food Waste and Their Implications for Sustainable Policy Development. *Resour. Conserv. Recycling*, **2016,** *106*, 110–123.
223. Timoneda, J.; Rodríguez-Fernández, L.; Zaragozá, R.; Marín, M. P.; Cabezuelo, M. T.; Torres, L.; Barber, T. Vitamin A Deficiency and the Lung. *Nutrients* **2018,** *10* (9), 1132–1136.
224. Toloza, F. J. K.; Motahari, H.; Maraka, S. Consequences of Severe Iodine Deficiency in Pregnancy: Evidence in Humans. *Frontiers in Endocrinology*, **2020,** *11*, 409–415.
225. Trienekens, J.; Zuurbier, P. Quality and Safety Standards in the Food Industry, Developments and Challenges. *Int. J. Prod. Econ.* **2008,** *113* (1), 107–122.
226. Tsagkaris, A. S.; Tzegkas, S. G.; Danezis, G. P. Nanomaterials in Food Packaging: State of the Art and Analysis. *J. Food Sci. Technol.* **2018,** *55* (8), 2862–2870.
227. Uddin, F. Clays, Nanoclays, and Montmorillonite Minerals. *Metallurgical Mater. Trans A* **2008,** *39* (12), 2804–2814.
228. Uttamchandani, I. False Shortage: Food Wastage in India and the Need for Resource Management Reforms. *Int. J. Soc. Sci. Econ. Res.* **2019,** *4* (7), 4898–4903. <ijsser.org/more2019.php?id=372>. (accessed Aug. 2, 2021).
229. Van Boekel, M.; Fogliano, V.; Pellegrini, N.; Stanton, C.; Scholz, G.; Lalljie, S.; Eisenbrand, G. A Review on the Beneficial Aspects of Food Processing. *Mol. Nutr. Food Res.* **2010,** *54* (9), 1215–1247.
230. Van Houdt, R.; Michiels, C. W. Biofilm Formation and the Food Industry, a Focus on the Bacterial Outer Surface. *J. Appl. Microbiol.* **2010,** *109* (4), 1117–1131.
231. Varela-Santos, E.; Ochoa-Martinez, A.; Tabilo-Munizaga, G.; Reyes, J. E.; Pérez-Won, M.; Briones-Labarca, V.; Morales-Castro, J. Effect of High Hydrostatic Pressure (HHP) Processing on Physicochemical Properties, Bioactive Compounds and Shelf-Life of Pomegranate Juice. *Innov. Food Sci. Emerg. Technol.* **2012,** *13*, 13–22.
232. Vasile, C.; Sivertsvik, M.; Miteluţ, A. C.; Brebu, M. A.; Stoleru, E.; Rosnes, J. T.; Popa, M. E. Comparative Analysis of the Composition and Active Property Evaluation of Certain Essential Oils to Assess Their Potential Applications in Active Food Packaging. *Materials* **2017,** *10* (1), 45–50.
233. Vazquez-Armenta, F. J.; Bernal-Mercado, A. T.; Tapia-Rodriguez, M. R.; Gonzalez-Aguilar, G. A.; Lopez-Zavala, A. A.; Martinez-Tellez, M. A.; Ayala-Zavala, J. F. Quercetin Reduces Adhesion and Inhibits Biofilm Development by Listeria Monocytogenes by Reducing the Amounts of Extracellular Proteins. *Food Control* **2018,** *90*, 266–273.
234. Vega-Gálvez, A.; Giovagnoli, C.; Pérez-Won, M.; Reyes, J. E.; Vergara, J.; Miranda, M.; Di-Scala, K. Application of High Hydrostatic Pressure to Aloe Vera Gel: Microbial Inactivation and Evaluation of Quality Parameters. *Innov. Food Sci. Emerg. Technol.* **2012,** *13*, 57–63.
235. Vetrano, D. L.; Zucchelli, A.; Marconi, E.; Levi, M.; Pegoraro, V.; Cataldo, N.; Lapi, F. Predictors of Iron-Deficiency Anemia in Primary Care Older Adults: A Real-World European Multi-Country Longitudinal Study. *Aging Clin. Exp. Res.* **2020,** ***2020***, 1–6.
236. Von Grebmer, K.; Saltzman, A.; Birol, E.; Wiesman, D.; Prasai, N.; Yin, S.; Sonntag, A. Global Hunger Index: The Challenge of Hidden Hunger. *IFPRI Books*, 2014; p 56. http://dx.doi.org/10.2499/9780896299580. (accessed Aug 2, 2021).
237. Wang, C. Y.; Huang, H. W.; Hsu, C. P.; Yang, B. B. Recent Advances in Food Processing Using High Hydrostatic Pressure Technology. *Crit. Rev. Food Sci. Nutr.* **2016,** *56* (4), 527–540.
238. Warner, M. J.; Kamran, M. T. Anemia, iron deficiency. *Stat Pearls* [online update], 2020. https://www.ncbi.nlm.nih.gov/books/NBK448065/ (accessed Aug 2, 2021).

239. Webb, A.; McQuaid, R.; Rand, S. Employment in the Informal Economy: Implications of the COVID-19 Pandemic. *Int. J. Sociol. Soc. Policy* **2020** (Dec); online. https://www.emerald.com/insight/content/doi/10.1108/IJSSP-08-2020-0371/full/html (accessed Aug 2, 2021).
240. Wells, C. R.; Sah, P.; Moghadas, S. M.; Pandey, A.; Shoukat, A.; Wang, Y.; Galvani, A. P. Impact of International Travel and Border Control Measures on the Global Spread of the Novel 2019 Coronavirus Outbreak. *Proc. Natl. Acad. Sci.* **2020,** *117* (13), 7504–7509.
241. Wilson, M. Bacterial Biofilms and Human Disease. *Sci. Progress* **2001,** *84* (3), 235–254.
242. Wójciak, K. M.; Dolatowski, Z. J. Oxidative Stability of Fermented Meat Products. *Acta Scientiarum Polonorum Technologia Alimentaria (Food Sciences Technology)* **2012,** *11* (2), 99–109.<food.actapol.net>
243. World Health Organization (WHO). UNICEF/WHO/The World Bank Group Joint Child Malnutrition Estimates: Levels and Trends in Child Malnutrition: Key Findings, 2016. https://www.who.int/publications/i/item/9789240025257 (accessed Aug 2, 2021).
244. World Health Organization (WHO). UNICEF/WHO/The World Bank Group Joint Child Malnutrition Estimates: Levels and Trends in Child Malnutrition: Key Findings, 2020. https://www.who.int/publications/i/item/9789240025257 (accessed Aug 2, 2021).
245. Wu, Y.; Zhou, Y.; Huang, H.; Chen, X.; Leng, Y.; Lai, W.; Xiong, Y. Engineered Gold Nanoparticles as Multicolor Labels for Simultaneous Multi-Mycotoxin Detection on the Immunochromatographic Test Strip Nanosensor. *Sens. Actuat. B: Chem.* **2020,** *316*, E-article 128107, p 8.
246. Wu, Z.; Huang, X.; Li, Y. C.; Xiao, H.; Wang, X. Novel Chitosan Films with Laponite Immobilized Ag Nanoparticles for Active Food Packaging. *Carbohydr. Polym.* **2018,** *199*, 210–218.
247. Wyrwa, J.; Barska, A. Innovations in the Food Packaging Market: Active Packaging. *Eur. Food Res. Technol.* **2017,** *243* (10), 1681–1692.
248. Yan, X.; Sedykh, A.; Wang, W.; Yan, B.; Zhu, H. Construction of a Web-Based Nanomaterial Database by Big Data Curation and Modeling Friendly Nanostructure Annotations. *Nat. Commun.* **2020,** *11* (1), 1–10.
249. Yetter, R. A.; Risha, G. A.; Son, S. F. Metal Particle Combustion and Nanotechnology. *Proc. Combustion Inst.* **2009,** *32* (2), 1819–1838.
250. Youssef, A. M.; El-Sayed, S. M. Bionanocomposites Materials for Food Packaging Applications: Concepts and Future Outlook. *Carbohydr. Polym.* **2018,** *193*, 19–27.
251. Zaidi, S.; Misba, L.; Khan, A. U. Nano-Therapeutics: A Revolution in Infection Control in Post Antibiotic Era. *Nanomed. Nanotechnol. Biol. Med.* **2017,** *13* (7), 2281–2301.
252. Zimmermann, M. B. Iodine Deficiency. *Endocrine Rev.* **2009,** *30* (4), 376–408.
253. Zimmermann, M. B. The Role of Iodine in Human Growth and Development. *Semin. Cell Dev. Biol.* **2011,** *22* (6), 645–652.
254. Zimmermann, M. B.; Boelaert, K. Iodine Deficiency and Thyroid Disorders. *Lancet Diab. Endocrinol.* **2015,** *3* (4), 286–295.
255. Zimmermann, M. B.; Hurrell, R. F. Nutritional Iron Deficiency. *Lancet*, **2007,** *370* (9586), 511–520.

PART III

ASSESSMENT OF QUALITY OF FRUITS AND VEGETABLES

CHAPTER 11

NONDESTRUCTIVE METHODS FOR QUALITY ASSESSMENT OF FRUITS AND VEGETABLES DURING STORAGE

MOHIT SHARMA and KIRTY PANT

ABSTRACT

Quality of fruits and vegetables (FAVs) is the primary fundamental parameter for consumer acceptability. Globally different foodborne illnesses have been identified in several whole and fresh-cut FAVs. These outbreaks have raised the concern to reassess the quality of these FAVs to investigate all critical hazards that arise during handling, processing, production, and storage, which ultimately endure the burden on the expense and value of the food. With the surge of expectations concerning the quality of food commodities, the demand for authentic, active, fast techniques is also progressing. Recently, numerous nondestructive approaches have been investigated to evaluate the quality and characteristics of FAVs. This chapter summarizes several nondestructive techniques to determine the quality aspects of FAVs.

11.1 INTRODUCTION

The representation of food quality (Figure 11.1) is constituted of all the characteristics, properties, likely attributes, perception, psychological, and notational value of food.[45] To some extent, it also includes convenience, safety and authenticity and safety of food. This food quality phenomenon requires dedicated techniques to explore the micro- and macro-investigation of food quality in a nondestructive way.

Quality Control in Fruit and Vegetable Processing: Methods and Strategies. Megh R. Goyal, & Faizan Ahmad (Eds.)

FIGURE 11.1 Various features of food quality.

FAVs are valuable reservoir of antioxidants, vitamins, minerals, and numerous bioactive compounds in our daily diet. The storage and transportation of food are considered a crucial food quality factors for selling, purchasing, and retailing FAV. Consequently, it is necessary to assure the quality of FAV before conveying them to the market.

The quality of both FAV can be assessed based on many external determinants, such as color intensity, size dimension, shape characteristics, odor, surface attributes, etc., whereas internal constituents such as total soluble solids, total carotenoid content, total phenolic content, pH, titrable acidity, reducing sugars, total sugars, DPPH assay, microbiological value, gas transmission rates, water repellence, and permeability designate the quality of various FAV.[35]

Mostly all of FAV pass through ripening and spoilage process by the natural means of biochemical and enzymatic reactions that take place internally in the metabolism system of fruits and vegetables (FAVs), which leads to weight loss, skin changes, and color changes. However, early ripening and spoilage are not desirable in both fresh and cut FAVs. Nowadays, consumers demand food that is sensorial and nutritional sound without compromising any food safety risks.

Fruits (such as apples, grapes, citrus, etc.) are mostly prone to mold attack, which leads to different defects, such as blue rot, brown rot, Rhizopus soft rot, green mold rot, grey rot, and pink rot. Unusual bacteria- and virus-related spoilage also takes place in FAV, which can provoke defects, such as specks, vascular rings, off-taste, off-odor, and spots on the surface or skin of FAV.[28] Due to the lack of information, cost factors and limited resources available to farmers and researchers have raised uncertainties of deterioration and inadequate quality of FAVs.

The food quality evolves during postharvesting and storage to entail innovative techniques for exploring the food quality. Until now, scientists have discovered different nondestructive methods, which are compact, cost-effective, flexible, and well-defined devices to assess the quality of FAVs during storage.

This chapter focuses on image analysis, near-infrared spectroscopy (NIRS), NMR, MRI, X-ray tomography, and computerized tomography. Further, the chapter also highlights future perspectives and limitations of nondestructive techniques.

11.2 NONDESTRUCTIVE TECHNOLOGY: AN OVERVIEW

Nondestructive technology comes under those techniques, which can be used to assess the quality of food with the aid of nondestructive devices by manifesting the data in both qualitative and quantitative ways and without causing any injury or damage to the attributes of the food. Nondestructive technology is not merely limited to detect the internal and external defects of food and food materials but also can predict the future probabilities and possibilities of shelf-life and characteristic behavior evaluation of various FAVs (such as microstructure analysis, thermal properties, moisture entrapment, and variation in structure of various fruits during drying and freezing, etc.).[1] Based on the food quality, nondestructive devices are of the following types:

- Nondestructive instruments are uniquely designed for research and development activities. These are usually costly and are used for high accuracy and repeatability measurement of food products.
- Nondestructive instruments are usually developed for sort analysis, ranking, and rating purposes of FAV in industries.

- Compact devices, which are commonly less costly, are available in hand plied designs. These devices are created for quality controllers and food processors. New manageable tools are also available for consumers.

11.3 SIGNIFICANCE OF NONDESTRUCTIVE TECHNOLOGY FOR QUALITY ASSESSMENT

- Nondestructive technologies can be utilized for both quantitative and qualitative investigations of FAVs.
- These techniques can be used for the exposure of unwanted blemishes and discontinuities in both the FAV without destroying the internal attributes.
- These can also be used for the disclosure of any extraneous foreign substances.
- These techniques can be used for the revealing of any irregularity in the shape and size dimensions, such as length, thickness, depth, width, and diameter.
- Nondestructive technology can be used for the measurement of different properties of FAV, such as optical, dynamic, electromagnetic, chemical properties, etc.
- Nowadays, these can also be adopted for the safety and quality assurance of FAV and detection of unwanted bacteria and pathogens, such as *Escherichia coli, pseudomonas spp., coliforms, Cladosporium spp,* etc.[27]
- These techniques are beneficial in the investigation of ripening and maturity stages of FAV.[57]
- These techniques can also be used to measure the freshness index and decay index of FAV during storage.[48]

11.4 NONDESTRUCTIVE SYSTEMS AND THEIR APPLICATION FOR QUALITY ASSESSMENT OF FRUITS AND VEGETABLES

11.4.1 OPTICAL METHODS

11.4.1.1 IMAGE ANALYSIS

Image analysis of any sample comprises of image acquisition, segmentation, object selection, measurement analysis, statistical analysis, and stereological interpretation (Figure 11.2). Image acquisition relies on an image obtained from the equipment in use and the imaging technique. The image can either be in analogue form or digital form. The image can be attained from magnetic resonance imagers, scanners, monochrome cameras, or digital cameras, which are sometimes used along with different color and grayscale filters to eliminate the unwanted noise and artefacts.

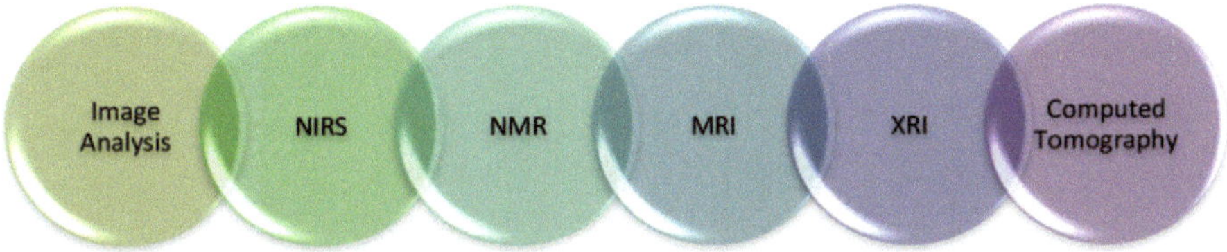

FIGURE 11.2 Applications of nondestructive systems for quality assessment of fruits and vegetables.

The collected images can be segmented based on structural dimensions, color, brightness, sharpness, temperature, and elemental attributes. Several types of algorithms and software are also employed for the segmentation of area of specific interest. Threshold, edge-based, and origin-based methods are generally used for object selection. Threshold, edge-based, and origin-based methods rely on intensity control and pixel separation, respectively.

After segmentation, measurement analysis could be done based on size (perimeter, length, area, curl ratio, roundness ratio, and extent ratio), shape (compaction and elongation), boundaries (curvatures and descriptors), distance between objects and object count, etc. Particle size analysis of the particles can be done based on measuring the surface diameter, volume diameter, free-falling diameter, perimeter diameter, martin, and projected area.[2] Various types of imaging methods along with their benefits and limitations are listed in Table 11.1.

11.4.1.2 NEAR INFRARED SPECTROSCOPY (NIRS)

NIRS is a highly stable, harmless analytical technique in which usually no specimen preparation is needed. NIRS operates on the phenomenon of absorption of waves at a specific wavelength. Generally for NIRS, the wavelength spectrum lies between 780 and 2500 nm.[12] When the specimen interacts with a near-infrared light source, the sample absorbs, transmits, and reflects radiation; and electrons of lower orbital excites to a higher orbital

TABLE 11.1 Benefits and Limitations of Different Imaging Methods.

Imaging method	Benefits	Limitations
Fluorescence	• Top-quality images due to limited noise	• Lack of linear intensity ratios
	• Fluorophore labeled images	• Issue of autofluorescence
Hyperspectral	• Advantageous for spatial and spectral differentiation of various similar resembling objects	• High-cost instrument setup
	• Beneficial for characterizing the attributes of objects	• Limited information yield
Laser backscattering	• Reasonable technique	• Limited resolution of images
	• Simple to operate	• Risk of loss of quality of images is high due to observation errors
Thermal	• Able to collect data of moving objects	• Viable environment impact on thermal images
	• Able to collect data from hard to reach areas, which are not attainable in other methods	• Challenging image interpretation due to variable temperatures
X-ray	• In-depth analysis	• Impact of X-ray radiations to the specimen and surrounding is still questionable
	• High penetration ability	

state. Due to the excitation of electrons, varied types of stretching, rotation, and bending appear in the specimen at different wavelength values, such as O–H stretching, C–H stretching, N–H stretching, etc. Various stretching and bending inference combinations justify the changes in the composition of FAVs. The O–H stretching and bending combinations imply the moisture content changes in FAVs. Whereas C–H bonds indicate the changes in the proportion of oils, carbohydrates, and fats in FA, and N–H stretching represents changes in protein compositions.[3]

The detector usually is adjusted according to the requirement of data observation. Tungsten and halogen lamps are normally adopted for the light source. When detectors are adjusted in reflectance mode, the reflected light or radiations are frequently captured and result in information about the external structure, such as texture, size, shape, color values, and blemishes. The detectors are generally attached to computers to convert radiations to electric signals for data processing. Due to the advancement in the information technology; several image analysis data analysis and statistical software systems have emerged to remove noise by employing various filters and data preprocessors models. Sometimes detectors are also adjusted in trans-reflectance and transmittance mode to detect the deeper information on the surface of FAVs. The main disadvantage of the NIRS system is the lack of spatial arrangement.[3] Mango, pear, peach, melons, grapes, strawberry, and blueberry samples were analyzed for ripening stages, maturity index, dry matter analysis, chlorophyll content, titrable acidity, sugar content, anthocyanin content, and firmness with the help of the NIRS technique.[5,70,71]

11.4.2 ELECTROMAGNETIC METHODS

11.4.2.1 NUCLEAR MAGNETIC RESONANCE

Nuclear magnetic resonance (NMR) is an exclusive technique, which determines the magnetic characteristics of spinning associated with the physical and chemical characteristics of matter.[53] During this process, the nucleus evolves in the magnetic moment, which significantly engrosses radiations of particular frequencies under an external magnetic field.[18] A nucleus comprising an odd number of nucleons (e.g., 1H and 13C) is essential to attain a net magnetization moment.[39] However, the 1H produces the highest NMR signal and provides outstanding contrast among diverse structures; thus, most of the researches have been using 1H NMR for solid food quality assessment, due to the existence of ample hydrogen atoms in the product.[20,39,41]

During the process, NMR signals are carried as electromagnetic energy, which is perceived and received by the detectors; further these signals are directed to the processer (CPU) and transformed in the form of an illustration via data computation.[18] NMR images possess individual pixel value to demonstrate the signal intensity in the evaluated matter, this further is linked with the resonance density and relaxation time, that is, T1 and T2.[18] NMR parameters (such as, relaxation time and proton density, various food properties) can be measured, such as chemical composition and structural quantifications.[50] Whereas several food features (such as chemical composition, cell structure and cell-size, cellular content viscosity, and magnetic exposure) affect the proton relaxation.[20,65]

Consequently, alteration in the interior qualities (such as firmness, browning, ripeness, decaying of tissues, bruising, mealiness, water core), woolly categorization, or circulation of water within the cellular structure and chemical composition (such as water, sugar, oil, pectin, starch, etc.) of fresh produce could be effortlessly observed through NMR.[4,9,11,17,36,49,75] While the defected and damaged structures inside the FAVs are identified by the darker region because of their low proton density.[20,42,65,75]

The operation of the NMR system in the food sector primarily is utilized to inspect various quality features, such as moisture, oil, carbohydrate, protein, etc. Utilizing the variations in the dispersal of NMR water-proton transverse relaxation times, apples under freeze-drying possessed lower water content in tissues as compared to fluidized-bed drying. However, NMR results indicated its damaged membrane integrity that leads to breakdown of cell-wall.[18] As compared to monochrome camera images, NMR images provide various benefits, such as 3D analysis of samples, clarity, particularly among fat and connective tissue.[6]

11.4.2.2 MAGNETIC RESONANCE IMAGING

Magnetic resonance imaging (MRI) is a nonintrusive and nondestructive spectroscopic method.[59] This significantly depends on the interaction between electromagnetic radiation and atomic particles (electrons and the nucleus) of matter.[53] Protons (chiefly 1H nuclei) are targeted, especially those which are existing in the water of FAVs. During this process, protons spin within the body and allied themselves under influence of powerful magnetic field conditions. Through regulating the proton dynamics subsequently, data can be attained related to the spatial distribution of proton density, relaxation parameters values, and self-diffusion factors of the specimen. Particular self-diffusional including relaxation characteristics give complimentary information, and improved contrast is also associated with proton mobility.[74]

Food commodity exhibits a huge variety of compositional complexity, skeletal complications, heterogeneity, and divergence, which describe its quality excellence and safety perspectives. Noninvasive methods comprise of products to get quantitative data evidence related to the spatial arrangement.[50] MRI technique is being already established technologically and commercially in the medical field, laboratory testing, and now recently in the food processing industries.[18]

MRI is considered appropriate for biological materials, primarily in water but similarly in oil, fat, or salt also, and is capable to differentiate these compounds. Moreover, MRI is profound toward various quality aspects, which possibly affect the product and selectively those parameters that influence the mobility or water concentration, for example, internal browning. However, a nondestructive feature of MRI makes it fascinating for scanning the FAV but also for regulating product quality over time.[20,24]

MRI process has relatively low image acquisition speed in comparison to other techniques, such as X-ray imaging. MRI mechanism possesses high sensitivity toward water molecules; however, it is predominantly employed to examine relative transformations or alterations occurrences in water concentration rather than quantification[51,78]; subsequently, the process is not straightforward due to other factors, such as temperature variations and proton containing compounds.

MRI technique has been successfully established to determine numerous quality characteristics of fresh FAV and the disclosure of internal checks, such as voids/pits,[68] worm destruction/holes,[16] wounding,[54] water core ailment in apples,[19] interior browning in apples,[19,24] browning of apples or peaches,[52] core crackup in pears,[42,43] freezing injury in citrus berry, or zucchini squash[33] to check tissue structure and water status of interior properties or fungal inhibition.[18] It is being utilized to check physical properties (such as shape, size, volume), which are interrelated with soluble solids, firmness, or acid content.[20,46]

MRI facilitates easy differentiation between immature and ripe FAV, for example, by determining oil content in avocado[16] or free moisture content in pineapple or tomatoes.[16] Additionally, the detection of pathogen infection has also been achieved successfully.[31] Due to the high cost and relatively low acquisition speed of MRI equipment as compared to the base cost, of large volume of FAVs, industrial implementation of MRI is in a nascent stage.[67]

Several types of research aimed to develop economical, precise, and quick processing equipment that is able to attain pragmatic outcomes. The primary aim of research studies is on the hardware systems, namely, to develop or use cheaper, quick, low-field, wide-bore MRI scanners[34,56] with smaller mobile (1D or 2D) systems,[22,23] and faster pulse sequences. These developments can enhance previous system potentials so that MRI is comparatively worthwhile tool for nondestructive quality assessment of raw agricultural products.

11.4.3 DYNAMIC METHODS

11.4.3.1 X-RAYS IMAGING

The nondestructive and instinctive quality assessment of agricultural commodities has considerably become the paramount solution to improve quality standards and reduce production expenses.[79] X-rays imaging is one of the highly significant and imminent technologies used in medical diagnostics and other perspectives of inspections. Besides that, the purpose of the X-ray imaging technique for characteristic inspection of farming products has proposed significant advantages. It also complements current evaluation techniques.[14]

However, the usage of X-rays imaging for quality assessment of agricultural products is not yet exploited to a full extent and it is still in a nascent stage. The application of X-ray imaging technology is for inspection of interior qualities parameters, such as nutrition, texture (crispiness, juiciness, firmness), and an internal cavity, rotten, pest infestation, water core, spongy tissues that are hard to determine by normal visual appearance. X-ray imaging is a well-recognized technique to access high attenuation and delicate kinds of stuff; therefore, it has been utilized for various inspection purposes within the agronomical and food industries.

X-rays are electromagnetic emissions with a wavelength of 0.01–10 nm, and with relative energies from 0.12 to 12 keV. The energy and penetration capacity of an X-ray significantly depends on its wavelength, where the shorter wavelength possesses higher energy and penetration potential and visa-versa. Soft X-rays are being recognized for their lower penetration potential and capability to disclose the internal density alteration in farming commodities. Neethirajan et al. [58] stated that these simple X-rays technique processing takes about 3–5 s to build an X-ray picture. Hence, X-ray imaging-based systems are operated as nondestructive techniques. Therefore, it is extensively utilized for research purposes to evaluate quality classification, quality components in FAV, and to detect interior defects in agricultural commodities.

The X-rays are released and reach within the fruit or vegetable and eventually fall on an X-ray scintillator (having a light coat of luminescent substance) to captivate the X-ray picture after processing. Yang et al.[82] checked the various prospects of using X-ray to determine internal defects and injuries of various fruits. Njorge et al.[61] have found internal defects, such as rind puffing in oranges by X-ray method. The results revealed that this technique is a valuable tool for the nondestructive examination of interior damages of fruits, which cannot be detected by visual appearance.

Cristiane et al. [21] developed digital images of a product via X-ray spectrum ranging from 18 to 20 keV, where the initial images were examined by using a digital image processing system for differentiation of speed, peel, pulp, and damaged region in mangoes. The X-ray processing accompanies the Law of Absorption: the strength of the estimated light subsequently advancing through each specimen remains mainly reliant on the primary penetration power, density, stiffness of the absorbent material and its immersion coefficient. FAVs have tremendous amount of water content, which causes the maximum portion of X-rays absorption. Various interior disorders with adverse impact on quality (such as astringent pit, water nucleus, cork spot, and brown nucleus for apple, seed germination, black rot, membranous stain, bruises, hollow heart, and reasonably black core for potato and freeze damage for citrus) would be demonstrable by X-ray technique.[79] This process is mainly reliant on the tissue density but not on the chemical composition of the material, which makes its application limited.[77]

Three main techniques are used for quality control: the two-dimensional radiography; the line scanning process generates a picture meanwhile the X-rays pass by the material; and Roentgen computed tomography (CT). This technique creates a three-dimensional image, which enables to compute the information from the two-dimensional image. Commercial practices with X-ray imaging help to detect the hollow heart in potato tubers,[62] the moisture content in apples,[76] density alteration throughout the ripening of tomatoes,[10] and insect infestation of pistachio nuts.[38]

Reyes et al.[66] analyzed the mango fruit seed and weevil in mango fruit via estimation of distinct characteristic of weevil-infested seeds. The prospective and potential of utilizing X-rays imaging is to identify internal damages due to pests or physical impingement in qualitative evaluation of FAVs. Yacob et al.[81] compared X-ray imaging technique with MRI systems applied on agricultural products for postharvest nondestructive inspection. Their study concluded that the X-ray is a conceivable technique to execute nondestructive investigation, since it is more acceptable and less expensive as compared to the MRI process.[79]

11.4.3.2 X-RAY COMPUTED TOMOGRAPHY

X-ray CT system primarily depends on a scientific algorithm to produce a tomogram (3-D illustration) of numerous radiographs of an object taken at various locations and angles.[69] It is also known as a highly impressive method due to its nondestructive visualization abilities and interior quality assessment of an object.[40] CT scanner has a 1 mm pixel resolution[59] and can produce 2D and 3D pictures (tomograms) from the acquired information for the evaluation of atomic and physiological details.[64,75] However, X-ray reckoned tomography technology is a well-recognized for cross-sectional evaluation of an object by combining a portable X-ray reservoir with an indicator equipment to accumulate information of a thin projected section of a specimen. Every attribute identification manner of an object is manual, though the image processing algorithm system has been testified as an image enhancer to facilitate better readability. The approach of X-ray tomography has additionally been searched for the measurement of the specific chemical composition of agricultural products.[40]

Initially, X-ray CT research studies were aimed to evaluate mellowness and ripeness-associated parameters of the fruit,[76] but a decent relationship was only observed between X-ray absorption and moisture content, or density as a derivation therefrom. Also, effective applications of X-ray CT can inscribe the detection of interior damages.[59] Lammertyn et al.[42,43] have used X-ray tomography to explore significant growth of core disruption defects in pears, perceiving both cavity formation and browning of tissue at high-resolution power. Hansen et al.[32] have observed that apples and cherries were able to identify codling moth causing feeding tunnels with the help of X-ray CT.

Ogawa et al.[63] detected particular nonmetallic substances around various liquids and food materials via processing of medical X-ray CT. In this study, authors evaluated the detection constraints for extraneous substances; and inferred the size of external materials was lesser than the physical resolution of the X-ray CT scanner, where tiny fragments might be recognized as extra particles but were not able to properly predict the characteristics of the materials.

Donis-Gonzalez et al.[26] used a CT at the initial stage from the inline sorting system to checkout defected chestnuts with holes and spoiled tissues. Moreover, the development must be based on the improvement in hardware system for better rebuilding of image, and picture processing models to attain appropriate, quick, and reasonable inline CT operations for quality appraisal[25] of goods. Lammertyn et al.[42] applied X-ray CT system to differentiate among wholesome and affected brownish segment of pears. Authors observed that the most despicable pixel intensity value for wholesome tissue was more powerful than the highest pixel intensity value of the brown tissue; and they stated that one threshold power was adequate to distinct both types of tissues.

Lim and Barigou[47] studied the applications of CT as a nondestructive system to assess the quality parameters of agricultural products by defining images, visualization, and examination of the 3D cellular microstructure by various CT scans. Ogawa et al.[63] mentioned that food substances own different CT numbers depending on the food type because of their inhomogeneity; when the CT value of extraneous substances fell inside the range than that particle was not noticeable even after possessing a size greater as compared to the physical interpretation of the X-ray CT scanner.

Barcelon et al.[8] studied CT processing of peaches and developed a histogram of the peach image, which demonstrated the rate of the CT value of peach slice under observation, where histogram depicting a fresh peach possessed lower attenuation frequency ranging from −300 to −1000 CT numbers. In contrast, a significant increase in attenuation rate of mentioned CT number range was observed in peaches under 2 weeks of ripening phase. This assured the drier region and existence of voids that were perhaps occurred due to moisture loss toward the peripheral layer of the fruit. Several researchers have further established the correlation between physicochemical contents and CT numbers.

Barcelon et al.[7,8] applied the X-ray CT method to examine the interior variations related to the maturation stages of mangoes. They appraised mangoes on various parameters such as X-ray penetration, moisture measurement, density, pH, and acidity values. The authors stated that the X-ray CT image displayed noticeable characteristics of the interior structural variations among ripened and fresh mangoes.[40]

11.5 MEASUREMENT OF EXTERIOR AND INTERIOR QUALITY PARAMETERS OF FAV DURING STORAGE

Excellence can be determined by the assessment of exterior and interior quality attributes of fresh agricultural commodities. Whereas the nondestructive techniques signify a crucial function to check the specific quality as well as safety of food material optimize postharvest processes by eliminating inter-product variability; the exterior quality parameters of FAV (such as texture, shape, measurements) are very crucial in terms of acceptance by consumers, as well as also indirectly influence the postharvesting processing and storage operation. Evaluation of the interior quality attributes of fresh farming goods focuses on total carotenoids, total phenolic content, pH, reducing sugars, total sugars, gas transmission rates, water repellence, permeability, TSS, moisture, nutritional composition, and other defects/damages.

Hernandez-Sanchez et al.[35] observed that it is essential to develop quality and safe product after processing, as it directly affects the postharvesting processing and the shelf-life of FAV.

Measurement of firmness is one of the most common practices to check the degree of ripeness and maturity, and to know the best time for harvesting, transportation, marketing, and consumption.[15] Riped fruits are delicate and susceptible to microbial decay, physiological deterioration, and mechanical damage; and ultrasonic[13] and electromagnetic[4,49] techniques are commonly applied to evaluate these parameters.

Shumin et al.[72] used X-ray and machine scan to observe the CT value of apples dropped from various heights. They found the destructive apples to possess a decrease in CT estimation with an increase in shelf-life. NIR spectroscopy must utilized effectively to evaluate the soluble solids from numerous fruits, for example, kiwifruit,[55] mandarin,[37] apple,[44] peach,[73] cherry[49] and melon and pineapple.[29] The application of hyperspectral imaging was evaluated in fresh FAV to inspect defects in cherries,[30] bitter pit, and bruises in apples.[60,80] However, MRI has huge potential for online processing operations, such as: sorting, grading or quality assessment of fresh agricultural produce.[67] However, more research is needed to exploit these techniques for the betterment of the postharvest processing and storage systems to improve the quality and safety of agricultural products by reducing postharvesting losses.

11.6 FUTURE PERSPECTIVES AND CHALLENGES

We must analyze the current scenario of the several nondestructive imaging technologies to evaluate food quality and safety aspects. In previous decades, various nondestructive imaging methods were evolved and have become appreciated tools for the investigation of quality and assurance of food. However, these present trends designate significant utilization of nondestructive methods to regulate production-cost and product quality and safety in several aspects. In this chapter, the authors discussed some technical challenges and future scope for these technologies in the food industries.

Primarily, nondestructive techniques are capable of online application in the food sector. However, many of these methods are in their initial stage due to high initial equipment cost. Therefore, numerous studies have been focused to develop economical, low-cost, and quick processing equipment to attain of logical outcomes. The principal focus is particularly on the device systems (i.e., cheaper and fast) and automation of the process for rapid detection of huge amounts of products successfully. Although to meet other essential requirements (such as overcome the food safety challenges, meet the consumer expectations, and improved equipment performance), all these approaches would reduce the complications in these practices. Besides, improvement in algorithms processing speed and improvement in image processing features is necessary to acquire real-time-based results. Several techniques, for example, soft X-ray technique, MRI, and NMR are yet under their experiment phases for fruit analysis, which may be considered on a huge level for quality detection of fruit and vegetable industries in the upcoming future; and therefore further efforts in investigation and growth in these methods are the demand of an era.

11.7 SUMMARY

The optical imaging, spectroscopic methods, X-ray computer tomography, etc., are being used for the quality assessment of fresh agricultural commodities, such as FAVs. These techniques provide real-time, online, in-field information and precise, accurate, inexpensive, rapid, reproducible measuring method to achieve the correct classification of a food material. Nondestructive methods are applied to assess both external and internal quality attributes, such as shape, size, color, texture, degree of ripeness, firmness, maturity, presence of any defect, etc. The detection of interior quality characteristics critically relies on the accuracy of an instrument to imitate the perception of humans.

KEYWORDS

- **color measurement system**
- **food quality**
- **NMR**
- **shelf-life**
- **X-ray tomography**

REFERENCES

1. Abasi, S.; Minaei, S.; Jamshidi, B.; Fathi, D. Dedicated Non-Destructive Devices for Food Quality Measurement: A Review. *Trends Food Sci. Technol.* **2018,** *78*, 197–205.
2. Aguilera, J. M.; Stanley, D. W., Eds. *Microstructural Principles of Food Processing and Engineering*, 2nd ed.; Springer Science and Business Media: Gaithersburg, MD, 1999; p 432.
3. Amodio, M. L.; Chaudhry, M. M. A.; Colelli, G. The Use of Non-Destructive Techniques to Assess the Nutritional Content of Fruits and Vegetables. In: *Fruit and Vegetable Phytochemicals*; Yahia, E. M., Ed.; John Wiley and Sons, Ltd: New York, 2017; pp 763–780.
4. Arendse, E.; Fawole, O. A.; Magwaza, L. S.; Opara, U. L. Non-Destructive Prediction of Internal and External Quality Attributes of Fruit with Thick Rind: A Review. *J. Food Eng.* **2018,** *217*, 11–23.
5. Bai, W.; Yoshimura, N.; Takayanagi, M. Quantitative Analysis of Ingredients of Blueberry Fruits by Near Infrared Spectroscopy. *J. Near Infrared Spectrosc.* **2014,** *22*, 357–365.
6. Ballerini, L.; Hogberg, A.; Borgefors, G.; Bylund, A. C.; Lindgard, A.; Lundstrom, K.; Rakotonirainy, O.; Soussi, B. Segmentation Technique to Determine Fat Content in NMR Images of Beef Meat. *IEEE Trans. Nucl. Sci.* **2002,** *49*, 195–199.
7. Barcelon, E. G.; Tojo, S.; Watanabe, K. Nondestructive Ripening Assessment of Mango Using an X-Ray Computed Tomography. *Agric. Eng. J.* **2000,** *9*, 73–80.
8. Barcelon, E. G.; Tojo, S.; Watanabe, K. X-Ray Computed Tomography for Internal Quality Evaluation of Peaches. *J. Agric. Eng. Res.* **1999,** *73*, 323–330.
9. Bizzani, M.; Flores, D. W. M.; Colnago, L. A.; Ferreira, M. D. Non-Invasive Spectroscopic Methods to Estimate Orange Firmness, Peel Thickness, and Total Pectin Content. *Microchem. J.* **2017,** *133*, 168–174.
10. Brecht, J. K.; Shewfelt, R. L.; Garner, J. C.; Tollner, E. W. Using X-Ray-Computed Tomography to Nondestructively Determine Maturity of Green Tomatoes. *HortScience* **1991,** *26*, 45–47.
11. Butz, P., Hofmann, C.; Tauscher, B. Recent Developments in Noninvasive Techniques for Fresh Fruit and Vegetable Internal Quality Analysis. *J. Food Sci.* **2005,** *70*, R131–R141.
12. Cakmak, H. Assessment of Fresh Fruit and Vegetable Quality with Non-Destructive Methods. In *Food Quality and Shelf-life*; Galanakis, C. M., Ed.; Apple Academic Press: London, 2019; pp 303–331.
13. Cao, S.; Hu, Z.; Pang, B.; Wang, H.; Xie, H.; Wu, F. Effect of Ultrasound Treatment on Fruit Decay and Quality Maintenance in Strawberry After Harvest. *Food Control* **2010,** *21*, 529–532.
14. Casasent, D. A.; Sipe, M. A.; Schatzki, T. F.; Keagy, P. M.; Lee, L. C. Neural Net Classification of X-Ray Pistachio Nut Data. *LWT-Food Sci. Technol.* **1998,** *31*, 122–128.
15. Chauhan, O. P.; Lakshmi, S.; Pandey, A. K.; Ravi, N.; Gopalan, N.; Sharma R. K. Non-Destructive Quality Monitoring of Fresh Fruits and Vegetables. *Defense Life Sci. J.* **2017,** *20*, 103–110.
16. Chen, P.; McCarthy, M. J.; Kauten, R. NMR for Internal Quality Evaluation of Fruits and Vegetables. *Trans. ASABE* **1989,** *32*, 1747–1753.
17. Chen, P.; Sun, Z. A Review of Non-Destructive Methods for Quality Evaluation and Sorting of Agricultural Products. *J. Agric. Eng. Res.* **1991,** *49*, 85–98.

18. Chen, Q.; Zhang, C.; Zhao, J.; Ouyang, Q. Recent Advances in Emerging Imaging Techniques for Non-Destructive Detection of Food Quality and Safety. *TrAC: Trends Analy. Chem.* **2013,** *52*, 261–274.
19. Cho, B. K.; Chayaprasert, W.; Stroshine, R. L. Effects of Internal Browning and Watercore on Low Field (5.4 MHz) Proton Magnetic Resonance Measurements of T2 Values of Whole Apples. *Postharvest Biol. Technol.* **2008,** *47*, 81–89.
20. Clark, C. J.; Hockings, P. D.; Joyce, D. C.; Mazucco, R. A. Application of Magnetic Resonance Imaging to Pre- and Post-Harvest Studies of Fruits and Vegetables. *Postharvest Biol. Technol.* **1997,** *11*, 1–21.
21. Cristiane de Queiroz, O.; Rubemar, S. F.; Estrada, J. Digital Image Processing Applied to Inspection of Internal Disorders in Mangoes CV Tommy Atkins. Unpublished Paper at *International Nuclear Atlantic Conference* (INAC) 2005 Santos, SP, Brazil, August 28 to September 2, **2005,** *2005*; p 8.
22. Danieli, E.; Berdel, K.; Perlo, J.; Michaeli, W.; Masberg, U.; Blumich, B.; Casanova, F. Determining Object Boundaries from MR Images with Sub-Pixel Resolution: Towards In-Line Inspection with a Mobile Tomograph. *J. Magnet. Resonance* **2010,** *207*, 53–58.
23. Danieli, E.; Mauler, J.; Perlo, J.; Blümich, B.; Casanova, F. Mobile Sensor for High Resolution NMR Spectroscopy and Imaging. *J. Magnetic* Resonance **2009,** *198*, 80–87.
24. Defraeye, T.; Lehmann, V.; Gross, D.; Holat, C.; Herremans, E.; Verboven, P.; Verlinden, B. E.; Nicolai, B. M. Application of MRI for Tissue Characterization of Apple. *Postharvest Biol. Technol.* **2013,** *75*, 96–105.
25. Donis-Gonzalez, I. R.; Guyer, D. E.; Pease, A. Application of Response Surface Methodology to Systematically Optimize Image Quality in Computer Tomography: A Case Study Using Fresh Chestnuts. *Comput. Electron. Agric.* **2012,** *87*, 94–107.
26. Donis-Gonzalez, I. R.; Guyer, D. E.; Pease, A.; Fulbright, D. W. Relation of Computerized Tomography Hounsfield Unit Measurements and Internal Components of Fresh Chestnuts. *Postharvest Biol. Technol.* **2012,** *87*, 74–82.
27. El-Mesery, H. S.; Mao, H.; Abomohra, A. E. F. Applications of Non-Destructive Technologies for Agricultural and Food Products Quality Inspection. *Sensors* **2019,** *19*, 846.
28. Erkmen, O.; Bozoglu, T. F. Spoilage of Vegetables and Fruits. In *Food Microbology Principles and Practice*; Erkmen, O., Bozoglu, T. F., Eds.; John Wiley and Sons, Ltd.: West Sussex, U. K., 2016; pp 337–363.
29. Guthrie, J. A.; Walsh, K. B. Non-Invasive Assessment of Pineapple and Mango Fruit Quality Using Near Infra-Red Spectroscopy. *Aust. J. Exp. Agric.* **1997,** *37*, 253–263.
30. Guyer, D.; Yang, X. K. Use of Genetic Artificial Neural Networks and Spectral Imaging for Defect Detection on Cherries. *Comput. Electron. Agric.* **2000,** *29*, 179–194.
31. Hall, L. D.; Evans, S. D.; Nott, K. P. Measurement of Textural Changes of Food by MRI Relaxometry. *Magnet. Resonance Imag.* **1998,** *16*, 485–492.
32. Hansen, J. D.; Schlaman, D. W.; Haff, R. P.; Yee W. L. Potential Postharvest Use of Radiography to Detect Internal Pests in Deciduous Tree Fruits. *J. Entomol. Sci.* **2005,** *40*, 255–262.
33. Hernández-Sánchez, N.; Barreiro, P.; Ruiz-Altisent, M.; Ruiz-Cabello, J.; Fernandez-Valle, M. E. Detection of Freeze Injury in Oranges by Magnetic Resonance Imaging of Moving Samples. *Appl. Magnet Resonance* **2004,** *26*, 431–445.
34. Hernández-Sánchez, N.; Hills, B. P.; Barreiro, P.; Marigheto, N. An NMR Study on Internal Browning in Pears. *Postharvest Biol. Technol.* **2007,** *44*, 260–270.

35. Hernandez-Sanchez, N.; Moreda, G. P.; Herrero-Langreo, A.; Melado-Herreros, A. Assessment of Internal and External Quality of Fruits and Vegetables. In *Imaging Technologies and Data processing for Food Engineers: Food Engineering Series*; Sozer, N., Ed.; Springer: Switzerland, **2016**; pp 269–309.
36. Hills, B. P. Applications of Low-Field NMR to Food Science. In *Annual Reports on NMR Spectroscopy*; Webb, G. A., Ed.; Academic Press: Norwich, **2006**; pp 177–230.
37. Kawano, S.; Fujiwara, T.; Iwamoto, M. Nondestructive Determination of Sugar Content in Satsuma Mandarin Using Near-Infrared (NIR) Transmittance. *J. Japanese Soc. Hort. Sci.* **1993,** *62*, 465–470.
38. Keagy, P. M.; Parvin, B.; Schatzki, T. F. Machine Recognition of Navel Orange Worm Damage in X-Ray Images of Pistachio Nuts. *LWT-Food Sci. Technol.* **1996,** *29*, 140–145.
39. Kirtil, E.; Oztop, M. H. Nuclear Magnetic Resonance Relaxometry and Magnetic Resonance Imaging and Applications in Food Science and Processing. *Food Eng. Rev.* **2016,** *8*, 1–22.
40. Kotwaliwale, N.; Singh, K.; Kalne, A.; Jha, S. N.; Seth, N.; Kar, A. X-Ray Imaging Methods for Internal Quality Evaluation of Agricultural Produce. *J. Food Sci. Technol.* **2014,** *51*, 1–15.
41. Kuperman, V., Ed., *Magnetic Resonance Imaging: Physical Principles and Applications*, 3rd ed.; Academic Press: San Diego, 2000; p 182.
42. Lammertyn, J, Dresselaers, T.; Van-Hecke, P.; Jancsók, P.; Wevers, M.; Nicolai, B. M. Analysis of the Time Course of Core Breakdown in pears by Means of MRI and X-Ray CT. *Postharvest Biol. Technol.* **2003,** *29*, 19–28.
43. Lammertyn, J.; Dresselaers, T.; Van-Hecke, P.; Jancsók, P.; Wevers, M.; Nicolai, B. M. MRI and X-Ray CT Study of Spatial Distribution of Core Breakdown in Pears. *Magnet. Resonance Imag.* **2003,** *21*, 805–815.
44. Lammertyn, J.; Nicolai, B.; Ooms, K.; DeSmedt, V.; De Baerdemaeker, J. Non-Destructive Measurement of Acidity, Soluble Solids, and Firmness of Jonagold Apples Using NIR Spectroscopy. *Trans. ASAE*, **1998,** *41*, 1089–1094.
45. Leitzmaan, C. Food Quality: Definition and Holistic View. In *Safeguarding Food Quality*; Sommer, H., Petersen B., Wittke P. V., Eds.; Springer: Berlin- Heidelberg, 1993; pp 3–15.
46. Létal, J.; Jirák, D.; Suderlová, L.; Hájek, M. MRI Texture Analysis of MR Images of Apples During Ripening and Storage. *LWT-Food Sci. Technol.* **2003,** *36*, 719–727.
47. Lim, K. S.; Barigou, M. X-Ray Micro-Computed Tomography of Cellular Food Products. *Food Res. Int.* **2004,** *37*, 1001–1012.
48. Lorente, D.; Blasco, J.; Serrano, A. J.; Soria-Olivas, E.; Aleixos, N.; Gomez-Sanchis, J. Comparison of ROC Feature Selection Method for Detection of Decay in Citrus Fruit Using Hyperspectral Images. *Food Bioprocess Technol.* **2012,** *6*, 3613–3619.
49. Lu, Y.; Lu, R. Non-Destructive Defect Detection of Apples by Spectroscopic and Imaging Technologies: A Review. *Trans. ASABE* **2017,** *60*, 1765–1790.
50. Mariette, F. *Relaxation RMN: un couplage indispensable pour l'etude des produits alimentaires* (Relaxation RMN: An Essential Coupling for the Study of Food Products). *Comptes Rendus Chimie* (Chem. Rep.) **2004,** *7*, 221–232.
51. Mariette, F.; Collewet, G.; Davenel, A.; Lucas, T.; Musse, M. Quantitative MRI in Food Science and Food Engineering. *eMag Res.* **2012,** *1*, 205–214.
52. Marigheto, N.; Venturi, L.; Hills, B. Two-Dimensional NMR Relaxation Studies of Apple Quality. *Postharvest Biol. Technol.* **2008,** *48*, 331–340.

53. McCarthy, M. J., Ed. *Magnetic Resonance Imaging in Foods*; Springer: New York, 2012; p 110.
54. McCarthy, M. J.; Zion, B.; Chen, P.; Ablett, S.; Darke, A. H.; Lillford, P. J. Diamagnetic Susceptibility Changes in Apple Tissue After Bruising. *J. Sci. Food Agric.* **1995,** *67*, 13–20.
55. McGlone, V. A.; Kawano, S. Firmness, Dry-Matter and Soluble-Solids Assessment of Postharvest Kiwifruit by NIR Spectroscopy. *Postharvest Biol. Technol.* **1998,** *13*, 131–141.
56. Milczarek, R. R.; McCarthy, M. J. Low-Field MR Sensors for Fruit Inspection. In *Magnetic Resonance Microscopy, Spatially Resolved NMR Techniques and Applications*; Codd, S., Seymour, J., Eds.; Wiley-VCH: Weinheim-Germany, 2009; pp 289–299.
57. Naderi-Boldaji, M.; Fazeliyan-Dehkordi, M.; Ahmad, M. S.; Ghasemi-Varnamkhasti, M. Dielectric Power Spectroscopy as a Potential Technique for the Non-Destructive Measurement of Sugar Concentration in Sugarcane. *Biosyst. Eng.* **2015,** *140*, 1–10.
58. Neethirajan, S.; Jayas, D. S.; Karunakaran, C. Dual Energy X-Ray Image Analysis for Classifying Vitreousness in Durum Wheat. *Postharvest Biol. Technol.* **2007,** *45*, 381–384.
59. Nicolaï, B. M.; Defraeye, T.; De-Ketelaere, B.; Herremans, E.; Hertog, M. L.; Saeys, W.; Torricelli, A.; Vandendriessche, T.; Verboven, P. Nondestructive Measurement of Fruit and Vegetable Quality. *Annu. Rev. Food Sci. Technol.* **2014,** *5*, 285–312.
60. Nicolaï, B. M.; Lotze, E.; Peirs, A.; Scheerlinck, N.; Theron, K. I. Non-Destructive Measurement of Bitter Pit in Apple Fruit Using NIR Hyperspectral Imaging. *Postharvest Biol. Technol.* **2006,** *40*, 1–6.
61. Njorje, J. B.; Ninomiya, K.; Kondo, N.; Toita, H. Automated Fruit Grading System Using Image Processing. *2002 Proc. 41st SICE Annu. Conf. Osaka–Japan*, **2002,** *2*, 1346–1351.
62. Nylund, R. E.; Lutz, J. M. Separation of Hollow Heart Potato Tubers by Means of Size Grading, Specific Gravity, and X-Ray Examination. *Am. Potato J.* **1950,** *27*, 214–222.
63. Ogawa, Y.; Morita, K.; Tanaka, S.; Setoguchi, M.; Thai, C. N. Application of X-ray CT for Detection of Physical Foreign Materials in Foods. *Trans. ASAE* **1998,** *41*, 157–162.
64. Okochi, T.; Hoshino, Y.; Fujii, H.; Mitsutani, T. Non-Destructive Measurements for Japanese Oak and Japanese Beech Using Micro-Focus X-Ray Computed Tomography. *Dendro Cronologia* (Tree Chronol.), **2007,** *24*, 155–164.
65. Patel, K. K.; Khan, M. A.; Kar, A. Recent Developments in Applications of MRI Techniques for Foods and Agricultural Produce—An Overview. *J. Food Sci. Technol.* **2015,** *52*, 1–26.
66. Reyes, M. U.; Paull, R. E.; Gautz, L. D.; Armstrong, J. W.; Follett, P. A. Non-Destructive Inspection of Mango (*Mangifera indica* L.) Fruit with Soft X-Ray Imaging. *Acta Hort.* **2000,** *509*, 787–792.
67. Ruiz-Altisent, M.; Ruiz-Garcia, L.; Moreda, G. P.; Renfu, L.; Hernandez-Sanchez, N, Correa, E. C.; Diezma, B.; Ramos, J. G. Sensors for Product Characterization and Quality of Specialty Crops: Review. *Comput. Electron. Agric.* **2010,** *74*, 176–194.
68. Saito, K.; Miki, T.; Hayashi, S.; Kajikawa, H.; Shimada, M.; kawate, Y.; Nishizawa, T.; Ikegaya, D.; Kimura, N.; Takabatake, K.; Sugiura, N.; Suzuki, M. Application of Magnetic Resonance Imaging to Non-Destructive Void Detection in Watermelon. *Cryogenics* **1996,** *36*, 1027–1031.
69. Salvo, L.; Cloetens, P.; Maire, E.; Zabler, S.; Blandin, J. J.; Buffere, J. Y.; Ludwig, W.; Boller, E.; Bellet, D.; Josserond, C. X-Ray Micro-Tomography an Attractive Characterization Technique in Materials Science. *Nucl. Instr. Methods Phys. Res. Sect. B: Beam Interact. Mater. Atoms* **2003,** *200*, 273–286.

70. Shah, S. A. S; Zeb, A.; Qureshi, W. S.; Arslan M.; Malik A. U.; Alasmary, W.; Alanazi, E. Towards Fruit Maturity Estimation Using NIR Spectroscopy. *Infrared Phys. Technol.* **2020,** *111*, 103479.
71. Shen, F.; Zhang, B.; Cao, C.; Jiang, X. On-Line Discrimination of Storage Shelf-Life and Prediction of Post-Harvest Quality for Strawberry Fruit by Visible and Near Infrared Spectroscopy. *J. Food Process Eng.* **2018,** *41*, E-article 12866, pp 7.
72. Shumin, X.; Yong, Y.; Jun, W. Study on CT Value of Damaged Apple. *Trans. Chinese Soc. Agric. Mach.* **2006,** *37*, 83–85.
73. Slaughter, S. Non-Destructive Determination of Internal Quality in Peaches and Nectarines. *Trans. ASAE* **1995,** *38*, 617–623.
74. Smith, N. B.; Webb, A., Eds. *Introduction to Medical Imaging: Physics, Engineering and Clinical Applications*; Cambridge University Press: Cambridge, 2010; p 296.
75. Sonego, L.; Ben-Arie, R.; Raynal, J.; Pech, J. C. Biochemical and Physical Evaluation of Textural Characteristics of Nectarines Exhibiting Woolly Breakdown: NMR Imaging, X-Ray Computed Tomography and Pectin Composition. *Postharvest Biol. Technol.* **1995,** *5*, 187–198.
76. Tollner, E. W., Hung, Y. C., Upchurch, B. L., Prussia, S. E. Relating X-ray absorption to density and water content in apples. *Transactions of the ASAE*, **1992,** *35*, 1921–1928.
77. Tollner, E. W.; Ramseur, E. L.; Murphy, C. Techniques and Approaches for Documenting Plant Root Development with X-Ray Computed Tomography. *Tomography Soil-Water-Root Processes* **1994,** *36*, 115–133.
78. VanAs, H.; van Duynhoven, J. MRI of Plants and Foods. *J. Magnetic Resonance* **2013,** *229*, 25–34.
79. Veena, T.; Chidanand, D. V.; Alagusundaram, K. Soft X-Ray Imaging System for Improving Quality Classification of Fruits and Vegetables: A Review. *Int. J. Agric. Sci. Res.* **2015,** *5*, 25–36.
80. Xing, J.; Saeys, W.; De Baerdemaeker, J. Combination of Chemometric Tools and Image Processing for Bruise Detection on Apples. *Comput. Electron. Agric.* **2007,** *56*, 1–13.
81. Yacob, Y.; Ahmad, H.; Saad, P.; Aliana, R.; Raof, A.; Ismail, S. A Comparison Between X-ray and MRI in Postharvest Non-Destructive Detection Method. Unpublished Paper at *International Conference on Information Technology and Multimedia at UNITEN (ICIMU'05),* Malaysia; **2005**; p 9.
82. Yang, E. C.; Yang, M. M.; Liao, L. H.; Wu, W. Y.; Chen, T. W.; Chen, T. M.; Lin. T.T.; Jiang, J. A. Non-Destructive Quarantine Technique-Potential Application of Using X-Ray Images to Detect Early Infestations Caused by Oriental Fruit Fly (*Bactroceradorsalis*) in Fruit. *Formosan Entomol.* **2006,** *26*, 171–186.

CHAPTER 12

MECHANICAL, OPTICAL, AND ACOUSTIC METHODS: NONINVASIVE QUALITY ASSESSMENT OF FRUITS AND VEGETABLES

TEJASWI BOYAPATI, PAYEL GHOSH, and SANDEEP SINGH RANA

ABSTRACT

Fresh fruits and vegetables (FAVs) provide essential micronutrients in developed and developing countries. Most of the FAVs are seasonal and perishable in nature. A huge loss is being observed for this food sector irrespective of the production. FAVs are mainly deteriorated due to biological and mechanical damages. For consumer safety and quality assurance, several nondestructive techniques have gained popularity to check the quality of the fresh FAV, such as mechanical methods (e-nose, e-tongue), optical methods (image analysis, thermal imaging), and dynamic methods (acoustic).

12.1 INTRODUCTION

These days, the safety and quality of agro-food products are an increasing concern among the consumers.[1] The 90% of total horticultural production accounts for fresh fruits and vegetable (FAV).[2] In order to determine the consistency and sort of agricultural produce and products, researchers have developed several methods. However, for agricultural products, there is no established quality standard and various researchers interpret quality differently.[3] The quality concept represents high quality, a superior

Quality Control in Fruit and Vegetable Processing: Methods and Strategies. Megh R. Goyal, & Faizan Ahmad (Eds.)

degree. The concept of quality can be defined as the acceptance of product characteristics by the user.[4]

Agricultural institutions are looking for advanced methods for detecting defects, freshness, grading based on color, shape, scale, etc., to understand the quality and in turn create a better value-added product from the product for commercializing the fresh produce.[2] The consistency of the commodity depends on chemical qualities, food makeup, external deficiencies, or physical parameters. There is a higher market value for delicious and nutritious FAV.[2]

Products are tested to determine quality on the basis of their physicochemical properties.[2] The physical properties of the product are used to determine the consistency of the product. It is also important that the consistency of FAVs should be measured using nondestructive techniques, and it is dependent on physical properties.[4]

This has contributed to a need for production and research in product quality determination after harvest.[1] A brief definition is referred to as nondestructive monitoring for the qualitative and quantitative steps of agricultural products and refined foods that have been tested without any process injury, including physical, chemical, temperature, and mechanical damage.[1]

Nondestructive measurement devices are commonly used in engineering study, which include the monitoring of raw materials at the time of receipt in the factory, the production of new goods and the shelf-life testing of the product sample, the rapid examination of intermediate or final items in the field of quality management or quality control.[4]

The understanding of postharvest loss through nondestructive assessment is more important in reducing postharvest losses for a food technologist.[1] Advances in science, infrastructure, computer systems, and their deployment are witnessing a decline in postharvest losses and in new approaches for nondestructive identification of goods. The use of these strategies for agriculture and food processing is also on the rise.[4]

Nondestructive approaches should have no physical or chemical effects on the product and the product that has undergone this analysis should satisfy the consumer.[1] Compared to these advanced technologies, the conventional methods for product analysis were too costly, ineffective, complex, time-consuming, and stressful. These are not readily measured by the human eye and cannot be detected with regard to consistency detection factors, such as flavor, chemical aspects, nutrient number, cloth, and physical injuries by the narrow band that raises the electromagnetic spectrum errors.[1] Several nondestructive techniques are indicated in Table 12.1.

TABLE 12.1 Nondestructive Techniques for Determining Consistency Characteristics of Agriculture and Food Products.

Technique	Measurable components	References
CT and X-ray imaging	Maturity, damaged internal tissues, core depth, and volume	[1]
E-nose	Smell or aroma	[1]
E-tongue	Acidity/ sour, sweet, salt, umami, bitterness	[1]
Sound waves	Maturity, damaged internal tissues, core depth, texture, firmness, and volume	[2]
Ultrasound and Sound	Stiffness, viscosity and elasticity, cavity, density, sweetness	[1]
Vibrational excitation	Firmness, viscosity and elasticity, the degree of maturity	[1]

This chapter highlights the basic principle of mechanical (E-nose, E-tongue), optical (image analysis, thermal imaging (TI)), and dynamic methods (acoustic) methods. This chapter also explores the application of these methods to identify the quality of fresh FAV in a nondestructive way.

12.2 QUALITY ANALYSIS OF FAV

Consistency testing of fruit and vegetables using an image processing technique requires five steps, namely, imaging preprocessing, image segmentation, extraction, and attribute classification.[5]

There are plenty of important vitamins and vegetables in FAV. The economic value of foods extracted from plants depends on their value. The quality covers a variety of aspects, including exterior appearance, nutritional aspects, health-related compound presence, safety, and protection.[6]

Quality-related characteristics must be measured to evaluate and validate accuracy. Sensory characteristics, nutrient values, chemical elements, mechanical properties, physical properties, and defects include product consistency. Instrumental tests are often preferred to sensory measures in analysis and market scenarios because they minimize decision disparities between persons and may give researchers, industry, and consumers a common language. Basically, electromagnetic properties (often optical) refer to shape, texture mechanical, and flavor properties (taste and aroma).

Instruments can bring human decisions in direct contact with the way in which people evaluate and mathematically integrate the substance to understand the fundamental characteristics. Quality may only be assessed by humans but measures that quantify quality are important for research and evaluation.[7]

12.3 MECHANICAL NONDESTRUCTIVE METHODS FOR ANALYSIS

12.3.1 E-NOSE

The E-nose method is used to calculate characteristics, such as sugar, acidity, etc.[1] It is an electronic sensing system to detect flavors and odors. This was developed to detect and identify odor, vapors, and gases.[4] This helps to activate the human olfactory system to work and consists of a chemical sensor and an olfactory system.

12.3.1.1 WORKING PRINCIPLE

An electronic nose system includes a detector, a database management unit, a digital pattern recognition, and an index library software.[2]

The distribution of samples makes it possible to generate the headspace of a sample (volatile compounds), which is the fraction to be analyzed. This headspace is then pumped into the nose detection system by the computer. To guarantee constant operating conditions.[8] the sample delivery system is important. The "reactive" component of the instrument is the detection device, which consists of a sensor collection. The sensors respond when are in contact with volatile compounds, which implies that they experience a change in electrical properties.[8] In most E-noses, each sensor is sensitive to various chemical substances. To assess the maturity of the fruit, the storage shelf-life, and the freshness prediction, an electronic nose has been used. By inducing chemical or physical changes, the sensor detects the components, which in turn senses the generated signal.

E-sensor arrays are known for use in gas sensing. Organic polymers, metal oxides, quartz crystal microbalance, and gas chromatography (GC) are the sensors in use. GC is often used with mass spectroscopy to detect different compounds.[2] along with MS.

12.3.1.2 APPLICATIONS OF E-NOSE METHOD

Different industries and R&D laboratories use E-Nose (both in quality control laboratories and in In-process and in production departments). This involves verifying raw materials consistency, medium- and final-goods, batch-wise examination, storage control, contamination detection, spoilage, adulteration, meat and meat products freshness, and so on. The identification of harmful microbes in foods and many more could also be a potential application (Table 12.2).

TABLE 12.2 E-Nose Applications in Different Agro-Food Produce.

Fruit/vegetable	Details	References
Citrus fruits	• Aromatic volatile substances are identified	[9]
Melon	• Aromatic compounds, volatile esters, etc., are identified	[9]
Peaches	• On the basis of TSS and firmness, freshness is predicted	[9]
Pineapple	• Aromatic flavor compounds	[11]
	• Shelf-life	[12]
Tomato	• Analysis of shelf-life	[9]
	• Effects of harvesting and postharvest handling procedures	[10]

12.3.1.3 FUTURE SCOPE

E nose method has been used to predict shelf-life, volatile compounds of many FAV. There is also a future scope for work on the detection of pesticide residues, the postharvest detection of spoilage, and the detection of other gasses released during the storage of the product.

12.3.2 E- TONGUE

Taste is a vital organoleptic property referring to oral administration acceptability.[13] The sense of taste comprises of five fundamental tastes: sour, salt content/salt, umami, bitter, and sweet.[14] Human panel members are the key tool for assessing the flavor profile of food items and new formulations. The use of sensory people on the panel is difficult and controversial due to nutrient toxicity or contaminants. E-tongue is a nondestructive technique used to measure characteristics, such as sugar and acidity. A taste-sensing system is an analytical sensory array capable of detecting specific substances using different artificial membranes and electrochemical techniques.[13] This has been developed for automatic taste and flavor detection.

According to IUPAC, "An electronic tongue is a multisensory system consisting of several low-selective sensors and cross-sensitivity in solution to different species and an appropriate pattern recognition and/or multivariate calibration method for data processing".[15] To enable olfactory systems to work, e-tongue consists of a sensor, a data processing device, a program for the detection of digital patterns and a repository.

12.3.2.1 WORKING PRINCIPLE

The core elements inside the electronic taste sensing device connected to the arm, the sample board, the amplification unit, and the data analysis unit are a variety of different sensor types.[13] Many of the chemical sensor options reported in e-tongue work include voltammetry (rare metal electrodes[16]), composite sensors,[17] and polymer conductors,[18] potentiometric sensors[19] (lipid membrane electrodes[14] chalcogenide glass electrodes[15] and PVC plasticized membrane electrodes[20]). Among all the above is the easiest and most natural calculation. Due to their availability, Chalcogenide glass sensors were selected.

12.3.2.2 APPLICATIONS

The aim of E-tongue is to be used in R&D to develop new products and detect wine spoilage by odors, detect fraud, and adulteration,[21] for the analysis of flavor, aging in beverages or wine,[13] to distinguish between multiple levels of hardness in mineral water,[22] identifying the spoilage of milk,[23] to distinguish coffee brands on the basis of their origin,[24] and quantify the taste and organic substances in food materials[13] and much more. E-tongue application in different agro-food produce is mentioned in Table 12.3.

12.3.2.3 FUTURE SCOPE

The E-tongue can be used for tasting other agricultural products and processed foods. This can be applied in the wine industry, for the detection of traces of organic substances in foods.[13] This reduces the risk of exposure of the human panel to contaminants and toxins if present in the product being tastes tested.

TABLE 12.3 E-Tongue Applications in Different Agro-Food Produce.

Fruit/vegetable	Observations	References
Black tea	The acceptance was assessed	[14]
Grapes	The degree of fermentation in barrels	[21]
Honeydew	The experiment gave a conclusion that on ripening the sourness is suppressed. This is sweeter	[19]
Pineapple juice	The experiment gave a conclusion that on ripening the sourness is suppressed	[19]
Wine	The development of off-flavors and odors during aging	[21]

12.4 OPTICAL NONDESTRUCTIVE METHODS FOR ANALYSIS

An electro-optic technique that uses optical and electronic properties is one of the effective and productive methods for estimating the quality and handling of agro-food commodities.[3] The fruit has different absorption qualities and reflective properties under light with different rays because of the variations in internal components.[4] The fruit's optical properties are dependent on reflection, transmission, absorption, fluorescence, or light scattering.[2] By integrating these properties with the optical detection devices, nondestructive measurement of fruit content can be achieved.[4]

Image analysis is a fairly large spectrum of nondestructive quality analysis. Traditional computer vision has been used to provide limited information. However, there are currently several imaging systems that can be used for detailed quality analysis. Various studies describe the amount of contusions and their effect by many criteria, such as postharvesting and pre-have variables including variety, state of maturity and temperature, and atmospheric storage conditions.[25]

There are currently a number of imaging techniques, including biospeckle tomography, fluorescence imaging, structural illumination reflectance imaging, hyperspectral/multispectral imaging (H/MSI), X-ray imaging (XRI), magnetic resonance imaging, and TI.[26] Optical imaging techniques application in different agro-food produce is given in Table 12.4.

TABLE 12.4 Application of Optical Imaging Techniques in Different Agro-Food Produce.

Fruit/vegetable	Technology used	References
Apples and pears	Temporal history of speckle patterns	[28]
Papaya	Light intensity	[27]
Potatoes	Bio speckle activity	[29]

Fluorescence imaging is a particular imaging technique that works on the principle when any organic body emits particular fluorescence excited by electromagnetic radiation.[30] Brightness of the images changes due to the change in chlorophyll content. The wavelength varies from 685 to 730 nm.[31]

12.4.1 X-RAY TOMOGRAPHY

The X-ray imagery is based on the rate difference of X-ray penetration of multiple tissues.[32] We may obtain a grayscale strength picture due to the penetration capability difference.[32] Using several images taken from various angles, a three-dimensional representation of the object can be obtained through reconstruction. By using the difference in the penetration of X-rays into normal product tissues, known as X-ray computed tomography, swelling and physical damage to the product can be detected.[33] The application of XRI in different agro-food produce is given in Table 12.5.

TABLE 12.5 Application of X-Ray Imaging in Different Agro-Food Produce.

Fruit/vegetable	Remarks	References
Apple	Tissues inside the apple were analyzed for bruise detection	[37]
Eggplant	Browning or chemical reactions and tissue distribution/ porosity is studied	[32]
Pear	Browning or chemical reactions and cavities in the fruit	[32]
Pomegranate	Physical and chemical properties of fruits are evaluated	[36]
Turnip	The evenness of the tissues is studied	[32]

By measuring the bruise volume based on the difference in the penetration of X-rays,[34] CT was used to determine the consistency of FAV. The drawbacks of using this X-ray image are that the subject must be rotated to achieve the complete image of the product and the imaging must be performed from all sides.[35]

12.4.2 HYPERSPECTRAL IMAGING/ MULTISPECTRAL IMAGING

This technique combines both imaging and spectroscopy for noninvasive quality analysis of horticultural produce.[38] Hyperspectral imaging (HSI) is being used as a reliable source for analyzing both the internal and external characteristics of the produce.[39] This system for imaging is based on the working principle of spectroscopy.[40]

The wavelength typically ranges from 400 to 1000 nm or wider in.[40] However, MSI only includes fixed wavelengths.[41] Then photographs of this area can be used to process the inspection with main component analysis.[41] Key components of the hyperspectral imagery system are presented in Figure 12.1.

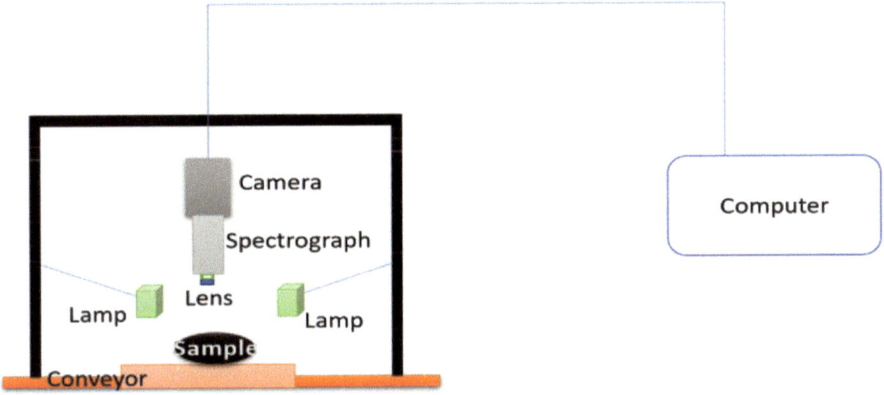

FIGURE 12.1 Key components of the hyperspectral imagery system.

This imaging can be used in apples, pears, and potatoes not only to detect wounds but also to inspect insect infestation, fungal infestation, surface area residues of pesticide. This imaging system also detects these infections and chemical properties, such as TSS, titratable acidity, etc.[42]

In HSI, detailed information on the classification and segmentation spectral space models is given, including precise and simultaneous study of various chemicals and potential for diseases and defects in farm produce.[43] The application of HSI/multispectral imaging in different agro-food produce is given in Table 12.6.

12.4.3 THERMAL IMAGING

Every material emits the infrared radiation that is measured in the TI, the infrared for a particular subject of interest.[45] The physical and chemical properties are particular to different materials and have differences in thermal diffusivities based on which the thermo-grams are generated.[49] Unlike other spectral imaging techniques, this TI does not require any external source of light, but, it needs an environment to support prolonged heating or cooling of the produce.[50] Primary elements of the TI system are given in Figure 12.2.

TABLE 12.6 Application of Hyperspectral Imaging/Multispectral Imaging in Different Agro-Food Produce.

Fruit/vegetable	Remarks	References
Apple	Bruise detection, soil contamination, and fungal infection	[44]
Kiwi	Bruise and insect infestation	[46]
Mangoes	Firmness	[47]
Mushrooms	Damage detection	[46]
Spinach	Bacterial infestation	[45]
Tomatoes	Sweetness evaluation	[48]

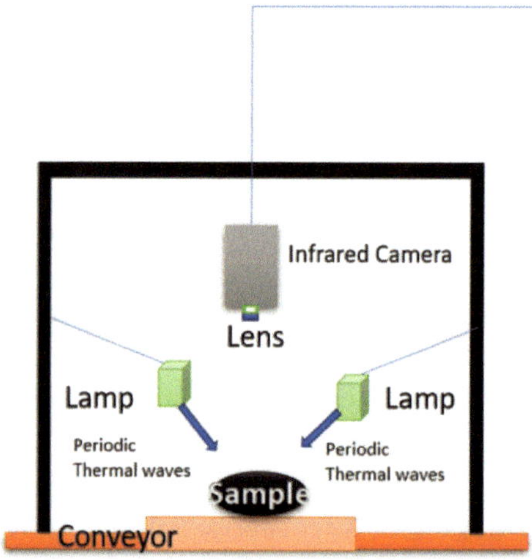

FIGURE 12.2 Primary components of the thermal imaging system.

TI is of two types: active thermography and passive thermography.[51] The difference in the thermal diffusivity is used to differentiate the bruised tissue from the healthy tissue of the produce.[37] The major advantage of this TI is that even the minute defect can be detected and can be used for dark fruits.[52] Furthermore, applications of TI in different agro-food produce are given in Table 12.7.

TABLE 12.7 Application of Thermal Imaging in Different Agro-Food Produce.

Fruit/vegetable	Remarks	References
Apple	Temperature of the surface and texture of the tissue	[53]
Guava	Assessment of texture, color, and TSS	[53]
Mango	Evaluated maturity index	[53]
Palm	The maturity and ripening are evaluated	[53]

12.5 DYNAMIC NONINVASIVE METHODS FOR ANALYSIS OF FAV

12.5.1 ACOUSTIC ANALYSIS

Acoustic analysis is an intriguing method for determining the consistency of the food and fruit. It is a quick, effective, economical, and nondestructive.[54] The device for assessing food consistency and structure diagram is given in Figure 12.3.

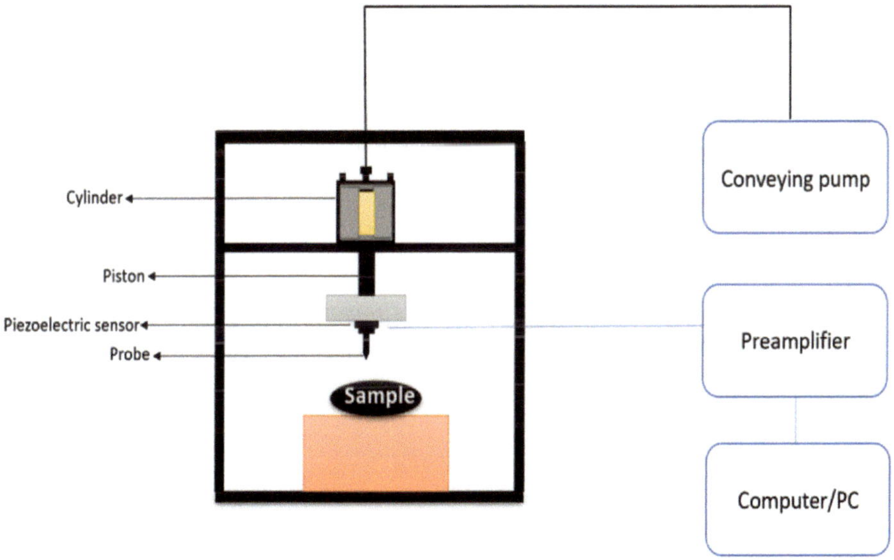

FIGURE 12.3 Device for assessing food consistency and structure diagram.

The implementation of acoustic analyses is simply a dispersion and projection of sound waves. Acoustic signals are sent from the transducer to the target and then dispersed, scattered, or mirrored from the target.[54] Sound measurement by food can be calculated destructively and without destruction (passive and active method). Sound generation, reflection, and other phase oscillation techniques are classified as acoustic approaches.[55] This is a new approach based on the mechanical and acoustic reactions of its pulp, and a structure analyzer coupled with an ultrasound instrument is used.[56] During the postharvest season, parts of the process shifts were evaluated for the elasticity index and structure indices of two mulberry cultivars.[57]

Applications of this method include fruit ripening estimation, processing, classification, and separation of agricultural products, where the physical properties of the crop are calculated.[54,58] Applications of acoustic method in different agro-food produce are given in Table 12.8.

TABLE 12.8 Application of Acoustic in Different Agro-Food Produce.

Fruit	Remarks	References
Apple	An acoustically evaluated disintegration index for tissue characterization of the true level of damage	[59]
Grapes	The textures of the flesh are classified as crisp, noncrystal, or medium depending on the energy strength from 10 Hz to 32 kHz	[59]
Mango	The best method to evaluate variations during mango ripening is its durability with the penetrometer	[61]
Pea	Establish a nondestructive strength sensing methodology for nonspherical fruit based on acoustic response volume	[62]
Potato	A hollow heart in potato has been done to identify (Spunta variety)	[60]

12.6 APPLICATIONS OF NONINVASIVE METHODS IN FOOD PROCESSING

a. **In the field of reception**
   - Management of raw materials.
   - Eliminates loss of product.
   - Increases the consistency of the commodity viewed by the customer.

b. **Regulation of the method**
   - On-line and off-line measurements are possible.
   - Standard instrument configuration shall be used in the off-line framework.

c. **Analysis of intermediate and final products**
   - Screening a large range of samples.
d. **Material production and testing of storage**
   - Over time, follow the properties of a single product for shelf-life extension.
e. **Research studies**
   - For making any changes in the process of quality control or quality analysis.

12.7 BENEFITS OF NONINVASIVE ANALYSIS

Sensitivity and selectivity have been increased because human error can be reduced to a minimum.[63] Low concentrations of toxic substances can be easily detected if they are present in the product. As the aromas of fruits reflect the nature of the fragrance of the fruit, the quality of the fruit can be detected without damaging the product.[11] The E-nose can be used in the manufacturing industry and by suppliers for quality management.[12]

This method is able to carry out multicomponent quantitative analyses of the formulated products and can also discriminated between near-liquid forms, that is, integral qualitative analysis.[64] There is also an important application for electronic nose systems to monitor polluted water and air pollution early in urban and rural areas that can avoid detrimental environmental impacts, such as smog formation or elimination.[20] These nondestructive approaches are potential solutions to time-consuming and costly computational procedures that require precise and rapid research and it does not require skilled staff.[20] It is less time-consuming and the requirement of the sample is small.

12.8 FUTURE SCOPE OF NONINVASIVE ANALYSIS

The goal of this chapter was to study the nondestructive approaches for the scientific detection of the quality of agricultural products and foods. Compared to other instrumental research and chemical analytical methods, the nondestructive detection of food content offers a distinctive advantage. Conventional chemical analysis techniques have several drawbacks, such as lengthy, complex, and high cost. The conclusions of this study indicate that the use of nondestructive approaches is of greater interest to the food industry.

In the future, the E-tongue method can be used for tasting other agricultural products and processed foods. This can be applied in the wine industry for the detection of traces of organic substances in foods.[13] This reduces the risk of exposure of the human panel to contaminants and toxins, if present in the product being tested.

Through designing more effective and reliable systems (such as vision techniques for machines, the future uses nondestructive technology for food and farm quality assessment) would improve the estimation of the inner quality of fruits. Although the expense of purchasing these instruments is currently high, yet it is possible to buy through research funds by numerous laboratories and gather them as a single research point for multiple researchers in a particular area. It is often done by actively seeking means of designing small systems that were equally effective but cost-efficient; therefore, conveniently available. In comparison to other contemporary approaches, the acoustic approach offers knowledge on the essence of the mechanisms of destruction and helps the onset of these phenomena to be observed before the full material stress occurs. The advantages of nondestructive acoustic systems include fast analysis in the food and agricultural sector. Researchers from different disciplines must take part in this since the scientific problems vary from science to information technology, food, and farming science. Through these joint measures, devices can be made available that are better tailored to the identification of food and agricultural items. Most manufacturers are currently producing machines for various research areas, but essentially with the same system setup.

12.9 SUMMARY

This chapter focuses on machinery and the identification of images, the technological future of the food and agro-industry. Scale, color, form, texture, and defects are the most important consistency attributes of an agricultural product. Image processing solutions for authentic, fair, and nondestructive tests are used to supplement manual food assessments. The quality inspection focused on machines consists of four main steps: detection, classification, isolation of components, and classification. It also lowers the human panel's likelihood of exposure to contaminants and toxins.

The future use of nondestructive technology in food and farm quality assessment would improve the evaluation of the internal quality of the fruit by designing systems that are more efficient and reliable, such as vision techniques for machines. The cooperative efforts should bring in place resources that are specifically intended to recognize food and agricultural goods.

KEYWORDS

- **acoustic**
- **fruits and vegetables**
- **image analysis**
- **nondestructive techniques**
- **quality analysis**
- **X-ray**

REFERENCES

1. Abbott, J. A. Quality Measurement of Fruits and Vegetables. *Postharvest Biol. Technol.* **1999**, *15*, 207–25.
2. Abdul Aziz, A. H.; Ali, A. Y.; Farook, R.; Adom, A. H.; Ahmad, M. N.; Mahat, N. I. Simple Implementation of an Electronic Tongue for Taste Assessments of Food and Beverage Products. *Sens. Transducers* **2011**, *132*, 136–150.
3. Aboonajmi M; Faridi H. Non-Destructive Quality Assessment of Agro-Food Products. Unpublished Paper Presentation at 3rd Iranian International NDT Conference, Olympic Hotel, Tehran, Iran; 21–22 February **2016**; pages 9.
4. Aboonajmi, M.; Jahangiri, M.; Hassan-Beygi, S. R. Review on Application of Acoustic Analysis in Quality Evaluation of Agro-Food Products. *J. Food Process Preserv.* **2015**, *39*, 3175–3188.
5. Adamiak, A.; Zdunek, A.; Kurenda, A.; Rutkowski, K. Application of the Biospeckle Method for Monitoring Bull's Eye Rot Development and Quality Changes of Apples Subjected to Various Storage Methods-Preliminary Studies. *Sensors* **2012**, *12*, 3215–3227.
6. Arendse, E.; Fawole, O. A.; Magwaza, L. S.; Opara, U. L. Non-Destructive Prediction of Internal and External Quality Attributes of Fruit with Thick Rind: A review. *J Food Eng. Elsevier B.V.* **2018**, *217*, 11–23.
7. Baietto, M.; Wilson, A. D. Electronic-Nose Applications for fruit Identification, Ripeness and Quality Grading. *Sensors (Switzerland)* **2015**, *15*(1), 899–931.
8. Bhargava, A.; Bansal, A. FAV Quality Evaluation Using Computer Vision: A review. *J King Saud Univ-Comput Inf Science*; **2018**; online;https://doi.org/10.1016/j.jksuci.2018.06.002 (accessed Aug 2, 2021).
9. Chauhan, O. P.; Lakshmi, S.; Pandey, A. K.; Ravi, N.; Gopalan, N.; Sharma, R. K. Non-Destructive Quality Monitoring of Fresh FAV. *Def. Life Sci. J.* **2017**, *2*, 103–110.

10. Che, W.; Sun, L.; Zhang, Q.; Tan, W.; Ye, D.; Zhang, D. Pixel Based Bruise Region Extraction of Apple Using Vis-NIR Hyperspectral imaging. *Comput Electron Agric.* **2018**, *146*, 12–21. https://doi.org/10.1016/j.compag.2018.01.013 (accessed Aug 2, 2021).
11. Chen, P.; Sun, Z. A Review of Non-Destructive Methods for Quality Evaluation and Sorting of Agricultural Products. *J. Agric. Eng. Res.* **1991**, *49*, 85–98.
12. Chen, Q.; Zhang, C.; Zhao, J.; Ouyang, Q. Recent Advances in Emerging Imaging Techniques for Non-Destructive Detection of Food Quality and Safety. *TrAC—Trends Anal. Chem.* **2013**, *52*, 261–274;http://dx.doi.org/10.1016/j.trac.2013.09.007 (accessed Aug 2, 2021).
13. Chiu, Y. C.; Chou, X. L.; Grift, T. E.; Chen, M. T. Automated Detection of Mechanically Induced Bruise Areas in Golden Delicious Apples Using Fluorescence Imagery. *Trans. ASABE* **2015**, *58*, 215–225.
14. Costa, F.; Cappellin, L.; Longhi, S.; Guerra, W.; Magnago, P.; Porro, D. Assessment of Apple Fruit Texture by a Combined Acoustic-Mechanical Profiling Strategy. *Postharvest Biol. Technol.* **2011**, *61*, 21–28.
15. Darras, B. T.; Ryan, M. M.; De Vivo, D. C. *Food Texture and Viscosity: Concept and Measurement-Preface*. New York: Elsevier Science & Technology Books;**2015**; pages 421.
16. Dian, R.; Li, S. Fang, L.; Wei, Q. Multispectral and Hyperspectral Image Fusion with Spatial-Spectral Sparse Representation. *Inf. Fusion* **2019**, *49*, 262–270; online;https://doi.org/10.1016/j.inffus.2018.11.012 (accessed Aug 2, 2021).
17. Diels, E.; van Dael, M.; Keresztes, J.; Vanmaercke, S.; Verboven, P.; Nicolai, B. Assessment Of Bruise Volumes in Apples Using X-Ray Computed Tomography. *Postharvest BiolTechnol.* **2017,** *128*, 24–32; online;http://dx.doi.org/10.1016/j.postharvbio.2017.01.013 (accessed Aug 2, 2021).
18. Doosti-Irani, O.; Golzarian, M. R.; Aghkhani, M. H.; Sadrnia, H.; Doosti-Irani, M. Development of Multiple Regression Model to Estimate the Apple's Bruise Depth Using Thermal Maps. *Postharvest Biol. Technol.* **2016**, *116*, 75–79; online;http://dx.doi.org/10.1016/j.postharvbio.2015.12.024 (accessed Aug 2, 2021).
19. Du, Z.; Zeng, X.; Li, X.; Ding, X.; Cao, J.; Jiang, W. Recent Advances in Imaging Techniques for Bruise Detection in FAV. *Trends Food Sci. Technol.* **2020**, *99*, 133–141. https://doi.org/10.1016/j.tifs.2020.02.024 (accessed Aug 2, 2021).
20. El Hadi, M. A. M.; Zhang, F. J.; Wu, F. F.; Zhou, C. H.; Tao, J. Advances in Fruit Aroma Volatile Research. *Molecules* **2013**, *18*, 8200–8229.
21. El-Mesery, H. S.; Mao, H.; Abomohra, A. E. F. Applications of Non-Destructive Technologies for Agricultural and Food Products Quality Inspection. *Sensors (Switzerland)*, **2019**, *19*, 1–23.
22. Elbatawi, I. E. An Acoustic Impact Method to Detect Hollow Heart of Potato Tubers. *Biosyst. Eng.* **2008**, *100*, 206–213.
23. Fukunaga, T.; Toko, K.; Mori, S.; Nakabayashi, Y.; Kanda, M. Quantification of Taste of Coffee Using Sensor with Global Selectivity. *Sens. Mater.* **1996**, *8*, 47–56.
24. Gao, H.; Zhu, F.; Cai, J. Review of Non-Destructive Detection for Fruit Quality. *IFIP Adv. Inf. Commun. Technol.* **2010**, *317*, 133–140.
25. Gardner, J. W. Review of Conventional Electronic Noses and Their Possible Application to the Detection of Explosives. *Electron. Noses Sens. Detect. Explos.* **2004**, *13*(8), 1–28.
26. Gowen, A. A.; Tiwari, B. K.; Cullen, P. J.; McDonnell, K.; O'Donnell, C. P. Applications of Thermal Imaging in Food Quality and Safety Assessment. *Trends Food Sci. Technol.* **2010**, *21*, 190–200. http://dx.doi.org/10.1016/j.tifs.2009.12.002 (accessed Aug 2, 2021).

27. Grimi, N.; Mamouni, F.; Lebovka, N.; Vorobiev, E.; Vaxelaire, J. Acoustic Impulse Response in Apple Tissues Treated by Pulsed Electric Field. *Biosyst. Eng.* **2010**, *105*, 266–272.
28. Haff, R. P.; Toyofuku, N. X-Ray Detection of Defects and Contaminants in the Food Industry. *Sens. Instrum. Food Qual. Saf.* **2008,** *2*, 262–73.
29. Hussain, A.; Pu, H.; Sun, D. W. Innovative Non-Destructive Imaging Techniques for Ripening and Maturity of Fruits: Review of Recent Applications. *Trends Food Sci Technol.* **2018**, *72*, 144–152. https://doi.org/10.1016/j.tifs.2017.12.010 (accessed Aug 2, 2021).
30. Hussein, Z. Bruise Damage Susceptibility of Pomegranates. *Int. J. Fruit Sci.* **2020**, *20* (Suppl. 2), online. https://doi.org/10.1080/15538362.2019.1709602 (accessed Feb 20, 2021).
31. Latha, R. S.; Lakshmi, P. K. Electronic Tongue: An Analytical Gustatory Tool. *J Adv. Pharm. Technol. Res.* **2012**, *3*, 3–8.
32. Lawo, D.; Neifer, T.; Esau, M.; Vonholdt, S.; Stevens, G. From Farms to Fridges: A Consumer-Oriented Design Approach to Sustainable Food Traceability. *Sustain Prod Consum.* **2021,** *27*, 282–297. https://doi.org/10.1016/j.spc.2020.11.007 (accessed Aug 2, 2021).
33. Li, J.; Chen, L.; Huang, W. Detection of Early Bruises on Peaches (*Amygdalus persica*) Using Hyperspectral Imaging Coupled with Improved Watershed Segmentation Algorithm. *Postharvest Biol. Technol.* **2018,** *135*, 104–113. http://dx.doi.org/10.1016/j.postharvbio.2017.09.007 (accessed Aug 2, 2021).
34. Li, J.; Luo, W.; Wang, Z.; Fan, S. Early Detection of Decay on Apples Using Hyperspectral Reflectance Imaging Combining Both Principal Component Analysis and Improved Watershed Segmentation Method. *Postharvest Biol. Technol.* **2019**, *149*, 235–246. https://doi.org/10.1016/j.postharvbio.2018.12.007 (accessed Aug 2, 2021).
35. Lu, Y.; Lu, R. Development of a Multispectral Structured Illumination Reflectance Imaging System and its Application to Bruise Detection of Apples. *Trans ASABE* **2017,** *60*, 1379–1389.
36. Lü, Q.; Tang, M. J.; Cai, J. R.; Zhao, J. W.; Vittayapadung, S. Vis/NIR Hyperspectral Imaging for Detection of Hidden Bruises on Kiwifruits. *Czech J. Food Sci.* **2011,** *29*, 595–602.
37. Mahesh S.; Jayas D.S.; Paliwal J.; White N.D.G. Hyperspectral Imaging to Classify and Monitor Quality of Agricultural Materials. *J. Stored Prod. Res.* **2015**, *61*, 17–26. http://dx.doi.org/10.1016/j.jspr.2015.01.006 (accessed Aug 2, 2021).
38. Maldague, X.; Galmiche, F.; Ziadi, A. Advances in Pulsed Phase Thermography. *Infrared Phys. Technol.* **2002**, *43*, 175–181.
39. Maul, F.; Sargent, S.; Huber, D.; Balaban, M.; Luzuriaga, D.; Baldwin, E. Non-Destructive Quality Screening of Tomato Fruit Using E-Nose Technology. *Postharvest Biol. Technol.* **1997**, *110*, 188–194.
40. Nugraha, B.; Verboven, P.; Janssen, S.; Wang, Z.; Nicolaï, B. M. Non-Destructive Porosity Mapping of Fruit and Vegetables Using X-Ray CT. *Postharvest Biol. Technol.* **2019**, *150*, 80–88. https://doi.org/10.1016/j.postharvbio.2018.12.016 (accessed Aug 2, 2021).
41. Padda, M. S.; do Amarante, C. V. T.; Garcia, R. M.; Slaughter, D. C.; Mitcham, E. J. Methods to Analyze Physicochemical Changes During Mango Ripening: A Multivariate Approach. *Postharvest Biol. Technol.* **2011**, *62*, 267–274. http://dx.doi.org/10.1016/j.postharvbio.2011.06.002 (accessed Aug 2, 2021).

42. Passoni, I.; Dai Pra, A.; Rabal, H.; Trivi, M.; Arizaga, R. Dynamic Speckle Processing Using Wavelets-Based Entropy. *Opt. Commun.* **2005,** *246*, 219–228.
43. Pioggia, G.; Di Francesco, F.; Marchetti, A.; Ferro, M.; Leardi, R.; Ahluwalia, A. Composite Sensor Array Impedentiometric Electronic Tongue. Part II. Discrimination of Basic Tastes. *Biosens. Bioelectron.* **2007,** *22*, 2624–2628.
44. Podrazka, M.; Báczyńska, E.; Kundys, M.; Jeleń, P. S.; Nery, E. W. Electronic Tongue: A Tool for All Tastes. *Biosensors* **2017**, *8*, 1–24.
45. Riul, A.; Dos Santos, D. S.; Wohnrath, K.; Di Tommazo, R.; Carvalho, A. C. P.L.F.; Fonseca, F. J. Artificial Taste Sensor: Efficient Combination of Sensors Made from Langmuir-Blodgett films of Conducting Polymers and a Ruthenium Complex and Self-Assembled Films of an Azobenzene-Containing Polymer. *Langmuir* **2002**, *18*, 239–245.
46. Rodríguez-Méndez, M. L.; De Saja, J. A.; González-Antón, R.; García-Hernández, C.; Medina-Plaza C.; García-Cabezón, C. Electronic Nose and E-Tongue in Wine Industry. *Front. Bioeng. Biotechnol.* **2016**, *4*, 1–12.
47. Röck, F.; Barsan, N.; Weimar, U. Electronic Nose: Current Status and Future Trends. *Chem. Rev.* **2008**, *108*, 705–725.
48. Sanz, T.; Primo-Martín, C.; van Vliet, T. Characterization of Crispness of French Fries by Fracture and Acoustic Measurements, Effect of Pre-Frying and Final Frying Times. *Food Res. Int.* **2007,** *40*, 63–70.
49. Scharwies, J. D.; Grimm, E.; Knoche, M. Russeting and Relative Growth Rate Are Positively Related in Pears. *Hort. Sci.* **2014**, *49*, 746–749.
50. Shah, R.; Gujarathi, P.; Unadkat, P.; Mulik, R.; Bandal, A. N. Quality Analysis of FAV Using Machine Learning Techniques. *Int. J. Res. Eng. Sci. Manage. (IJRESM)* **2018**, *1* (4), 91–95.
51. Sim, M. Y. M.; Shya, T. J.; Ahmad, M. N.; Shakaff, A. Y. M.; Othman, A. R.; Hitam, M. S. Monitoring of Milk Quality with Disposable Taste Sensor. *Sensors* **2003**, *3*, 340–349.
52. Siripatrawan, U.; Makino, Y.; Kawagoe, Y.; Oshita, S. Rapid Detection of *Escherichia Coli* Contamination in Packaged Fresh Spinach Using Hyperspectral Imaging. *Talanta* **2011,** *85*, 276–281. http://dx.doi.org/10.1016/j.talanta.2011.03.061 (accessed Aug 2, 2021).
53. Payel, G.; Sunil, E. C. K. Quality Analysis of Pomegranate by Non-Destructive Soft X-Ray Method. *J. Food Process Technol.* **2014,** *5*(1), 341–346.
54. Tahara, Y.; Toko, K. Electronic Tongues-a Review. *IEEE Sens. J.* **2013**, *13*, 3001–3011.
55. Taniwaki, M.; Hanada, T.; Sakurai, N. Postharvest Quality Evaluation of Fuyu And Taishuu Persimmons Using a Nondestructive Vibrational Method and an Acoustic Vibration Technique. *Postharvest Biol. Technol.* **2009**, *51*, 80–85.
56. Toko, K. Taste Sensor. *Meas. Sci. Technol.* **1998**, *9*, 1919–1936.
57. Torri, L.; Sinelli, N.; Limbo, S. Shelf-life Evaluation of Fresh-Cut Pineapple by Using an Electronic Nose. *Postharvest Biol. Technol.* **2010**, *56*, 239–245. http://dx.doi.org/10.1016/j.postharvbio.2010.01.012 (accessed Aug 2, 2021).
58. Varith, J.; Hyde, G. M.; Baritelle, A. L.; Fellman, J. K.; Sattabongkot, T. Non-Contact Bruise Detection in Apples by Thermal Imaging. *Innov. Food Sci. Emerg. Technol.* **2003,** *4*, 211–218.
59. Varith, J.; Noochuay, C.; Netsawang, P.; Hirunstitporn, B.; Janin, S.; Krairiksh, M. Design of Multimode-Circular Microwave Cavity for Agri-Food Processing. In *Asia-Pacific Microwave Conf (APMC) Proc.* 2007; pp 20–23.
60. Vega, F.; Torres, M. C. Automatic Detection of Bruises in Fruit Using Biospeckle Techniques. *Symp. Signals Images Artif. Vis—2013 (IEEE)*, **2013**, *2013*, 1–5.

61. Vlasov, Y.; Legin, A.; Rudnitskaya, A.; Di Natale, C.; D'Amico, A. Nonspecific Sensor Arrays (Electronic Tongue) for Chemical Analysis of Liquids (IUPAC Technical Report). *Pure Appl. Chem.* **2005**, *77*, 1965–1983.
62. Vlasov, Y. G.; Legin, A. V.; Rudnitskaya, A. M.; D'Amico, A.; Di Natale, C. Electronic Tongue—New Analytical Tool for Liquid Analysis on the Basis of Non-Specific Sensors and Methods of Pattern Recognition. *Sens. Actuat. B Chem.* **2000**, *65*, 235–236.
63. Yang, Z.; Wang, Z.; Yuan, W.; Li, C.; Jing, X.; Han, H. Classification of Wolfberry from Different Geographical Origins by Using Electronic Tongue and Deep Learning Algorithm. *IFAC Papers OnLine* **2019**, *52*, 397–402. https://doi.org/10.1016/j.ifacol.2019.12.592 (accessed Aug 2, 2021).
64. Zhang, L.; McCarthy, M. J. Measurement and Evaluation of Tomato Maturity Using Magnetic Resonance Imaging. *Postharvest Biol. Technol.* **2012**, *67*, 37–43. http://dx.doi.org/10.1016/j.postharvbio.2011.12.004 (accessed Aug 2, 2021).

PART IV

POTENTIAL USE OF FRUIT AND VEGETABLE WASTE

CHAPTER 13

FRUIT AND VEGETABLE WASTE: A VIABLE FEEDSTOCK FOR BIOENERGY

OGECHUKWU BOSE CHUKWUMA, MOHD RAFATULLAH, MOHAMMAD FAISAL UMAR, HUSNUL AZAN TAJARUDIN, and NORLI ISMAIL

ABSTRACT

Fruits and vegetables waste (FVW) is a significant contributor to the composition of compostable organic materials in municipal solid wastes (MSW). Organic waste makes up around half of the total MSW generated. FVWs are produced throughout the production process, from harvesting to fork. As bioenergy gains prominence as a green replacement to fossils, viable, renewable feedstock for its production is still challenging issue. With their high organic and moisture content, FVW is a cheap and renewable source for several streams of biorefinery. First-generation biofuels come from food crops, starch, and sugar. With the global population on a steady rise, food production will need to be increased by over 50% to meet the demands. This calls into question of its sustainability as the possible food competition that will put further pressure on food demand. FVW falls under second-generation feedstocks as they are rich lignocellulosic biomass with potential for biorefinery. However, their use in the waste-to-energy approach has several challenges, including the recalcitrant nature of the biomass and the high-cost of pretreatment. However, this group of biomass requires further characterization and evaluation to maximize its utilization to yield value-added products and to aid in overcoming the cost and technological challenges in their wide spread adoption. Research to help ensure all parts of

Quality Control in Fruit and Vegetable Processing: Methods and Strategies. Megh R. Goyal, & Faizan Ahmad (Eds.)

the biomass are used in the biorefinery process to help improve the cost of biorefinery process suitability driven by this class of feedstock.

13.1 INTRODUCTION

Energy is an indicator of social and economic development of a country or a community. Despite being in a technologically advanced times, over two billion people of the world's population cannot access advanced energy sources and still use traditional methods, such as, burning wood and coal to meet their daily energy needs. These continue to pose environmental and health problems, and hence, need more sustainable and greener alternatives.[73] Because fossil resources are finite, therefore, the issue of them being exhaustible is no longer as distant as humanity would prefer. This realization has led to find alternatives to secure the future of our energy needs.[44]

Bioenergy is an emerging and feasible energy source tasked with meeting emerging and pressing energy demands that have come about as a result of fossil fuels' damaging effects and their imminent scarcity.[71] With the steady incline of the global population[20,60] leading to enormous demands for fuels, the scientific community is pressed to find solutions to meet these demands.[21]] Interestingly, bioenergy is fed by a diverse range of readily available feedstocks, most of which are wastes, thus increasing their appeal.[71]

Lignocellulosic biomass largely made up of crop residues, energy crops, animal manures, and municipal solid wastes (MSW) can be converted and then utilized for bioenergy.[20] Fruits and vegetables wastes (FVWs) are very prominent part of lignocellulosic biomasses that are used for biofuel production.[62] FVWs are also vital source of raw materials for producing natural and synthetic products and chemicals.[25] FVW generated from any part of the food supply chain (production, distribution, transportation, processing, and retailing) can either be solid or semisolid except for waste effluents from processing plants.

Vegetable waste refers to biodegradable biomass available in huge quantities across the entire food production line from harvesting, transportation, and storage to marketing and processing to fork.[58] It is a carbohydrate-rich biomass, highly degradable with high moisture content, and thus holds great appeal for sustainable bioenergy production. On the other hand, fruit waste refers to the pulp, seeds, and peels, and they likely form up to half (50%) of the total harvest/product of fruits. They contain high amounts of cellulose and hemicellulose, but their lignin and protein contents are quite low.[77]

Fruits and vegetables are significant part of our diet. Thus, as the population increases, their demand is also expected to increase. Losses of FVW in the supply chain have been estimated to be 60%, making it the highest among all food types.[63] The useful part, after they have been processed, may be less than the waste by-product thus leaving adverse effects on the economics of growing fruits. FVW disposal further adds cost to processors; and when improperly disposed of, it leads to major environmental problems.[66] The vast amounts of FVW being generated globally cause severe issues as the existing conventional disposal methods (landfills, incineration, composting) leave much to be desired. Landfills result in methane being released into the atmosphere, which is more than 20 times more potent than carbon dioxide as a greenhouse gas (GHG).[44] They also cause pollution as they are usually disposed of and discarded in water bodies and on land/soil, where they pose environmental and health concerns and emit huge amounts of greenhouse gases.[52]

Managing this group of solid organic waste is a huge problem for waste management and government agencies as they are linked with the spread of diseases, bad smells in communities, contamination of water sources, and rodent infestation, etc. Asides from affecting the environment and health, the real cost of managing this waste will begin to impact the standard of living of the people in the communities, where these are generated.[15] Sustainable waste management practices are necessary to manage the useful organic fraction in FVW to successfully mitigate the existing problems, from GHG emissions to soil and water pollutions.[73]

FVWs are particularly high in organic content, which makes them attractive in the biological employment of microorganisms to both manage waste and produce bioenergy. This has become very desirable as their high organic and water content produces bad odors and lots of leachate in the course of their collection, transportation, and disposal at the MSW landfills[23,41] This option is being preferred because the other options come with pitfalls, the highest being they do not produce energy as output. Incineration of FVW is discouraged by its high moisture feature, while disposal by landfill produces GHG and leachate, both environmental issues.[51]

FVWs are the largest contributors to the kitchen waste. The processing of fruits and vegetables produces peels, which can be used to feed livestock or as fertilizer. FVWs are also used to produce essential oils, food additives, enzymes, biofuels, bioplastics, and other products.[10] Using FVW to produce renewable energies like biohydrogen is a welcome sustainable solution to the economic and environmental problems, and the production and disposal of this group of MSW pose issues on daily basis.[37]

In a study in 2014, FVW was used to produce energy from anaerobic digestion ensuring that FVW was the only substrate used, and the findings revealed savings in terms of cost.[65] While FVW for two-stage anaerobic degradation.[14] was used to produce biogas and it was observed that the process was more stable and it produced higher amounts of biogas.

This chapter focuses on (1) how FVW can be harnessed to produce useful products and (2) existing challenges and future prospective of this biomass in the biorefinery.

13.2 CHARACTERIZATION OF FRUIT AND VEGETABLE WASTE

FVW belongs to a group of biomasses that characterization is required to use it as a raw material and utilize it maximally. Waste is characterized based on its physical, chemical, and biological attributes and these attributes affect the performance of the biomass in the digestion process. They also determine the recovery strategy and processing method used in the waste to energy (WTE) conversion process.

The physical characterization of FVW refers to moisture, ash, total solid, volatile solid, color, odor, pH, and temperature. When waste has high moisture content levels, it degrades rapidly in comparison to those with low moisture content. Moisture loss is linked to a waste biomass's suitability to be converted and then used as a solid fuel. A material's ability to lose moisture will determine how combustible it will be and will be used as a fuel.[15] FVWs with pH around 5.5–6.5 degrades easily, as putrefaction starts early and continues without any restrictions. However, when FVWs have pH 6–7, decomposition is at a much slower rate than those with lower pH values. Reducing the biomass size to finer particles and optimizing temperature to higher temperatures have been used in studies to overcome the challenges of this increased pH.

FVW can be degraded in either the psychrophilic range (15–25°C), mesophilic range (35–37°C), or thermophilic range (50–60°C). Studies have reported that thermophilic and mesophilic ranges are more favorable, with the thermophilic conditions being able to degrade various kinds of FVWs as it has high metabolic rates and is less specific when it comes to substrates.[15]

Chemical characterization refers to carbon, nitrogen, total organic carbon, starch, cellulose, hemicellulose, phosphorus, and protein. Carbon–Nitrogen (C/N) ratio is the interaction of nitrogen and carbon and a low value means that the biomass is rich in protein while a high value means the biomass is amino deficient. The 25:1 is the optimal value for high biogas yields with a higher ratio indicating nitrogen will be rapidly depleted, and yields will be low.[35] These chemical characteristics of FVW affect the conversion of the biomass via anaerobic digestion.[51]

Biological characterization includes chemical oxygen demand (COD) and biological oxygen demand (BOD), commonly used as pollution indicators.[71] and as indicators of how much hydrolyzable content is present in FVW biomass. Together, they show how putrescible FVW is as the lesser the BOD and COD ratio, the lesser the degradation potential of the waste and vice versa. COD indicates the potential of FVW for biogas production, while BOD determines the availability of oxygen, which microorganisms need to break down organic matter in FVW.[35]

There must be a balance between the microorganisms and organic biomass used in anaerobic digestion (AD), as the slightest imbalance will lead to the entire digestion being ineffective and having low yields.[51]

13.2.1 COMPONENTS OF THE FRUIT AND VEGETABLE WASTE

FVW comprises peels and twigs, which are mostly the nonedible parts of fruits and vegetables. The pomace comprises pulp, skin, seeds, and stem from fruits and vegetables.[8] FVW can also contain leaves, rinds cores, pits, twigs, and spoiled fruits and vegetables in addition to the aforementioned components.[80] The group of fruits and vegetables usually determines the nutrients and bioactive components it contains. Fruits and vegetables in the same group also share growing conditions, and this information can be useful in terms of handling and logistics in the supply chain.

The highest amount of FVW comes mostly from the pre-consumer stage of the supply chain and includes residues from harvest as well as peels and pomace. There is a whooping drop from 48% of FVW in the processing stage to 21% in the consumer stage.[80]

13.2.2 LIGNOCELLULOSE COMPOSITION IN FRUIT AND VEGETABLE WASTE

Lignocellulosic composition in feed stocks (animal manure and crop residues) is known to be higher than that in FVWs.[55] FVWs are rich in cellulose, hemicellulose, pectin, and minerals but have very low concentrations of lignin compared with other crop residues. FVW is characterized by low lipid and high cellulose content. They also have higher pH and C/N ratio than the other food wastes, but their lignocellulosic materials are more resistant to microbial attacks and impede acidification.[12] FVWs are known to be low in total solids (TS) and high in volatile solids (VS).[68] Fruit wastes have a higher TS, VS, and C/N ratio than the vegetable wastes. The TS of fruits is usually around 7.5–23% (wet weight basis), while TS for vegetables are between 3–11%. VS of fruits at 5–12% exceed that of vegetables at 2–9%. Then C/N ratio of fruit wastes is 19–53% higher than that of vegetables which is at 10–21%.[25]

FVW residue after processing contains pulp, peel, seeds, and stem. The main constituents of this waste are pectin that accounts for 1–13%, cellulose accounts for 7–44%, hemicellulose for 4–34%, lignin for 15–69%, and trace amounts of gums.[74] These components affect the value that any particular FVW can produce. They are the major factors that make up the cell wall saccharification potential of a lignocellulosic biomass feedstock, which drives its suitability in the biorefinery process. This in turn t affects the extent and efficiency of hydrolysis and other steps in the degradation process.[74]

FVWs have high moisture content, which alongside high VS (>90%) makes them very suitable choice for anaerobic digestion (AD) to produce organic products.[15] These properties also affect how the FVW is handled and processed, as the high VS may pose challenging issue for the AD process if monodigestion is used. FVW with a high percentage of VS and TS < 10% is processed via wet AD, while semidry AD is employed for FVW with 10–20% TS and dry AD is utilized for FVW with TS>20%. Temperature also plays a role in AD, as mesophilic temperature (30–39°C) encourages methanogenic activity that is a vital step in breaking down FVW. In contrast, breakdown under thermophilic conditions results in the production of various fatty acids that inhibit methanogenic activities by reducing the pH in the processing tanks.

13.3 FRUIT AND VEGETABLE WASTE AS FEEDSTOCK FOR BIOENERGY PRODUCTION

Miscanthus, switch grass, blue stem, yellow sweet clover, and alfalfa are some of the vegetable feedstocks that have been previously studied for bioenergy production. Switch grasses and miscanthus have garnered the most attention and have shown better prospects in terms of their yields and reduced production and cost requirements. Switchgrass is also drought tolerant as it adapts to different soil and climate.[3] Mango endocarp, banana peel, orange peel, and watermelon peels were studied for their potential usefulness in producing biochar, biooil, and biogas, and the results showed huge promise in producing useful products.[39] Mango, banana, lemon, orange, pineapple, tomato, pomegranate, grape vine, onion, potato, cabbage, coriander, cauliflower, turnip, radish, carrot, sugar beet, garden pea, sorghum, and Napier grass were used as feedstock for methane production. High methane yield was observed after fermentation.[32]

Globally, environmental issues, such as global warming, depletion of natural fossils, and other natural resources have fueled sustainable solutions to the energy challenges. Hence, producing bioenergy from biomass is gaining popularity as it has various positive environmental and economic implications.[38,72] Biofuels are fuels from lignocellulosic solid biomass that have gone through various chemical and biological processes to yield useful products. Biofuels are grouped into four generations based on their raw materials/biomass type and technology, such as first-, second-, third-, and fourth-generation biofuels.[43] They include solid fuels, liquid fuels, and various biogas products. They are also grouped as conventional biofuels, which refer to only first-generation biofuels and advanced biofuels which explore the second, the third, and the fourth-generations of biofuel.[34]

13.3.1 FIRST-GENERATION FUEL

First-generation biofuels are produced using food-based crops, such as sugar cane, wheat, cassava, and corn, etc. Bioethanol and biodiesel are the most common products of first-generation biofuels. They account for 96% of the global worldwide production of biofuels.[61] They represent the initial shift from the use of fossils to a more eco-friendly option in meeting energy demands.[28]

While the first-generation biofuels are already seeing global commercialization and have mature markets and technologies, their sustainability is still a major issue because the source of its feedstock competes directly with food crops for land, which places more pressure on land use and has more negative effects on the climate and the environment at large.[33] It also causes upsurge in food prices worldwide, which has seen cereal and sugar prices continue to grow at an alarming rate.[49]

13.3.2 SECOND-GENERATION FUEL

The second generation biofuel inculcates lignocellulosic waste and biomass, which are generally nonfood and nonedible biomasses.[4] FVW form part of agricultural wastes that is used in producing second-generation biofuels.

The biochemical routes of this production process produce bioethanol, natural gas, biohydrogen, biodiesel, biogas, and biooils. All biofuels are capable of supplementing or acting as replacements for fossils in a green way.[33,55] Using lignocellulosic biomass for bioenergy production is encouraged, as it does not compete with food production.[17] Lignin residue thus generated can be used for power generation and fuel additives. In the second-generation process, a reuse design exists and encourages the conversion of the entire biomass to useful products.[79]

The third- and fourth-generation biofuels use microorganisms as their feedstocks. However, the fourth-generation fuels use genetically modified algae with high lipid content as its feedstock.[61]

13.4 BIOENERGY PRODUCTS

Lignocellulosic waste represents a cheap, highly available, renewable resource that is used for bioenergy production. The use of these wastes is a means of conservation of resources and a sustainable, environmentally safe alternative in the energy production.[5] Bioenergy is a key consideration in the search for short- and medium-term solutions to lower greenhouse gas (GHG) emissions and find replacements for fossil fuels. It is gaining in popularity and acceptability everyday.[26] The main biofuel products that are derived from lignocellulosic biomass include biogas, bioethanol, and biodiesel. However, electricity and heat are also by-products of biorefineries.[31]

13.4.1 BIOETHANOL

Bioethanol from lignocellulosic biomass accounts for 10–14% of the global energy supply and is a welcome effort in mitigating global pollution and energy crisis. FVWs are rich in high sugars and are thus suitable raw materials for producing bioethanol, which can be used as an industrial solvent and as a fuel supplement.[48] FVWs are chosen for bioethanol production because they are cheap and readily available. Bioethanol is obtained by the microbial action on FVW via fermentation in controlled and specific conditions. These conditions include separate enzymatic hydrolysis and fermentation (SHF), simultaneous saccharification and co-fermentation (SSCF), and simultaneous saccharification and fermentation (SSF). The SSF is the fermentation of solid biomass by microorganisms growing and acting on these biomasses in the absence of moisture. SSF method is preferred than other methods as its affinity for little or no moisture means that no foam is produced when fermentation is carried out, and this helps to keep other parameters (aeration, temperature, and pH) under control. SSF is also known to show higher ethanol yields than other fermentation methods.[50] Microorganisms that have been successfully used in bioethanol production include *Saccharomyces cerevisiae*,[24] microalgae,[30] *Saccharomyces cerevisiae* acting on potato peel,[6] and *Trichoderma reesei* and *Trichoderma viride* on MSW.[40] Bioethanol is presently the most common biofuel with wide acceptance in USA, Brazil, European Union, China, and India being the fore front producer.[38]

13.4.2 BIOMETHANE

Biomethane represents the gas produced by the biodegradation of FVW in a series of reactions. Biogas is a significant contributor to the global energy market.[35] and is a gas mixture that contains methane and carbon dioxide, and is a clean, renewable source of energy used to power homes and industries, as it supplies heat and electricity. It can also be used for transportation energy needs.[73] Biogas generally contains 50–75% of methane, 25–50% of carbon dioxide, and trace elements (water vapor, hydrogen sulfide, and ammonia). These fractions are dependent on the condition of processing and biomass feedstock. COD of the biomass affects the amount of methane produced.[13]

AD is biologically converting organic biomass to biogas in the absence of oxygen by a consortium of microbes in several metabolic steps. Treating FVW anaerobically results in high biomethane yields and involves three steps: hydrolysis, acidogenesis, and methanogenesis. All three stages involve bacteria to break down FVW biomass in the presence of little or no oxygen.[47,71] These three stages are the following:

- **Hydrolysis** involves the conversion of complex, insoluble polymers into simple sugars, amino acids. It is a result of hydrolytic enzymes produced by cellulolytic bacteria.[17] Temperature is a major factor in determining the effect of hydrolysis. While thermophilic treatment produces more in a shorter time, it is however less stable and requires more energy than to complete the process. It is observed that hydrolysis rate is more than double when compared to mesophilic rates.[13]
- **Acidogenesis** involves acetogenic bacteria (acid-formers) converting the by-products of hydrolysis into organic acids, carbon dioxide, and hydrogen.

- **Methanogenesis** takes place by the action of methanogens by either causing cleavage to the organic acid molecules to produce CO_2 and methane or by reducing CO_2 using hydrogen. The latter action produces higher yields of methane and is preferred when high hydrogen concentrations can be arranged.[48]

Anaerobic digestion can either be executed in a single-stage AD system or in a two-stage AD system. The latter is usually simple, while the former is used to produce multiple products simultaneously in a shorter time frame.[51] The two-stage AD system is more stable but is more expensive and requires a certain level of expertise and knowledge to operate.[65]

13.4.3 BIODIESEL

The interest in biodiesel is on the rise as it represents a greener option and supplement for fossil derived gasoline, and diesel. This environmental advantage, its higher energy density, and ease of handling have led to biodiesel being an attractive investment option globally.[11,75] Biodiesel is a yellowish liquid biofuel used in "home heating system" and power heavy duty automobiles, such as tractors, military carriers/trucks, and mining equipment.[67]

Biodiesel is basically alkyl ester that contains low aliphatic alcohols and high fatty acids, formed through transesterification (triglyceride reacts with alcohol forming esters and glycerol).[54,81]. Biodiesel production is maximized when biomass used contains high lipid content such as vegetable wastes. The biomass feedstock makes up over 70% of the production cost and subsequently the final biodiesel cost.[71] Biodiesel does not produce any output of carbon and is thus carbon neutral.[81]

Some microorganisms have been used to produce biodiesel, such as *Rhodotorula glutinis,*[59] *Cryptococcus curvatus* MTCC 2698,[16] members of *Mogibacteriaceae and Ruminococcaceae,*[19] *Rhodosporidium* sp., *Lipomyces* sp., and recombinant bacteria.[45] Major biodiesel producers are the European Union and United States, Brazil, Argentina, Malaysia, Indonesia, India, and China. Some Sub-Saharan and Southeast Asia countries are starting to increase their biodiesel production.[28]

13.4.4 BIOHYDROGEN

Hydrogen in many quarters is being seen as potential clean energy carrier that can be applied to replace existing fossils, giving full benefits in an environmentally sustainable system. Biohydrogen comes from microbial communities due to breaking down of organic matter from biomass, such as FVW.[76] Use of FVW can lower the production costs and thus make biohydrogen more affordable and acceptable. Feedstock for biohydrogen production must be carbohydrate-rich waste and must be deficient in nitrogen.[2] Vegetable waste is a good example. Microorganisms that produce biohydrogen are either bacteria (anaerobic, aerobic, or photosynthetic) or algae, and the use of cultures increases the yield of biohydrogen. *Enterobacter, Bacillus, Clostridium,*[53] *Thermotoga maritima,*[64] *Bifidobacterium,* and *Lactobacillus,*[1] green algae, cyanobacteria, purple nonsulfur bacteria have been used to produce biohydrogen successfully using FVW.[78]

Biohydrogen can be produced through several means, but dark fermentation is the most widely accepted method, because it has the flexibility of substrates and takes place in the absence of light, and does not require sterile conditions or oxygen.[36,64] FVWs have a higher yield when compared with other biomasses for the production of biohydrogen.[18,19] Biohydrogen output is predicted to grow 10% annually and accounts for over 8% of the total energy globally by 2025 due to rising industrial and chemical demands for clean energy.[64]

13.4.5 BIOCHAR

Biochar is a black.[70] carbonaceous by-product of pyrolysis of lignocellulosic biomass that is now being used in anaerobic digestion, among other applications.[42] FVW (peels) can be converted into biochar, a solid carbon product of thermal pyrolysis (heating in the absence of oxygen at 350–700°C). Biochar has been an effective application that includes[21,69] improved crop growth and soil amendment, heavy metal/contamination treatment, developing biocomposite, cancel out negative effects of waste, improve soil fertility, and reduce gas emissions.

Vegetable waste has been confirmed to be suitable for producing biochar with good agricultural properties.[57] Biochar can also be obtained through gasification, hydrothermal carbonization, and torrefaction. It contains lignin, hemicellulose and cellulose, and trace quantities of extractives, protein, pectin, and ash.[42] The use of biochar for soil fertility, plant growth, and agricultural purposes is gaining prominence, and biochar from FVW is already being applied as biofertilizer.[7]

13.5 CHALLENGES FOR ADOPTION OF FRUIT AND VEGETABLE WASTE IN BIOENERGY PRODUCTION

Despite the promise of using FVW for bioenergy production, the major challenge is the recalcitrant nature of the lignocellulosic biomass making pretreatment a necessity. Pretreatment is necessary for increased biodegradation of FVW biomass prior to hydrolysis. Pretreatment increases the digestibility of the FVW feedstock as the lignin barrier is broken, releasing cellulose for hydrolysis.[22] Pretreatment requires a huge amount of energy and chemicals and thus is an expensive part of the biorefinery process. However, it is unavoidable if one must reach the second-generation bioenergy goal.[55] In a bid to overcome the drawbacks of the first-generation fuels, second-generation fuels have several added steps (thermal, biological, enzymatic, or chemical), which raise the overall cost of second-generation biorefineries. They are thus more difficult to scale up from small-scale level to a more commercial scale.[4] At times, anaerobic digestion is plagued by low biodegradability as the microbial consortia may show a preference for a specific feedstock.[9]

Commercialization of biofuels has only been successful for first-generation as the others require improvement to make them more cost-effective on a large scale.[56] The infrastructural and technological demands may also be challenges to the use of such biomass in bioenergy. Bioenergy can at times be marred by low yields and slow production rates.[22] Also, the supply chain for sustainable bioenergy is still scanty and in its infancy in most developing countries. It is yet to be fully embraced as a business model and there is still room for growth and wider adoption.[29] It also has logistics and operational challenges (preprocessing, processing, storage, and transportation), making it less desirable for large-scale adoption. In addition, it still has issues with supply due to its seasonal nature, spatial distribution, and annual variation.[33]

The challenges of biomass' supply chain are very complex, and most of the times vary from biomass to biomass. Some quarters have proposed a uniform protocol along the entire supply chain to reduce cost and ensure more effective use of produced biomass. However, fruits are harvested and collected in a different way than the vegetables, they also have different processing methods making the proposed uniformed protocol not feasible and practicable.[46] Biorefinery technology also requires high implementation and high maintenance costs as well as huge expertise. Unlike the traditional forms of energy, adopting these more sustainable forms of energy production costs several times more, which most households cannot afford.[73] It also needs specialized expertise for its construction and maintenance, which is difficult to find. Especially in regions with huge poverty, which is matched with high illiteracy levels, making them unattractive for professionals with such skills.

13.6 FUTURE PROSPECTS

There is a projection that bioenergy's contribution to the global energy pile will increase from 20 to 30% by 2035. This has led to more countries embracing biomass from agricultural and forestry activities, which includes FVW. The FVW and its subsequent use in second-generation fuels is being hailed as a more preferred option, where land use and food supply concerns are almost nonexistent. First-generation biorefineries may look to replace their key substrates with lignocellulosic wastes, to help reduce environmental and food concerns and lower the cost, because the existing infrastructures can be used for biofuel production as opposed to purchasing new set-ups. FVWs represent a cheap renewable raw material, whose entire biomass components can be used to produce economically viable end products. Bioenergy is geared to contribute to social and economic development in communities globally.[27]

Construction of biorefineries should take into account logistic issues of using FVW to reduce the cost of moving these wastes and maximize profits and benefits. They should also be constructed with operations that allow different use beyond the original purpose.[56] The implementation of separating waste from source will help to improve the quality of FVW that is obtained from waste streams for AD purposes. This will wean out plastics and metals and reduce sorting costs and waste-handling costs for AD purposes.[51]

Consolidated bioprocessing (CBP) involves directly introducing microorganisms into the biorefinery system to ferment cellulose after it has broken down the lignocellulosic raw material. This process can lower the cost as no enzymes are added as the production of cellulosic ethanol and other useful by-products are dependent on the action of one microorganism.[50] Co-digestion, which means the processing of FVW with other kinds of waste like sludge and effluents, is being encouraged as it overcomes instability and loading challenges experiences with pure substrates of FVW. These other waste streams provide alkaline conditions and allow for more efficient use of the biorefinery equipment, and the cost for processing of multiple waste streams is shared and lowered.[80]

FVWs are pivotal in achieving a zero-waste economy, as their use by microorganisms yields various valuable end products. Recent advances in microbiology and biotechnology are already utilizing FVW to produce enzymes and other high-end products. Genetic engineering and recombinant DNA technology has some success in research to improve the quality and quantity of biological products. Commercially, genetically modified microorganisms and recombinant DNA technology has produced products that led to improved yields of 100-folds when applied in production. These research results have helped to overcome the existing challenges in the use of FVW as feedstock for bioenergy production.[52] Increased efforts from governments and private sector to consciously raise awareness on environmental challenges traditional energy production will go a long way in closing the knowledge and skill gaps. They also need to provide subsidies and programs for bioenergy companies and citizens to support and encourage the implementation of bioenergy technologies up to the rural communities.

13.7 SUMMARY

The use of FVW to produce energy has allowed for more focus on sustainable energy solutions, preferably those that tackle more than one environmental issue. These benefits include reducing aesthetic problems arising from landfills, indiscriminate dumping of FVW, waste handling and management costs, and other environmental and health issues that may arise from these wastes. Advances in technology and social education are required to see significant, improved adoption in the ways FVWs are handled as we try to shift from burning and landfilling to more sustainable options like biorefinery. Despite the challenges in their adoption in terms of the lignocellulose hurdle, the pros definitely outweigh the cons as their use can be linked to responsible consumption and production. Successfully adopting this group of waste will be a sustainable process that means fruits and vegetables produced are put to the most efficient use with less amounts being wasted. It will create jobs and spur economic growth, as economies will see alternative streams of energy and income. They are thus a welcome green, cost-effective approach with barely tapped into potentials for the future.

ACKNOWLEDGMENT

The authors express their appreciation to Universiti Sains Malaysia RUI grant (1001/PTEKIND/8011044) for the financial support and research facilities in this study.

KEYWORDS

- **anaerobic digestion**
- **bioenergy**
- **biofuels**
- **biomass**
- **biorefinery**
- **feedstock**
- **lignocellulose**

REFERENCES

1. Abubackar, H. N.; Keskin, T.; Arslan, K.; Vural, C.; Aksu, D.; Yavuzyılmaz, D. K.; Ozdemir, G.; Azbar, N. Effects of Size and Autoclavation of Fruit and Vegetable Wastes on Biohydrogen Production by Dark Dry Anaerobic Fermentation Under Mesophilic Condition. *Int. J. Hydrogen Energy* **2019,** *44*, 17767–17780. DOI:10.1016/j.ijhydene.2019.05.106.
2. Abubackar, H. N.; Keskin, T.; Yazgin, O.; Gunay, B.; Arslan, K.; Azbar, N. Biohydrogen Production from Autoclaved Fruit and Vegetable Wastes by Dry Fermentation Under Thermophilic Condition. *Int. J. Hydrogen Energy* **2019,** *44*, 18776–18784. DOI:10.1016/j.ijhydene.2018.12.068.
3. Ahorsu, R.; Medina, F.; Constantí, M. Significance and Challenges of Biomass as a Suitable Feedstock for Bioenergy and Biochemical Production: A Review. *Energies* **2018,** *11*, E-article 3366. DOI:10.3390/en11123366.
4. Alalwan, H. A.; Alminshid, A. H.; Aljaafari, H. A. S. Promising Evolution of Biofuel Generations. Subject Review. *Renew. Energy Focus* **2019,** *28*, 127–139. DOI:10.1016/j.ref.2018.12.006.
5. Anwar, Z.; Gulfraz, M.; Irshad, M. Agro-Industrial Lignocellulosic Biomass a Key to Unlock the Future Bio-Energy: A Brief Review. *J. Radiat. Res. Appl. Sci.* **2014,** *7*, 163–173. DOI:10.1016/j.jrras.2014.02.003.
6. Arapoglou, D.; Varzakas, T.; Vlyssides, A.; Israilides, C. Ethanol Production from Potato Peel Waste (PPW). *Waste Manage.* **2010,** *30*, 1898–1902. DOI:10.1016/j.wasman.2010.04.017.
7. Aziz, S.; Yaseen, L.; Jamal, A.; Farooq, U.; Qureshi, Z.; Tauseef, I.; Haleem, S.; Ali, M. Fabrication of Biochar from Organic Wastes and Its Effect on Wheat Growth and Soil Microflora. *Polish J. Environ. Stud.* **2020,** *29*, 1069–1076. DOI:10.15244/pjoes/99825.
8. Ben-Othman, S.; Jõudu, I.; Bhat, R. Bioactives from Agri-Food Wastes: Present Insights and Future Challenges. *Molecules*, **2020,** *25*, E-article 510. DOI:10.3390/molecules25030510.
9. Bhatia, S. K.; Mehariya, S.; Bhatia, R. K.; Kumar, M.; Pugazhendhi, A.; Awasthi, M. K.; Atabani, A. E.; Kumar, G.; Kim, W.; Seo, S. O. Wastewater Based Microalgal

Biorefinery for Bioenergy Production: Progress and Challenges. *Sci. Total Environ.* **2021,** *751*, E-article 141599. DOI:10.1016/j.scitotenv.2020.141599.
10. Bhatnagar, A.; Sillanpää, M.; Witek-Krowiak, A. Agricultural Waste Peels as Versatile Biomass for Water Purification—A Review. *Chem. Eng. J.* **2015,** *270*, 244–271. DOI:10.1016/j.cej.2015.01.135.
11. Bhuvaneswari, M.; Sivakumar, N. Bioethanol Production from Fruit and Vegetable Wastes. In *Bioprocessing for Biomolecules Production*; John Wiley & Sons, Ltd.: Chichester, UK, 2019; pp 417–427.
12. Bong, C. P. C.; Lim, L. Y.; Lee, C. T.; Klemeš, J. J.; Ho, C. S.; Ho, W. S. The Characterisation and Treatment of Food Waste for Improvement of Biogas Production During Anaerobic Digestion—A Review. *J. Clean. Prod.* **2018,** *172*, 1545–1558. DOI:10.1016/j.jclepro.2017.10.199.
13. Bouallagui, H.; Haouari, O.; Touhami, Y.; Ben Cheikh, R.; Marouani, L.; Hamdi, M. Effect of Temperature on the Performance of an Anaerobic Tubular Reactor Treating Fruit and Vegetable Waste. *Process Biochem.* **2004,** *39*, 2143–2148. DOI:10.1016/j.procbio.2003.11.022.
14. Bouallagui, H.; Torrijos, M.; Godon, J. J.; Moletta, R.; Ben Cheikh, R.; Touhami, Y.; Delgenes, J. P.; Hamdi, M. Two-Phases Anaerobic Digestion of Fruit and Vegetable Wastes: Bioreactors Performance. *Biochem. Eng. J.* **2004,** *21*, 193–197. DOI:10.1016/j.bej.2004.05.001.
15. Chatterjee, B.; Mazumder, D. New Approach of Characterizing Fruit and Vegetable Waste (FVW) to Ascertain Its Biological Stabilization via Two-Stage Anaerobic Digestion (AD). *Biomass Bioenergy* **2020,** *139*, E-article 105594. DOI:10.1016/j.biombioe.2020.105594
16. Chatterjee, S.; Mohan, S. V. Microbial Lipid Production by Cryptococcus Curvatus from Vegetable Waste Hydrolysate. *Bioresour. Technol.* **2018,** *254*, 284–289. DOI:10.1016/j.biortech.2018.01.079.
17. Chukwuma, O. B.; Rafatullah, M.; Tajarudin, H. A.; Ismail, N. Lignocellulolytic Enzymes in Biotechnological and Industrial Processes: A Review. *Sustainability* **2020,** *12*, E-article 7282. DOI:10.3390/su12187282.
18. Cieciura-Włoch, W.; Borowski, S. Biohydrogen Production From Wastes Of Plant and Animal Origin via Dark Fermentation. *J. Environ. Eng. Landsc. Manage.* **2019,** *27*, 101–113. DOI:10.3846/jeelm.2019.9806.
19. Cieciura-Włoch, W.; Borowski, S.; Otlewska, A. Biohydrogen Production from Fruit and Vegetable Waste, Sugar Beet Pulp and Corn Silage via Dark Fermentation. *Renew. Energy* **2020,** *153*, 1226–1237. DOI:10.1016/j.renene.2020.02.085.
20. Dar, R. A.; Parmar, M.; Dar, E. A.; Sani, R. K.; Phutela, U. G. Biomethanation of Agricultural Residues: Potential, Limitations and Possible Solutions. *Renew. Sustain. Energy Rev.* **2021,** *135*, E-article 110217. DOI:10.1016/j.rser.2020.110217.
21. De Bhowmick, G.; Sarmah, A. K.; Sen, R. Zero-Waste Algal Biorefinery for Bioenergy and Biochar: A Green Leap Towards Achieving Energy and Environmental Sustainability. *Sci. Total Environ.* **2019,** *650*, 2467–2482. DOI:10.1016/j.scitotenv.2018.10.002.
22. Dhanya, B. S.; Mishra, A.; Chandel, A. K.; Verma, M. L. Development of Sustainable Approaches for Converting the Organic Waste to Bioenergy. *Sci. Total Environ.* **2020,** *723*, E-article 138109. DOI:10.1016/j.scitotenv.2020.138109.
23. Edwiges, T.; Frare, L.; Mayer, B.; Lins, L.; Mi Triolo, J.; Flotats, X.; de Mendonça Costa, M. S. S. Influence of Chemical Composition on Biochemical Methane Potential

of Fruit and Vegetable Waste. *Waste Manag.* **2018,** *71*, 618–625. DOI:10.1016/j.wasman.2017.05.030.

24. Elemike, E. E.; Oseghale, O. C.; Okoye, A. C. Utilization of Cellulosic Cassava Waste for Bio-Ethanol Production. *J. Environ. Chem. Eng.* **2015,** *3*, 2797–2800. DOI:10.1016/j.jece.2015.10.021.
25. Esparza, I.; Jiménez-Moreno, N.; Bimbela, F.; Ancín-Azpilicueta, C.; Gandía, L. M. Fruit and Vegetable Waste Management: Conventional and Emerging Approaches. *J. Environ. Manage.* **2020,** *265*, E-article 110510. DOI:10.1016/j.jenvman.2020.110510.
26. Faaij, A. P. C. Bio-Energy in Europe: Changing Technology Choices. *Energy Policy* **2006,** *34*, 322–342. DOI:10.1016/j.enpol.2004.03.026.
27. Fazio, S.; Barbanti, L. Energy and Economic Assessments of Bio-Energy Systems Based on Annual and Perennial Crops for Temperate and Tropical Areas. *Renew. Energy* **2014,** *69*, 233–241. DOI:10.1016/j.renene.2014.03.045.
28. Gasparatos, A.; Stromberg, P.; Takeuchi, K. Sustainability Impacts of First-Generation Biofuels. *Anim. Front.* **2013,** *3*, 12–26. DOI:10.2527/af.2013-0011.
29. Ghosh, S. K. Biomass & Bio-Waste Supply Chain Sustainability for Bio-Energy and Bio-Fuel Production. *Procedia Environ. Sci.* **2016,** *31*, 31–39. DOI:10.1016/j.proenv.2016.02.005.
30. Gohain, M.; Hasin, M.; Eldiehy, K. S. H.; Bardhan, P.; Laskar, K.; Phukon, H.; Mandal, M.; Kalita, D.; Deka, D. Bio-Ethanol Production: A Route to Sustainability of Fuels Using Bio-Based Heterogeneous Catalyst Derived from Waste. *Process Saf. Environ. Prot.* **2021,** *146*, 190–200. DOI:10.1016/j.psep.2020.08.046.
31. González-García, S.; Bacenetti, J. Exploring the Production of Bio-Energy from Wood Biomass. Italian Case Study. *Sci. Total Environ.* **2019,** *647*, 158–168. DOI:10.1016/j.scitotenv.2018.07.295.
32. Gunaseelan, V. N. Biochemical Methane Potential of Fruits and Vegetable Solid Waste Feedstocks. *Biomass Bioenergy* **2004,** *26*, 389–399. DOI:10.1016/j.biombioe.2003.08.006.
33. Ho, D. P.; Ngo, H. H.; Guo, W. A Mini Review on Renewable Sources for Biofuel. *Bioresour. Technol.* **2014,** *169*, 742–749. DOI:10.1016/j.biortech.2014.07.022.
34. Janda, K.; Kristoufek, L.; Zilberman, D. Biofuels: Policies and Impacts. *Agric. Econ.* **2012,** *58*, 372–386. DOI:10.17221/124/2011-AGRICECON.
35. Jingura, R. M.; Kamusoko, R. Methods for Determination of Biomethane Potential of Feedstocks: A Review. *Biofuel Res. J.* **2017,** *4*, 573–586. DOI:10.18331/BRJ2017.4.2.3.
36. Keskin, T.; Abubackar, H. N.; Yazgin, O.; Gunay, B.; Azbar, N. Effect of Percolation Frequency on Biohydrogen Production from Fruit and Vegetable Wastes by Dry Fermentation. *Int. J. Hydrogen Energy* **2019,** *44*, 18767–18775. DOI:10.1016/j.ijhydene.2018.12.099.
37. Keskin, T.; Arslan, K.; NalakthAbubackar, H.; Vural, C.; Eroglu, D.; Karaalp, D.; Yanik, J.; Ozdemir, G.; Azbar, N. Determining the Effect of Trace Elements on Biohydrogen Production from Fruit and Vegetable Wastes. *Int. J. Hydrogen Energy* **2018,** *43*, 10666–10677. DOI:10.1016/j.ijhydene.2018.01.028.
38. Kumar, A.; Kumar, N.; Baredar, P.; Shukla, A. A Review on Biomass Energy Resources, Potential, Conversion and Policy in India. *Renew. Sustain. Energy Rev.* **2015,** *45*, 530–539. DOI:10.1016/j.rser.2015.02.007.
39. Lam, S. S.; Liew, R. K.; Lim, X. Y.; Ani, F. N.; Jusoh, A. Fruit Waste as Feedstock for Recovery by Pyrolysis Technique. *Int. Biodeterior. Biodegr.* **2016,** *113*, 325–333. DOI:10.1016/j.ibiod.2016.02.021.

40. Li, A.; Antizar-Ladislao, B.; Khraisheh, M. Bioconversion of Municipal Solid Waste to Glucose for Bio-Ethanol Production. *Bioprocess Biosyst. Eng.* **2007,** *30*, 189–196. DOI:10.1007/s00449-007-0114-3.
41. Lin, J.; Zuo, J.; Gan, L.; Li, P.; Liu, F.; Wang, K.; Chen, L.; Gan, H. Effects of Mixture Ratio on Anaerobic Co-Digestion with Fruit and Vegetable Waste and Food Waste of China. *J. Environ. Sci.* **2011,** *23*, 1403–1408. DOI:10.1016/S1001-0742(10)60572-4.
42. Masebinu, S. O.; Akinlabi, E. T.; Muzenda, E.; Aboyade, A. O. A Review of Biochar Properties and Their Roles in Mitigating Challenges with Anaerobic Digestion. *Renew. Sustain. Energy Rev.* **2019,** *103*, 291–307. DOI:10.1016/j.rser.2018.12.048.
43. Mat Aron, N. S.; Khoo, K. S.; Chew, K. W.; Show, P. L.; Chen, W.; Nguyen, T. H. P. Sustainability of the Four Generations of Biofuels—A Review. *Int. J. Energy Res.* **2020,** *44*, 9266–9282. DOI:10.1002/er.5557.
44. Matsakas, L.; Gao, Q.; Jansson, S.; Rova, U.; Christakopoulos, P. Green Conversion of Municipal Solid Wastes Into Fuels and Chemicals. *Electron. J. Biotechnol.* **2017,** *26*, 69–83. DOI:10.1016/j.ejbt.2017.01.004.
45. Meng, X.; Yang, J.; Xu, X.; Zhang, L.; Nie, Q.; Xian, M. Biodiesel Production from Oleaginous Microorganisms. *Renew. Energy* **2009,** *34*, 1–5. DOI:10.1016/j.renene.2008.04.014.
46. Mohanty, A. K.; Vivekanandhan, S.; Pin, J.-M.; Misra, M. Composites from Renewable and Sustainable Resources: Challenges and Innovations. *Science* **2018,** *362*, 536–542. DOI:10.1126/science.aat9072.
47. Molino, A.; Nanna, F.; Ding, Y.; Bikson, B.; Braccio, G. Biomethane Production by Anaerobic Digestion of Organic Waste. *Fuel* **2013,** *103*, 1003–1009. DOI:10.1016/j.fuel.2012.07.070.
48. Moneruzzaman Khandaker, M.; Aliyu Abdullahi, U.; Dogara Abdulrahman, M.; Afiza Badaluddin, N.; Suryati Mohd, K. Bio-Ethanol Production from Fruit and Vegetable Waste by Using Saccharomyces Cerevisiae. In *Bioethanol*; IntechOpen; online, 2020.
49. Müller, A.; Schmidhuber, J.; Hoogeveen, J.; Steduto, P. Some Insights in the Effect of Growing Bio-Energy Demand on Global Food Security and Natural Resources. Water *Policy* **2008,** *10*, 83–94. DOI:10.2166/wp.2008.053.
50. Mupondwa, E.; Li, X.; Tabil, L.; Sokhansanj, S.; Adapa, P. Status of Canada's Lignocellulosic Ethanol: Part II: Hydrolysis and Fermentation Technologies. *Renew. Sustain. Energy Rev.* **2017,** *79*, 1535–1555
51. Negri, C.; Ricci, M.; Zilio, M.; D'Imporzano, G.; Qiao, W.; Dong, R.; Adani, F. Anaerobic Digestion of Food Waste for Bio-Energy Production in China and Southeast Asia: A Review. *Renew. Sustain. Energy Rev.* **2020,** *133*, E-article 110138. DOI:10.1016/j.rser.2020.110138.
52. Panda, S. K.; Mishra, S. S.; Kayitesi, E.; Ray, R. C. Microbial-Processing of Fruit and Vegetable Wastes for Production of Vital Enzymes and Organic Acids: Biotechnology and Scopes. *Environ. Res.* **2016,** *146*, 161–172. DOI:10.1016/j.envres.2015.12.035.
53. Pascualone, M. J.; Gómez Costa, M. B.; Dalmasso, P. R. Fermentative Biohydrogen Production from a Novel Combination of Vermicompost as Inoculum and Mild Heat-Pretreated Fruit and Vegetable Waste. *Biofuel Res. J.* **2019,** *6*, 1046–1053. DOI:10.18331/BRJ2019.6.3.5.
54. Pattanaik, L.; Pattnaik, F.; Saxena, D. K.; Naik, S. N. Biofuels from Agricultural Wastes. In *Second and Third Generation of Feedstocks*; Elsevier: New York, 2019; pp 103–142.
55. Paudel, S. R.; Banjara, S. P.; Choi, O. K.; Park, K. Y.; Kim, Y. M.; Lee, J. W. Pretreatment of Agricultural Biomass for Anaerobic Digestion: Current State and Challenges. *Bioresour. Technol.* **2017,** *245*, 1194–1205. DOI:10.1016/j.biortech.2017.08.182.

56. Porqueras, E. M.; Rittmann, S.; Herwig, C. Biofuels and CO_2 Neutrality: An Opportunity. *Biofuels* **2012,** *3*, 413–426. DOI:10.4155/bfs.12.25.
57. Pradhan, S.; Abdelaal, A. H.; Mroue, K.; Al-Ansari, T.; Mackey, H. R.; McKay, G. Biochar from Vegetable Wastes: Agro-Environmental Characterization. *Biochar* **2020,** *2*, 439–453. DOI:10.1007/s42773-020-00069-9.
58. Ramanujam, B. V.; Kepadatan, R. A. Population in Pemalang Regency Year 2014. *Int. J. Emerg. Sci.* **2015,** *1*, 478–486.
59. Razaghi, A.; Karthikeyan, O. P.; Hao, H. T. N.; Heimann, K. Hydrolysis Treatments of Fruit and Vegetable Waste for Production of Biofuel Precursors. *Bioresour. Technol.* **2016,** *217*, 100–103. DOI:10.1016/j.biortech.2016.03.041.
60. Rena; Mohammed Bin Zacharia, K.; Yadav, S.; Machhirake, N. P.; Kim, S.-H.; Lee, B.-D.; Jeong, H.; Singh, L.; Kumar, S.; Kumar, R. Bio-Hydrogen and Bio-Methane Potential Analysis for Production of Bio-Hythane Using Various Agricultural Residues. *Bioresour. Technol.* **2020,** *309*, E-article 123297. DOI:10.1016/j.biortech.2020.123297.
61. Robak, K.; Balcerek, M. Current State-of-the-Art in Ethanol Production from Lignocellulosic Feedstocks. *Microbiol. Res.* **2020,** *240*, E-article 126534. DOI:10.1016/j.micres.2020.126534.
62. Rodríguez-Valderrama, S.; Escamilla-Alvarado, C.; Magnin, J.-P.; Rivas-García, P.; Valdez-Vazquez, I.; Ríos-Leal, E. Batch Biohydrogen Production from Dilute Acid Hydrolyzates of Fruits-and-Vegetables Wastes and Corn Stover as Co-Substrates. *Biomass Bioenergy* **2020,** *140*, E-article 105666. DOI:10.1016/j.biombioe.2020.105666.
63. Sagar, N. A.; Pareek, S.; Sharma, S.; Yahia, E. M.; Lobo, M. G. Fruit and Vegetable Waste: Bioactive Compounds, Their Extraction, and Possible Utilization. *Compr. Rev. Food Sci. Food Safety*, **2018,** *17*, 512–531. DOI:10.1111/1541-4337.12330.
64. Saidi, R.; Liebgott, P. P.; Gannoun, H.; Ben Gaida, L.; Miladi, B.; Hamdi, M.; Bouallagui, H.; Auria, R. Biohydrogen Production from Hyperthermophilic Anaerobic Digestion of Fruit and Vegetable Wastes in Seawater: Simplification of the Culture Medium of Thermotoga Maritima. *Waste Manag.* **2018,** *71*, 474–484. DOI:10.1016/j.wasman.2017.09.042.
65. Scano, E. A.; Asquer, C.; Pistis, A.; Ortu, L.; Demontis, V.; Cocco, D. Biogas from Anaerobic Digestion of Fruit and Vegetable Wastes: Experimental Results on Pilot-Scale and Preliminary Performance Evaluation of a Full-Scale Power Plant. *Energy Convers. Manag.* **2014,** *77*, 22–30. DOI:10.1016/j.enconman.2013.09.004.
66. Sharma, R.; Oberoi, H. S.; Dhillon, G. S. Fruit and Vegetable Processing Waste. In *Agro-Industrial Wastes as Feedstock for Enzyme Production*; Elsevier: New York, 2016; pp 23–59.
67. Shehu, I.; Akanbi, T. O.; Wyatt, V.; Aryee, A. N. A. Fruit, Nut, Cereal, and Vegetable Waste Valorization to Produce Biofuel. In *Byproducts from Agriculture and Fisheries*; John Wiley & Sons: New York, 2019; pp 665–684.
68. Shen, F.; Yuan, H.; Pang, Y.; Chen, S.; Zhu, B.; Zou, D.; Liu, Y.; Ma, J.; Yu, L.; Li, X. Performances of Anaerobic Co-Digestion of Fruit & Vegetable Waste (FVW) and Food Waste (FW): Single-Phase vs. Two-Phase. *Bioresour. Technol.* **2013,** *144*, 80–85. DOI:10.1016/j.biortech.2013.06.099.
69. Sial, T. A.; Khan, M. N.; Lan, Z.; Kumbhar, F.; Ying, Z.; Zhang, J.; Sun, D.; Li, X. Contrasting Effects of Banana Peels Waste and Its Biochar on Greenhouse Gas Emissions and Soil Biochemical Properties. *Process Saf. Environ. Prot.* **2019,** *122*, 366–377. DOI:10.1016/j.psep.2018.10.030.

70. Sial, T. A.; Lan, Z.; Khan, M. N.; Zhao, Y.; Kumbhar, F.; Liu, J.; Zhang, A.; Hill, R. L.; Lahori, A. H.; Memon, M. Evaluation of Orange Peel Waste and Its Biochar on Greenhouse Gas Emissions and Soil Biochemical Properties within a Loess Soil. *Waste Manag.* **2019,** *87*, 125–134. DOI:10.1016/j.wasman.2019.01.042.
71. Singh, A.; Kuila, A.; Adak, S.; Bishai, M.; Banerjee, R. Utilization of Vegetable Wastes for Bioenergy Generation. *Agric. Res.* **2012,** *1*, 213–222. DOI:10.1007/s40003-012-0030-x.
72. Soltanian, S.; Aghbashlo, M.; Almasi, F.; Hosseinzadeh-Bandbafha, H.; Nizami, A.-S.; Ok, Y. S.; Lam, S. S.; Tabatabaei, M. A Critical Review of the Effects of Pretreatment Methods on the Exergetic Aspects of Lignocellulosic Biofuels. *Energy Convers. Manag.* **2020,** *212*, E-article 112792. DOI:10.1016/j.enconman.2020.112792.
73. Surendra, K. C.; Takara, D.; Hashimoto, A. G.; Khanal, S. K. Biogas as a Sustainable Energy Source for Developing Countries: Opportunities and Challenges. *Renew. Sustain. Energy Rev.* **2014,** *31*, 846–859;doi:10.1016/j.rser.2013.12.015.
74. Szymańska-Chargot, M.; Chylińska, M.; Gdula, K.; Kozioł, A.; Zdunek, A. Isolation and Characterization of Cellulose from Different Fruit and Vegetable Pomaces. *Polymers* **2017,** *9*, E-article 495. DOI:10.3390/polym9100495.
75. Taghizadeh-Alisaraei, A.; Hosseini, S. H.; Ghobadian, B.; Motevali, A. Biofuel Production from Citrus Wastes: A Feasibility Study in Iran. *Renew. Sustain. Energy Rev.* **2017,** *69*, 1100–1112. DOI:10.1016/j.rser.2016.09.102.
76. Tenca, A.; Schievano, A.; Perazzolo, F.; Adani, F.; Oberti, R. Biohydrogen from Thermophilic Co-Fermentation of Swine Manure with Fruit and Vegetable Waste: Maximizing Stable Production Without pH Control. *Bioresour. Technol.* **2011,** *102*, 8582–8588. DOI:10.1016/j.biortech.2011.03.102.
77. Ugwuanyi, J. O.; McNeil, B.; Harvey, L. M. Production of Protein-Enriched Feed Using Agro-Industrial Residues as Substrates. In *Biotechnology for Agro-Industrial Residues Utilisation: Utilisation of Agro-Residues*; nee' Nigam, P., Pandey, A., Eds.; Springer: Netherlands—Dordrecht, 2009; pp 77–103.
78. Vendruscolo, F. Starch: A Potential Substrate for Biohydrogen Production. *Int. J. Energy Res.* **2015,** *39*, 293–302. DOI:10.1002/er.3224.
79. Wijaya, Y. P.; Putra, R. D. D.; Widyaya, V. T.; Ha, J.-M.; Suh, D. J.; Kim, C. S. Comparative Study on Two-Step Concentrated Acid Hydrolysis for the Extraction of Sugars from Lignocellulosic Biomass. *Bioresour. Technol.* **2014,** *164*, 221–231. DOI:10.1016/j.biortech.2014.04.084.
80. Xu, F.; Li, Y.; Ge, X.; Yang, L.; Li, Y. Anaerobic Digestion of Food Waste—Challenges and Opportunities. *Bioresour. Technol.* **2018,** *247*, 1047–1058. DOI:10.1016/j.biortech.2017.09.020.
81. Zhang, Z.; O'Hara, I. M.; Mundree, S.; Gao, B.; Ball, A. S.; Zhu, N.; Bai, Z.; Jin, B. Biofuels from Food Processing Wastes. *Curr. Opin. Biotechnol.* **2016,** *38*, 97–105. DOI:10.1016/j.copbio.2016.01.010.

CHAPTER 14

POTENTIAL OF FRUIT AND VEGETABLE WASTE: TREATMENT OF WASTEWATER

ASIMA SHAFI, FAIZAN AHMAD, and SADAF ZAIDI

ABSTRACT

The increase in the number of industries worldwide has greatly contaminated the water bodies due to the release of various pollutants in the form of toxic metals, which negatively affect living systems. There are a number of methods and techniques to remove toxic metals from water bodies, for example, precipitation, chemical oxidation and reduction, ultrafiltration, etc. However, these methods are not economically sound. Adsorption is yet another technique, which uses carbon to adsorb different kinds of pollution substances from wastewater. However, its use is also not common because of its high cost. Therefore, there is an urgent need of agro-based inexpensive commodities, which can fruitfully adsorb heavy metals or metal ions. Fruit and vegetable processing industries and households generate waste (e.g., peel and pomace in large quantity). They contain large number of cellulose content and other vital components, such as hemicelluloses, lipids, lignin, simple sugars, proteins, water, starch, and hydrocarbon. As a result, these agrowastes are potentially utilized in the treatment of highly polluted water. Fruit and vegetable wastes are endowed with high adsorption capacities and therefore are eco-friendly alternatives to the costly methods and are generally used methodologies in the wastewater remediation.

Quality Control in Fruit and Vegetable Processing: Methods and Strategies. Megh R. Goyal, & Faizan Ahmad (Eds.)

14.1 INTRODUCTION

Since last few decades, the water quality has been adversely affected and deteriorated worldwide due to various factors, such as lack of maintenance of water resources, unplanned urbanization, rapid industrialization, etc. The main sources include sanitary discharges, toxic wastes from industries and their dumping, agricultural field runoff leading to the deterioration of water, and severe health issues. The toxicity of heavy metal/metal ions present in water is hazardous to living beings. Different types of industrial sectors, such as tanneries, mining sectors, storage battery, alloy produce toxic and heavy metals.[32] Therefore, there is an urgent need of introducing eco-friendly methods to ensure the safety of our ecological system.

Various conventional technologies, including exchange of ions, chemical-based precipitation, chemical oxidation (CO) and chemical reduction (CR), and ultrafiltration (UF) are available to control or alleviate water pollution. Since, these methods possess complex procedures in the wastewater remediation. They are also highly costly and generate toxic sludge in the solution. On the other hand, adsorption is comparatively an effective and significant alternating method in treating the wastewater, because it is easy to operate, design simplicity, and it is highly convenient technique.[12] This method can also remove multitude of pollutants, thus, it is widely applicable in the treatment of wastewater.

Carbon in activated form is commonly used in eliminating various heavy and toxic metals from water and is considered as universal adsorbent in the treatment of effluents. However, due to its high cost, its widespread use in effluent-treatment is limited. Several low-cost adsorbents have been observed and their potential to remove toxic pollutants from various effluents have been examined. Agricultural (fruit and vegetable) wastes, due to their inimitable chemical composition, are economically feasible and eco-friendly. Their abundant availability, low-cost, and renewable nature makes them usable for redressing toxic metallic pollutants.

Hemicellulose, lipids, simple sugars, proteins, starch, water, and hydrocarbons are basic units of fruit and vegetable waste biomass. These contain a number of functional groups that lead to metal complexation to remove heavy metal or metal ions.[80] Fruit and vegetable peels and pomace are of significant importance among various agricultural wastes for wastewater treatment, since most of them are discarded.

This chapter highlights the global production of fruits and vegetable wastes. Further, this chapter focuses on the potential role of different fruits and vegetable wastes (peel and pomace) in the removal of heavy metals from wastewater. The chapter also includes discussions on the adsorption capacity of various fruit and vegetable wastes and their procedure in removing heavy toxic metals from the wastewater.

14.2 PRESENT STATUS OF WASTEWATER GENERATION: INDIA AND WORLDWIDE

The noticeable increment in water is predicted for industrial sector and energy-producing sector along with the agricultural sector, which globally abstracts 70% of water. The economy of a country generally depends on the treatment of its municipal and industrial effluents. In developing countries, only 8% of the effluent undergoes such treatment. It is estimated that around 80% of wastewater is discharged worldwide without undergoing any treatment.

Wastewater consists of domestic effluents, sanitary effluents, industrial effluents, municipal effluents, runoff from agriculture, horticulture, and fish-farming sectors.[69] In the developing countries, the generation of wastewater is one of the major challenges for informal growth and settlements of slums. The toxicity of pollutants, discharged from industries in water has negative impacts on human health as well as aquatic life.

FAO estimates 3928 km^3/year of worldwide freshwater withdrawal in its AQUASTAT database. Agricultural sector consumes 1716 km^3/year (44%) of this freshwater. The remaining 56%, that is, 2212 km^3/year is released as effluents that is discharged from municipal as well as industrial sectors and agricultural drainage (Figure 14.1).

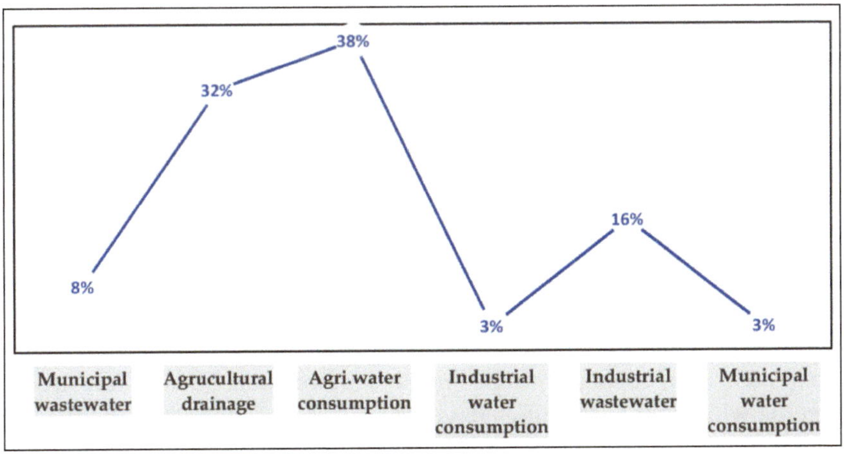

FIGURE 14.1 Worldwide wastewater production and consumption by major sectors.[49]

The constant increase in the volume of wastewater and sewage expansion pressurizes the available treatment facilities, and can also lead to poor performance. After collecting and treating wastewater, the final quality may be negatively influenced by poor maintenance and operation. Thus, much of the wastewater is not adequately treated, which eventually affects the quality and availability of water.

Domestic, commercial, industrial, and institutional sectors generate municipal wastewater in correspondence with existing community. Urban wastewater encompasses municipal wastewater and urban runoff. Throughout the world, urban areas are dealing with unusual challenges. Migration in conflicted areas, climatic change, and extreme and sudden events alter the urbanization patterns leading to unjust acts, as in some developed regions, poor community faces the same challenges similar to those belonging to developing regions. According to UN [2016], the demand for energy will increase to 40%; and for water it will increase to 50% worldwide in the year 2030.

Wastewater generated by domestic holds is usually devoid of health deteriorating and toxic components. However, commonly used medications may have long-term impacts even at low concentrations and thereby are highly concerned. In developing regions, hospitals and other sanitary establishments discharge untreated medical waste. The excessive use of antibiotics in animal husbandry and intensive farming techniques has resulted in high concentrations in municipal wastewater. It can lead to the additional peril of antimicrobial drug resistance.

14.3 GENERATION AND TREATMENT OF MUNICIPAL WASTEWATER IN METROPOLITAN CITIES OF INDIA

Around 15,644 million liter/day (ML/D) of sewage is generated from 35 metropolitan cities in India, out of which, only 51% of cities has treatment capacity, that means only 8040 ML/D of sewage is treated. Delhi has 30% of the total treatment capacity, which covers 2330 ML/D. Mumbai has 26% of total capacity, that is, 2130 ML/D. Therefore, combined treatment capacity of Delhi and Mumbai is 55% of the total treatment capacity of the metropolitan cities. The treatment capacity of various cities, such as Vadodara, Hyderabad, Ludhiana, Chennai, and Ahmadabad encounters the volume of generation. Cities such as Dhanbad and Delhi have more than 50% of the treatment capacity, while rest of the cities have less than 50% of the treatment capacity.

In India, water is provided by the Government Agencies that have become politically sensitive to overcome elections, that is why wastewater projects are not integrated with the projects of water supply. Also, wastewater projects are only employed as disposal units. Parallel increment in technologies and water crisis make people understand that wastewater treatment leads to recycle and reuse of water in various sectors, including agriculture, horticulture and can also be utilized in sludge compost to generate power.[71] Table 14.1 depicts the present status of generated, treated, and untreated sewage in India. A large number of industries also generate about 500 ML/D of effluents.[71] Table 14.2 shows the industry-wise generation of wastewater.

TABLE 14.1 Present Status of Sewage in India.

Sewage status	Quantity, ML/D
Generation of sewage	61,754
Sewage which is untreated	38,791
Treatment capacity of sewage	22,963

TABLE 14.2 Amount of Effluents Generated from Various Industries.

Industries generating wastewater	Total number of units	Quantity of wastewater generation (ML/D)
Chemical	27	97.79
Distillery	35	37
Food, Dairy and Beverage	22	6.5
Pulp and Paper	67	201.39
Sugar	67	96.69
Tannery	442	22.1
Textile, Bleaching, and Dyeing	63	11.39
Other	41	28.59
Total	**764 Units**	**501.45 ML/D**

14.4 TOXICITY OF HEAVY METALS

Heavy metal contamination in water is a global concern, particularly in countries having substandard economy. Constant growth of industrial sector and different activities of humans are the main causes that lead to the heavy metal contamination in water body. Nonbiodegradable nature of such toxic metals and their presence (beyond the permissible value) in the effluents, discharged from various industries, can deteriorate the whole living community (Table 14.3).[18] The most common toxic metals are phosphate, arsenic, fluoride, cadmium, iron, etc. Cadmium leads to toxicity in lungs, cardiovascular vessels, and kidney and lowers the count of red blood cells (anemia). Nickel can cause gastrointestinal cancer, brain and heart disease. Toxicity of iron in humans may cause sickness, cardiovascular collapse, and diarrhea. Main emphasis is on contamination caused by arsenic, because of its toxic and poisonous nature. Arsenic cannot be easily degraded but is only changed into different forms.[76]

14.5 PRODUCTION OF FRUIT AND VEGETABLE PROCESSING WASTES: GLOBAL SCENARIO

With the increasing worldwide population and changes in the human diets, the demand for various agricultural commodities has increased simultaneously. Among the agricultural materials, fruits and vegetables have major role in the diet of a human being.[74,83] Fruits are globally produced in significant amounts, for example, production of citrus fruits is about 124.729 million metric tons, banana production is about 114.079 million metric tons, apple production is about 84.629 million metric tons, grapes production is about 74.5 million metric tons, and pineapple production is about 25.429 million metric tons.[22] Similarly, vegetables also cover the myriad number of tons, such as potatoes production is 3820 million metric tons, tomatoes production is 171 million metric tons, brassicas production is 71.769 million metric tons, turnips and carrots production is 38.8 million metric tons, and peas production is 17.419 million metric tons.[22]

The term "fruit and vegetable waste" relates to the indigestible substances, which are removed or eliminated during collection, managing, and transporting and processing of fruits and vegetables.[19] Therefore, fruit and vegetable wastes can be considered as loss of fruits and vegetables. This waste can be generated at different food chain stages, that is, from farm to end user, which involves pre-consumer and postconsumer sites.[55] Horticulture sector also generates some fruit and vegetable waste. However, presently it is not of high concern. Since, they contain the materials to facilitate health-related problems and environmental problems, only if they are not used in an appropriate manner.[72]

In the world, one-third of the total food produced goes waste annually, which is about 1.3 billion metric tons,[23] among which, fruit and vegetable loss touches the maximum level, that is, 60%.[31]

During postharvest processes, about one-third of total worldwide production of fruits and vegetables is reduced and is not available to consumers. On the contrary, waste generated during postharvest processes is significant for

TABLE 14.3 Permissible Value of Toxic Metals and Their Negative Impacts on Living community.[80]

Toxic metal	Permissible limit of effluent discharged from industries (mg/L)			Permissible value given at international level (lg/L)		Impacts
	Inland surface water, IS:2490	To sewage disposal/ drainage pipeIS:3306	To irrigating water supply IS:3307	WHO	USEPA	
As	0.2	0.2	0.2	10	50	Leads to cancer.
Cd	2	1	Nil	3	5	Causes cancer, lung disease, breath-lessness, etc.
Cr	0.1	2	Nil	50	100	Neoplasm in lungs, dermatitis, etc.
Cu	3	3	Nil	Nil	1300	Irritation of nose, mouth, headache, stomach ache, dizziness, diarrhea.
Hg	0.01	0.01	Nil	1	2	Causes damage to kidneys and causes various skin diseases
Ni	3	3	Nil	Nil	Nil	Chronic bronchitis, reducing lung function, etc.
Pb	0.1	1	Nil	10	5	Anemia, muscle pain, joint pain, infertility, etc.
Zn	5	15	Nil	Nil	nil	Nausea, dizziness, chills, fidgetiness.

the consumer in developed countries. Postharvest loss ranges from 2 to 23% in the USA[40] compared with 9% in the United Kingdom.[25]

Buzby et al.[17] assessed $42.8 billion ($141/person) loss in fruits and vegetables in the US (2008). During fruits and vegetable processing, approximately, 55 MMT waste is generated in economically sophisticated regions of China, Philippines, India, and US.[23] Globally, losses in the agriculture sectors can be lowered by 47% and waste consumption by 86%, which implies that the worldwide amelioration is enormous in such regions.[44]

Along with household wastes, fruit and vegetable processing releases highest amount of waste in the world.[27] Approximately, 89 million tons of waste is generated by food processing sectors in European countries and it will increase by 40 times in the upcoming years. According to FAO, around 40% of total food is discarded as waste in India.[52,61] Indian Food Corporation (IFC) has also estimated that the amount of waste of the total production ranges from 10% to 15%, which amounts to about $4.39 billion and $10.6 billion of the total value of the produced food waste.[52] The Indian Ministry of Food Processing has suggested that fruit and vegetable loss ranges from 12 to 21 million tons. Figure 14.2 indicates the production of fruits and vegetables, their waste generation during processing.

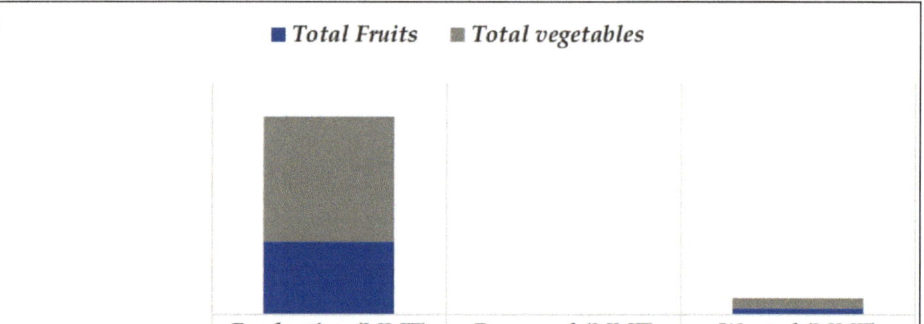

	Production (MMT)	Processed (MMT)	Wasted (MMT)
Total vegetables	865.8	0.5	70.2
Total Fruits	503.3	1.4	42.4

FIGURE 14.2 Fruit and vegetable commodities: their generation, their processing amount and generation of wasted quantity in million metric tons (MMT).[72]

In India, apple juice processing industry pomace production ranges from 35 to 40% by weight of reused apples. Although the processing volume can be more than 8 tons of dried bagasse per annum during the production of apple juice, yet most of the pomace is not presently utilized.[43]

During slicing of apples around 10.9% of pulp/seeds and 89.1% of finished products are produced. Around 8.49% of peels, 6.49% of seed residues, 32% of pulp residues and about 53% of the final product are generated during dicing of papaya. Similarly, 16% of peels and 84% of the final product are produced during the peeling of mandarins. Processing of pineapple generates about 9% of core, 14% of peels, 15% of pulp and 48% of the finished product. Mango processing industries generate 11% of peels, 13% of seed residues, 18% of pulp residues, and 58% of the finished product.[39] Furthermore, fruit and vegetable juice processing yields about 5.49 MMT of waste containing pomace also. Globally, grape and wine processing industries produce around 5–9 MMT annually of solid waste that consists of processed materials in the range of 20–30%.[72]

Fruits and vegetables during canning and freezing processes produce approximately 6 million metric tons of solid waste (comprising around 20–30% of stalk, leaves, and stems) yearly.[57] Characteristics of fruit and vegetable wastes are summarized in Table 14.4.

TABLE 14.4 Characteristics of Potential Fruit and Vegetable Waste.

Product	Waste characteristic	Amount of waste	Reference
Apples	Peels, Seeds, and Pomace	-	[30]
Bananas	Peel	35%	[30]
Citrus fruits	Peel, Seeds, etc.	50%	[30]
Dragon fruits	Seeds	(30–45) %	[20]
Durian	Seed and Skin.	(60–70) %	[79]
Grapes	Stem, Skin, and Seed	20%	[30]
Guava fruit	Core, Peels, and Seeds	10%	[30]
Jackfruit	Seeds, Rind	(50–70) %	[73]
Mango fruit	Stone and Peel	45%	[50]
Onions	Surrounded leaf	–	[30]
Potatoes	Peel	15%	[30]
Tomato	Core, Skin	20%	[30]

14.6 UTILIZATION OF PEELS AND POMACE AS ADSORBENTS: FRUITS AND VEGETABLES

Households produce large amount of fruit and vegetable wastes in the form of peels and pomace. Moreover, fruit and vegetable processing industries produce different types of waste in a significant amount during various processing operations, including peels and pomace, which are commonly discarded or used as feed for the farm animals. This negatively affects the environment and can lead to severe issues due to which there arises an essential management and better utilization of such wastes. Fruit and vegetable peels and pomace are organic, sustainable, and economic adsorbent sources and can fairly be utilized in removing several toxic metals and thereby can alleviate the water pollution. Moreover, they are renewable in nature, thus they can be employed as effective approach in the wastewater treatment.[12]

14.6.1 CITRUS FRUIT WASTE

The global yield and utilization of citrus fruits has increased tremendously during the past several years. A huge quantity of citrus waste is produced that comprises around 50% of the total fruit weight. However, they pollute the environment, as large amount of citrus fruit wastes is being burned for deposition. They are also used in the manufacture of feed for animals and in extracting various active biosubstances. To overcome this problem, they need to be managed and utilized properly to produce economic and effective products, which can be usable in different sectors without affecting the environment. Since, fruit and vegetable wastes have high adsorption capacity, therefore, much of the citrus fruit wastes can be used as bioadsorbents in the treatment of wastewater.

14.6.2 ORANGE FRUIT PEEL WASTE

Orange juice processing industries generate stacks of orange waste, which results in the environmental pollution due to the release of phenolic compounds. On the other hand, orange waste including peels can be used as a best approach for adsorbing various toxic polluting materials from the effluents as it consists of pectins, cellulosic and hemicellulosic compounds, lignin, chlorophyll pigments, and several hydrocarbons of low molecular weight. The orange peel waste from orange juice processing industries is also free of cost and its ability as an adsorbent in removing various toxic metal and metal ions from water has been studied by various researchers.

For example, Ajmal et al.[3] have observed the adsorption capacity of orange peel waste for the removal of metals (e.g., zinc, nickel, copper, lead, and chromium) from wastewater. The maximum sequestering of nickel was observed (pH = 6) followed by copper, lead, zinc, and chromium.

14.6.3 POMELO FRUIT PEEL WASTE

The adsorbing capacity of pomelo peel waste for lead removal from wastewater has been practically observed by the use of activated $ZnCl_2$ method.[46] through optimization of various adsorbing conditions, such as 5.29–6.49 of pH of wastewater at the time of starting the experiment, 10 g/L of dosage of adsorbent, 1.49 h of exposure time and concentration of 100 mg/L (t = 0) in the effluent at 30°C. This resulted in the removal of more amount of lead >90%. The adsorption of copper on the pomelo peel waste (pectinated as well as depectinated forms) has also been observed.[81] and it has been reported that the pectinated pomelo peel adsorbs around 19.69 mg/g of copper compared with 21.09 mg/g of depectinated form at various optimal provisions (concentration of 125 mg/L initially, 25°C of temperature, pH of 4 and equilibrium time of 60 min).

14.6.4 GRAPE FRUIT PEEL WASTE

Around 3.99 million metric tons of grapes are produced annually in both tropical and subtropical segments all over the globe, which implies abundant waste including peels and pomace. Grape peel wastes are composed of monomers and polymers, which are soluble and insoluble in water. The water dissolved parts consist of various units, fructose, glucose, xylose, and sucrose. On the other hand, the insoluble parts consist of cellulose, pectin, lignin, and hemicellulose units, which can fairly adsorb aquatic pollutants, because they contain carboxyl and hydroxyl functional groups.[12] Grape peels possess good adsorption capacity for removing uranium from water,[85] which is done by ion-exchange phenomenon, where anionic groups present in grape peel cell wall interchange with cations of uranium. The batch method has been employed for observing the adsorption of nickel and cadmium ions on grape peel wastes,[82] and it has been reported that grape peel waste adsorbs 46.1 mg/g of nickel and 42.1 mg/g of cadmium, respectively from wastewater.

14.6.5 BANANA PEEL WASTE

Approximately, 30–40% by weight of banana peel waste is generated, which is commonly utilized for feeding the animals, composting, and in producing several enzymes, pectins methane, proteins, etc.[77] Banana peel waste mainly consists of cellulosic and hemicellulosic units, chlorophyll pigments and other substances of low molecular weight. Since, carboxylic group and hydroxyl groups are present in the banana peels, they can adsorb toxic metals and other polluting organic compounds in large quantity from wastewater.

Cadmium and lead adsorption on the banana peels has been observed by using batch method.[9] It has been reported with the Langmuir method that 5.7 mg of cadmium and 2.2 mg of lead were absorbed by using 1 gram of banana peels. The removal of metals from the solution, synthesized in laboratory, by using chemically treated banana peels has also been observed.[8] It was found that the adsorption capacity of chemically treated banana peels for lead, nickel, zinc, copper, and cobalt removal is approximately 8, 6.9, 5.8, 4.8, and 2.5 mg/g, respectively. The banana peels also adsorb significant number of phenolic substances from water.[1] Physicochemical characteristics of banana peels are summarized in Table 14.5.

TABLE 14.5 Physicochemical Characteristics of Banana Peels.

Contents	Value
Aluminum (%)	1.049
Ash (%)	3.849
Carbon (%)	31.789
Chlorine (%)	3.219
Moisture (%)	13.549
Oxygen (%)	42.869
Potassium (%)	14.859
Silicon (%)	1.479
Sodium (%)	1.329
Volatile substance (%)	86.439
pH	6.6
Size of particle	<1

14.6.6 POMEGRANATE PEEL WASTE

The peel waste from pomegranate juice processing industries ranges from 5 to 15% of the net weight. Pomegranate peel consists of polyphenols, ellagitannin, gallic acids, etc. Pomegranate peels are utilized as cost-effective and sustainable adsorbents. Raw pomegranate peel, carbon-activated peels of pomegranate and chemically modified carbon-activated peels of pomegranate have been utilized in removing lead and copper from wastewater.[21]

The maximum adsorption of lead and copper on the peels has been reported at the optimum pH 5.59 and 5.79, respectively. Equilibrium time of 2 h has also been reported. The higher sorption capability of pomegranate peels has been reported for removing nickel, which is about 52 mg/g.[13]

Similarly, the adsorption capacity of pomegranate peels has been observed for various other metal ions, for example, copper, nickel, cadmium, zinc, where copper ions are adsorbed in abundant quantity and chromium ions are adsorbed in least quantity, while as zinc, cadmium and nickel ions are adsorbed in moderate quantity.

The ability of carbon-activated pomegranate peels has also been observed for adsorbing iron (Fe) ions from solutions.[51] The maximum adsorption capacity of carbon-activated pomegranate peels for removing iron ions was 18.5 mg/g at pH 5.99 and 1 g/L concentration of adsorbent at 29°C.

14.6.7 GARLIC PEEL WASTE

Garlic peel waste can be used as a better option in the remediation of wastewater. The availability of garlic peels as adsorbent for removing lead, copper, and nickel has been evaluated.[45] Garlic peel waste has remarkably higher adsorption capability for lead, that is, 209 mg/g than the copper and nickel. The chelation mechanism occurs between lead and oxygen atoms present on the garlic peel surface due to which lead is eventually adsorbed on the garlic peels.

14.6.8 APPLE POMACE WASTE

Apple pomace consists of polyphenolic compounds, which contain hydroxyl groups.[16] The adsorption capacity of apple pomace waste can be efficiently increased by replacing hydroxyl groups with chelating groups, such as esters, xanthate, sulfur-containing groups.

Furthermore, apple pomace is also composed of pectins, cellulose, hemicellulose, polysaccharides and lignin, which are comprised of carboxylic, ketonic, amino and hydroxyl groups. These functional groups get easily bound with the metals present in wastewater. The apple pomace also chelates the metals up to a certain limit. However, it needs to be chemically modified or treated for having the better adsorbing affinity for metal removal from wastewater. Apple waste particularly its pomace has an inimitable chemical composition and it can be utilized as a strong adsorbent if it undergoes chemical modifications. The utilization of chemically modified apple pomace as the better adsorbent for removing toxic metals from wastewater has also been reported by Chand et al.[18]

14.6.9 CUCUMBER WASTE

Cucumber wastes particularly peels are utilized in the wastewater treatment. A negatively charged polysaccharide produced from seaweed (brown in color), known as alginic acid, is used to immobilize cucumber peel components in the form of globules or beads,[11] so that the toxic metals would be easily removed from wastewater. Alginic acid has been employed as the source for beads. It has been reported that cucumber peel beads remove all contaminating substances from the wastewater. However, the adsorption capacity of cucumber peel beads for cadmium, copper, lead, and nickel removal has been found in low percentage in comparison to nonionized contaminating synthesized wastewater. It was observed that the immobilization of cucumber peels on sodium alginate/alginic acid bead makes the peel potentially efficient adsorbent for removing toxic contaminants from the drinking water.[70]

Figure 14.3 shows that cucumber beaded peels adsorbed more than 50% of cadmium, mercury, leads, nickel and copper, while they adsorb <3% of chromium and arsenic. The cucumber beaded peels are effectively usable compared with other fruit and vegetable beaded peels, such as, apple peel, potato peel, banana peel, and orange-peel beads prepared and utilized by optimizing the conditions at the same level both chemically and physically.

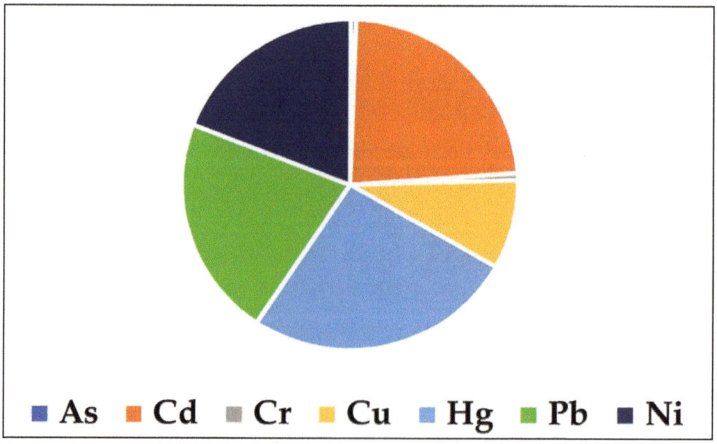

FIGURE 14.3 Adsorption of heavy metal ions by cucumber peel bead.

14.6.10 TOMATO PEEL WASTE

Tomato is the second most consumed agricultural commodity around the globe and approximately 30% of its processed form is commonly consumed. In various tomato processing industries, economically 90% of tomato peel is generated as waste. To overcome the problem of deposition of tomato waste, much of the tomato peel waste is utilized for wastewater treatment and thus becomes an alternative biotechnology method for water purification. Tomato peel waste contains pectins, carotenes, and phenolic components consisting of various functional groups, for example, amino, hydroxyl, and carboxylic functional groups, which potentially bind various toxic contaminants. The cost-effective and efficient availability and character of tomato peel waste-based adsorbent has been completely evaluated.[66]

The tomato peel waste-based adsorbent can be utilized for the extraction of multitude of contaminants from wastewater (Figure 14.4). The extraction of heavy metal ions (Pb^{2+}, Ni^{2+}, As^{3+} and Cr^{6+}) from wastewater has been investigated and analyzed to acknowledge the sorption technique.

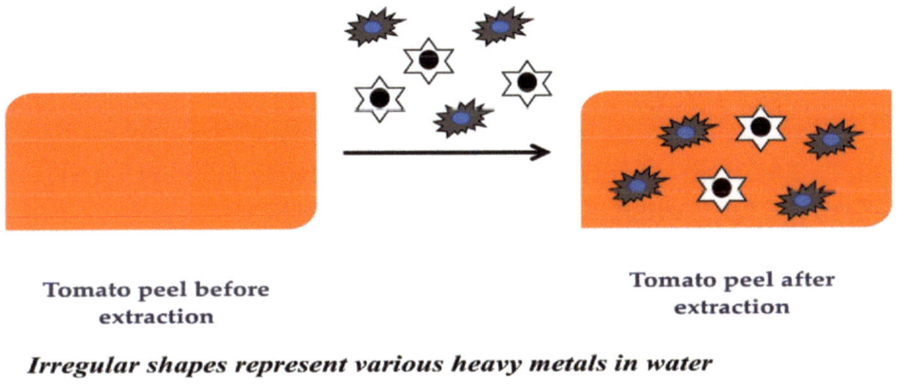

FIGURE 14.4 Extraction of heavy metals using tomato peel.

14.6.11 MISCELLANEOUS PEELS AS ADSORBENTS

The adsorption character of mandarin peel wastes for removing mercury (Hg in ionic form) from wastewater has been investigated. The investigation was done on three kinds of peels of mandarin: raw mandarin peels, pretreated mandarin peels with sodium hydroxide, and mandarin peels treated with carbon. The three types of mandarin peels exhibited different adsorption capacities for the removal of mercury ions from wastewater. For example, raw peels of mandarin exhibited approximately 19 mg/g of adsorption capacity, sodium hydroxide-treated mandarin peels exhibited 23.3 mg/g adsorption capacity and carbon-treated mandarin peels exhibited 34.8 mg/g adsorption capacity for the removal of mercury ions from water.[37]

It has been reported that the adsorption capacity of charcoal-treated potato peel for removing the copper ions from wastewater was approximately 0.39 mg/g.[7] The monolayer adsorption capability of melon peels for removing cadmium ions from water has been estimated as 82 mg/g.[33] Chemically treated watermelon peel adsorbent has been utilized for lead ions removal.

Iqbal et al.[38] have reported that mango peels can be potentially utilized as adsorbent for adsorbing nickel, copper, and zinc ions from wastewater discharged from electroplating industries. The maximum adsorption occurs at pH of 5–6. Litchi fruit peel also showed efficient adsorption capacity for chromium ions removal, while the maximum adsorption of chromium occurred at pH of 2 and the adsorption capability of litchi fruit peels for eliminating chromium ions was estimated as 101.1 mg/g. The viability and cost-effectiveness of peels of lemon, banana, and orange have been examined for the removal of toxic metals in their ionic forms, for example, copper, nickel, zinc, cadmium, and lead ions. Table 14.6 illustrates the adsorption capacities of peels and pomace of various fruits and vegetables.

14.7 FACTORS AFFECTING THE ADSORPTION OF PEELS AND POMACE OF VARIOUS FRUITS AND VEGETABLES

14.7.1 DOSAGE OF ADSORBENT

It is an important factor to estimate the adsorption capacity of fruit and vegetable peels and pomace adsorbents. The dosage of adsorbent determines the effectiveness of existing active sites present on the surfaces for the process of adsorption. Therefore, the withdrawal of adsorbate from wastewater linearly increases with the dosage of adsorbent.[42] The amount of adsorbent has a great impact on the adsorption of heavy metal ions from wastewater. Chand et al.[18] have studied the effect of apple pomace and chemically modified apple pomace doses for adsorbing cadmium at a concentration of ions 0.1 and 2.49 g per 50 mL, 0.01 and 0.1 g per 50 mL at optimal provisions, respectively. It was found that the adsorption capacity increases with the amount of adsorbent.

TABLE 14.6 Adsorbing Capacity of Different Peels and Pomace-Based Adsorbents of Fruits and Vegetables for Removing Toxic Metals.

Types of adsorbents	Toxic metal	Amount of metal adsorbed (mg/g)	References
Tomato peel	Ni^{2+}	8.71	[66]
Apple pomace	Cd^{2+}	4.45	[18]
Banana peels	Cd^{2+}	5.7	[9]
Cabbage waste	Pb^{2+}	60.57	[35]
Cassava peels	Cu^{2+} and Cd^{2+}	127.29 & 119.59, respectively	[34]
Cauliflower wastes	Pb^{2+}	47.63	[35]
Cauliflower wastes	Cd^{2+}	21.32	[35]
Chemically treated orange peels	Cu^{2+}	289	[24]
Depectinated Pomelo peels	Cu^{2+}	21.09	[81]
Garlic waste	Pb^{2+}	209	[45]
Gooseberries	Cu^{2+}	24.0	[67]
Grape pomace	Pb^{2+}, Cd^{2+}	312.5 & 135.2, respectively	[53]
Grapefruit peels	U(VI)	140.8	[85]
KCl treated orange peels	Pb^{2+}	141.8	[29]
KCl treated orange peels	Zn^{2+}	45.3	[29]
KCl treated orange peels	Ni^{2+}	49.14	[29]
KCl treated orange peels	Cd^{2+}	125.629	[29]
Lemon peels	Cobalt	22	[14]
Litchi peels	Chromium	101.1	[68]
Mango peels	Cu^{2+}	46.1	[38]
Mercerized garlic peels	Pb^{2+}	109	[47]
Modified cucumber peels	Cd^{2+}	58.2	[56]
Muskmelon peels	Pb^{2+}	0.8 mol/kg	[36]
Olive pomace waste	Ni^{2+}	14.8	[84]
Orange peels	As (III)	1.18	[41]
Peels of carrots	Cr^{3+}, Cr^{6+}	86.6 & 88.29, respectively	[15]
Pomegranate peels	Ni^{2+}	51.99	[13]
Pomelo peels	Cu^{2+}	19.69	[81]
Potato peel waste	Cu^{2+}	0.39	[7]
Pulp of sugar beet	Cu^{2+}	28.49	[4]
Thermally activated potato peels	Cu^{2+}	84.7	[28]
Zirconium-treated apple peel wastes	AsO_2^-	15.6	[48]

14.7.2 TIME OF CONTACT

The heterogeneous mixture of solids and gases often completes various stages of transfer of mass, out of which few stages are sluggish in a comparative manner. Hence, to ensure the attainment of equilibrium, it is mandatory to determine the time of contact. Adsorption of toxic metallic contaminants on fruit and vegetable peel and pomace-based adsorbents can be increased with increasing the time of contact. It can be attributed to the significant amount of metal removal from water by the adsorbent activities with the increasing time of contact.[54,60,64]

Nayak et al.[53] have depicted the effect of contact time on the adsorbing capacity of grape pomace. Results have revealed that in spite of the initial concentration of metallic ions, around 75–90% of metal was adsorbed within the first 120 min, followed by a slow falling rate and eventually the equilibrium at 270 min.

14.7.3 pH

It affects the adsorption of toxic metals by fruit and vegetable waste-based adsorbents. It influences the composition of homogenous mixture, usability of functional groups on the surface of fruit and vegetable peel and pomace and the adsorbate composition. With the variation of pH from acidic state to basic state, it facilitates the adsorption of positively charged metals. For positively charged metals, the low adsorption occurs at pH <7, because of excess protons competing with the cationic metals for the sites of adsorption. For instance, the functional groups on the tomato peels get ionized at high pH, which leads to the maximum adsorption of cations via electrostatic mechanism.[66] In the case of anionic pollutants, the low pH results in a higher adsorption since positive sites of peels can react with anionic pollutants. The removal of Cr (VI) and As (III) is maximum at pH less than 7, owing to electrostatic mechanism between the peel having cationic sites and the negatively charged ions.[66] The metal ions removal in percentage with variation of pH is shown in Figure 14.5.

14.7.4 INITIAL CONCENTRATION OF THE METAL

The adsorbing capability of fruit and vegetable peels and pomace-based adsorbent increases with the increase of initial concentration of toxic metals.

The analyzing materials in wastewater generate the propulsion due to which metals can be adsorbed by fruit and vegetable waste adsorbents. In the multi-component system, there is an efficient chemical reaction between lead and adsorbent material active sites. Moreover, the adsorption of garlic and onion wastes for metal is in the order[65]: Pb^{2+}> Sn^{2+}> Fe^{2+}> Hg^{2+}> As^{3+}> Cd^{2+}.

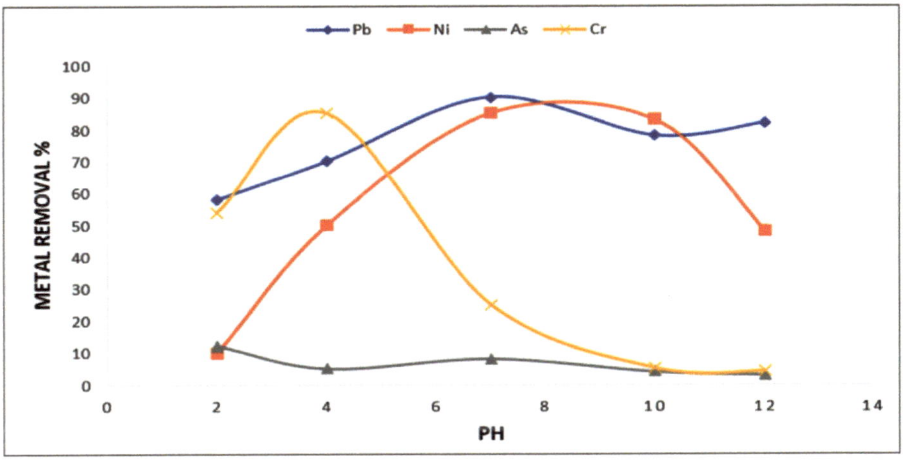

FIGURE 14.5 Effect of pH on the metal ion adsorption.

Diffusion of metals of more atomic weight generates more amounts of momentum energy that facilitates the adsorption of metals with the increment uncertainty of collision between metal and solid surface. Therefore, metals of more atomic weights are significantly and effectively adsorbed.

14.7.5 TEMPERATURE

Since temperature brings the change in the adsorption capacity of a particular adsorbent, therefore, it highly affects the adsorption of fruit and vegetable waste-based adsorbents. The extent of sorption increases with increasing temperature, since higher temperature increases the mobile nature in adsorbates and causes the decrement in the viscous nature of the homogenous mixture. For example, the maximum adsorbing capability of grape pomace for lead and cadmium ions is 312.5 and 135.2 mg/g, respectively at 25°C.[53] The adsorption of metal ions increases with the increasing temperature. This indicates that at higher temperature effective adsorption occurs, which is remarkable for adsorbing different types of pollutants from the wastewater.

14.7.6 PARTICLE SIZE/SURFACE AREA

Adsorption linearly increases with the increasing surface area. Surface area is indirectly proportional to size of particle, which generates a greater number of adsorption binding sites. It plays an effective role in the adsorption of toxic metals on fruit and vegetable peel and pomace-based adsorbents. The surface of fruit and vegetable wastes contain porous surface. Volume and diameter of pores determine the degree of diffusion of pollutants into the porous surface. If the pore diameter is larger than the adsorbate molecule diameter, then the molecule diffuses deeply into pores.[63]

Chand et al.[18] estimated that the surface area of untreated apple pomace and chemically treated apple pomace was 0.7129 m^2/g and 0.8690 m^2/g, respectively. Thus, 18% of the surface area of adsorbent has been enhanced after the chemical treatment. The particle size of untreated apple pomace and chemically treated apple pomace was 219.99 and 232.99 deci-nm, respectively.

14.8 MODIFICATION OF FRUIT AND VEGETABLE PEELS

The fruit and vegetable peels and pomace adsorbents need a modification before being used in heavy metal remediation, because they release soluble organic substances, which lead to the generation of more BOD as well as COD level thus decreasing the adsorption process that eventually depletes the oxygen level in water; and aquatic habitants are thus endangered. The proper selection of additives for modifying peels and pomace of fruits and vegetables ensures their efficient usability and re-usability. The effectiveness and efficiency of such adsorbents are enhanced by the application of available physicochemical treatment methods.[59]

To modify the size of the adsorbents, several physical methods are used, such as chopping, grinding, treatment by steam, microwave treatment, irradiation, drying (sublimation), and high pressure by using an autoclave. Similarly, in chemical modifications, adsorbents are treated with a myriad of chemicals for cleansing, enhancing the binding groups, eliminating the hampering groups, and graft copolymerization.[58]

Graft polymerization is done by applying the elevated energy radiation grafting, chemical initiation, and photochemical grafting. The circumstances for generating cost-effective adsorbents are required to be optimized after the surface modification. The treatment of banana peels with highly critical carbon dioxide brought noticeable variations in the lignocellulosic tissues.

However, any kind of considerable increment is not manifested by its capability for adsorption of metal/metal ions.[5]

Ali and Saeed[6] used un-treated banana peel wastes, banana peels hydrolyzed with alkaline solution, acid-hydrolyzed banana peel wastes, and banana peel wastes treated with bleach as adsorbents separately for washing off chromium (VI) and manganese (II) from wastewater. For chromium, untreated banana peel wastes exhibited 45% adsorption capacity, alkali and acidic-hydrolyzed banana peel wastes exhibited 87% and 67% adsorption capacities, respectively, while bleach-treated banana peel wastes exhibited adsorption capacity of 40%. Likewise for manganese, the final adsorption capacities of UT, Al, Ac, and B-banana peel wastes were 51%, 90%,74%, and 67%, respectively at optimum conditions.

It has also been reported that metal ion adsorbing capability of alkali-hydrolyzed banana peel is greatest, followed by acid-hydrolyzed and untreated banana peel, whereas bleach-treated banana peel had the lowest adsorption capacity for metal/metal ion removal from water, which indicates that the adsorption of metals is enhanced while chemically treated with banana peel wastes. Another example of modification is the application of apple pomace wastes, which are modified by using succinic anhydride and the phenomenon using simple-ring opening method. The chemically modified apple pomace wastes provide sufficient surface area for adsorbing Cd. Surface area of chemically treated apple pomace increases up to 18% and it is 50 times less quantity than the untreated apple pomace for removing cadmium from wastewater.[18]

In the same manner, HCl-treated cucumber peel waste exhibited the adsorption capacity of 85% for cadmium (II) at a metal concentration of 20 mg/L and pH of 5.[56]

14.9 REGENERATION OF PEEL-BASED ADSORBENTS

Regeneration of an adsorbent is an important step for lowering the cost of processing and metal/metal ion recovery at the end of adsorption. Regeneration is a phenomenon, in which fruits and vegetables wastes are washed off by treating with a specific solvent that leads to a concentrated metallic contaminant recovery. At industrial level, the most considerable application of adsorption is to sequester the heavy and toxic metals by utilizing specific adsorbents and regenerating such adsorbents simultaneously so that they can be used in further wastewater treatment operations.

The applicability of adsorbents depends on various factors and functional groups (active). There are many chemical compounds, which are used in different concentrations in desorbing toxic metals from fruit and vegetable peels, such as HNO_3, H_2SO_4, HCl. For example, alkaline solutions of sodium hydroxide, potassium hydroxide, and acidic solution of phosphoric acid are suitably employed in the desorption of As (arsenic) from agricultural waste-based adsorbents.[42]

Negi et al.[65] observed metal recovery, using 2% of HNO_3 in 100 mL concentrated solution in the sequence: iron (70.99%) > tin (66.19%) > lead (59.59%) > cadmium (59.19%) > arsenic (57.59%) > mercury (57.19%). When the metallic pollutants are retrieved, the biomaterials do not cause any degeneracy in the natural ecosystem as they are not degraded in the earth's crust and these materials can also be used in fertilizing and conditioning of the soil. Thus, the agricultural (fruit and vegetable) wastes can be utilized to purify the ecosystem as well as sewage, discharged from various sectors, from toxic metals. Therefore, recycled adsorbents can be efficiently used and their usefulness results in an affordable and economical devolution phenomenon.

14.10 FUTURE PERSPECTIVES

Since, adsorption is a physical and chemical process, which merely depends upon different parameters and its mechanisms. Thorough understanding of fruit and vegetable waste adsorption mechanisms remains a challenge. Fruit and vegetable peels and pomace waste adsorption requires further research to make it easier for understanding and applying in various effluent discharge sectors to sequester toxic metals from the effluents. Fruit and vegetable peels and pomace are observed for their utility in removing toxic metals, some organic compounds, and dyes from wastewater.

Exploration of utilization of fruit and vegetable peels and pomace for the removal of more organic and inorganic polluting materials is highly concerned and further needed. They are also required to be utilized for adsorbing contaminants (NH_3, CH_4, NO_X, CO_2, etc.).[26] To promote the utilization of adsorbents at large-scale level, further investigation is needed for characterizing fruit and vegetable peels and pomace-based adsorbents with maximum adsorption capacities for toxic metals. During the treatment process, fruit and vegetable peels release soluble organic compounds into water, which restrict their employment in high-level applications. Therefore, further study is required in this field.

Moreover, adsorbents can be altered by applying sustainable chemical materials to cease more deterioration of water bodies. This type of modification can lead to the increment in multitude of active sites, for binding metals, in the substance, thus improves the properties of exchange of ions and new active groups are formed, which favor the uptake of polluting materials.

Studies are further needed to understand the adsorption mechanism at the adsorbent–water boundary. The idea of utilizing fruit and vegetable peels and pomace-based adsorbents might be favorable in developing a sustainable (green) filtration technology for purifying toxic metal-contaminated water that may facilitate the communities having low economy to utilize this water for drinking purposes and to keep them from health ailments.[42] Preparing, optimizing and employing new technology of fruit and vegetable waste-based adsorbent for wastewater treatment at pilot as well as commercial stage by applying column adsorption method needs further research. It seems to be glistening future for new such adsorbents, which can be effective and economically feasible.

14.11 SUMMARY

In developing countries, wastewater treatment is not up to the mark, because of lacking an economy to facilitate and alleviate the pollution of water bodies. Economic level in developing countries is also not parallel with the increasing population and volume of wastewater. which consequently increases. This results in the generation of untreated wastewater and significant amount of this untreated wastewater is usually used for irrigating the fields by less-skilled farmers to improve the quality of the wastewater.

The utilization of waste peel-based adsorbents for eluting various metals from effluents offers various alluring capabilities, like the noticeable adsorption capacity for different polluting materials and these waste peel-based adsorbent also possesses low-cost. These are nontoxic and biologically more preferable. Metals having cationic nature are more effectively adsorbed on fruit and vegetable wastes. Moreover, different methods of surface modification, utilized by researchers are of beneficial use for enhancing the capacity of adsorption.

Treated or modified fruit and vegetable wastes show better adsorption capacity in comparison to the wastes, which are not treated or modified by any means. Heavy metallic polluting materials are efficiently removed by cellulose extracted and polyacrylic acid grafted fruit and vegetable wastes. Overall, fruit and vegetable waste-based adsorbents are collectively a better alternative for controlling water pollution over presently activated carbon technique, which is expensive. In addition to this, they can also contribute in minimizing whole waste production.

KEYWORDS

- **adsorbent**
- **fruit and vegetable waste**
- **heavy toxic metals**
- **peel and pomace**
- **regeneration**
- **wastewater treatment**
- **wastewater**

REFERENCES

1. Achak, M.; Hafidi, A.; Ouazzani, N.; Sayadi, S.; Mandi, L. Low-Cost Biosorbent (Banana Peel) for the Removal of Phenolic Compounds from Olive Mill Wastewater: Kinetic and Equilibrium Studies. *J. Hazard. Mater.* **2009,** *166*, 117–125.
2. Ahmad, T.; Danish, M. Prospects of Banana Waste Utilization in Wastewater Treatment: A Review. *J. Environ. Manage.* **2018,** *206*, 330–348.
3. Ajmal, M.; Rao, R. A. K.; Ahmad, R.; Ahmad, J. Adsorption Studies on *Citrus Reticulata* (Fruit Peel of Orange): Removal and Recovery of Ni (II) from Electroplating Wastewater. *J. Hazard. Mater.* **2000,** *79*, 117–131.
4. Aksu, Z.; İşoğlu, I. A. Removal of Copper (II) Ions from Aqueous Solution by Biosorption Onto Agricultural Waste Sugar Beet Pulp. *Process Biochem.* **2005,** *40*, 3031–3044.
5. Albarelli, J. Q.; Rabelo, R. B.; Santos, D. T.; Beppu, M. M.; Meireles, M. A. A. Effects of Supercritical Carbon Dioxide on Waste Banana Peels for Heavy Metal Removal. *J. Superscript Fluids* **2011,** *58*, 343–351.
6. Ashraf, A.; Khalid, S. Decontamination of Cr (VI) and Mn (II) from Aqueous Media by Untreated and Chemically Treated Banana Peel: A Comparative Study. *Desalination Water Treatment* **2015,** *53* (13), 3586–3591.
7. Aman, T. et al. Potato Peels as Solid Waste for the Removal of Heavy Metal Copper (II) from Wastewater/Industrial Effluent. *Colloids Surf. B: Bio Interf.* **2008,** *63*, 116–121.
8. Annadurai, G.; Juang, R. S.; Lee, D. J. Adsorption of Heavy Metals from Water Using Banana and Orange Peels. *Water Sci. Technol.* **2003,** *47*, 185–190.
9. Anwar, J.; Shafique, U.; Waheed, Z.; Salman, M.; Dar, A.; Anwar, S. Removal of Pb (II) and Cd (II) from Water by Adsorption on Peels of Banana. *Bioresour. Technol.* **2010,** *101*, 1752–1755.

10. Ayala-Zavala, J F.; Rosas-Domínguez C.; Vega-Vega V.; Gonzalez-Aguilar GA. Antioxidant Enrichment and Antimicrobial Protection of Fresh-Cut Fruits Using Their Own by Products: Looking for Integral Exploitation. *J. Food Sci.* **2010,** *75*, 175–81.
11. Banerjee, S.; Banerjee, A.; Sarkar, P. Statistical Optimization of Arsenic Biosorption by Microbial Enzyme via Ca Alginate Beads. *J. Environ. Sci. Health* **2018,** *53* (5), 436–442.
12. Bhatnagar, A.; Sillanpää, M.; Witek-Krowiak, A. Agricultural Waste Peels as Versatile Biomass for Water Purification: A Review. *Chem. Eng. J.* **2015,** *270*, 244–271.
13. Bhatnagar, A.; Minocha, A. K. Biosorption Optimization of Nickel Removal from Water Using Pomegranate Peel Waste. *Colloids Surfaces B* **2010,** *76*, 544–548.
14. Bhatnagar, A.; Minocha, A. K.; Sillanpaa, M. Adsorptive Removal of Cobalt from Aqueous Solution by Utilizing Lemon Peel as Biosorbent. *BioChem. Eng. J.* **2010,** *48*, 181–186.
15. Bhatti, H. N.; Nasir, A. W.; Hanif, M. A. Efficacy of Daucus Carrot Waste Biomass for the Removal of Chromium from Aqueous Solutions. *Desalination* **2010,** *253*, 78–87.
16. Bhushan, S.; Kalia, K.; Sharma, M.; Singh, B.; Ahuja, P. S. Processing of Apple Pomace for Bioactive Molecules. *Crit. Rev. Biotechnol.* **2008,** *28*, 285–296.
17. Buzby, J. C.; Hyman, J.; Stewart, H.; Wells, H. F. The Value of Retail- and Consumer-Level Fruit and Vegetable Losses in the United States. *J. Consum. Affairs Fall* **2015,** *45*, 492–515.
18. Chand, P.; Shil, A. K.; Sharma, M.; Pakade, Y. B. Improved Adsorption of Cadmium Ions from Aqueous Solution Using Chemically Modified Apple Pomace: Mechanism, Kinetics, and Thermodynamics. *Int. Biodeterior. Biodegrad.* **2014,** *90*, 8–16.
19. Chang, J. I.; Tsai, J. J.; Wu, K. H. Composting of Vegetable Waste. *Waste Manage. Resour.* **2006,** *24*, 354–362.
20. Cheok, C. Y.; Adzahan, N. M.; Rahman, R. A.; Abedin, N. H. Z.; Hussain, N.; Sulaiman, R.; Chong, G. H. Current Trends of Tropical Fruit Waste Utilization. *Crit. Rev. Food Sci. Nutr.* **2018,** *58* (3), 335–61.
21. El-Ashtoukhy, E. S. Z.; Amin, N. K.; Abdelwahab, O. Removal of Lead (II) and Copper (II) from Aqueous Solution Using Pomegranate Peel as a New Adsorbent. *Desalination* **2008,** *223*, 162–173.
22. *FAO Statistics Data.* 2014. www.fao.org/faostat/en data; Accessed on April 26, 2021.
23. FAO. *Definitional Framework of Food Losses and Waste*; FAO: Rome, Italy, 2014. http://www.alimentosargentinos.gob.ar/HomeAlimentos/ValoremoslosAlimentos/documentos/FLW_Definition_and_Scope_2014.pdf (accessed April 26, 2021).
24. Feng, N.; Guo, X.; Liang, S. Adsorption Study of Copper (II) by Chemically Treated Orange Peel. *J. Hazard. Mater.* **2009,** *164*, 1286–1292.
25. Garnett, T. Fruit and Vegetables and UK Greenhouse Emissions: Exploring the Relationship; Food and Climate Research Network, University of Surrey, 2006.
26. Giraldo, L.; Bastidas-Barranco, M.; Moreno-Pirajan, J. C. Preparation of Carbon Monoliths from Orange Peel for NOx Retention. *Orient. J. Chem.* **2014,** *30*, 1517–1528.
27. Gowe, C. Review on Potential Use of Fruit and Vegetables by-Products as a Valuable Source of Natural Food Additives. *Food Sci. Qual. Manage.* **2015,** *45*, 47–61.
28. Guechi, E. K.; Hamdaoui, O. Evaluation of Potato Peel as a Novel Adsorbent for the Removal of Cu (II) from Aqueous Solutions: Equilibrium, Kinetic, and Thermodynamic Studies. *Desalination Water Treatment* **2015,** *57* (23), 10677–10688.
29. Guo, X. Y.; Liang, S.; Tian, Q. H. Removal of Heavy Metal Ions from Aqueous Solutions by Adsorption Using Modified Orange Peel as Adsorbent. *Adv. Mater. Res.* **2011,** 236–240.

30. Gupta, K.; Joshi, V. K. (Eds.) *Postharvest Technology of Fruits and Vegetables: Handling, Processing, Fermentation and Waste Management*, 2nd ed.; Indus Publishing Company: New Delhi, 2002; p 1240.
31. Gustavsson, J.; Cederberg, C.; Sonesson, U.; van Otterdijk, R.; Meybeck, A. *Global Food Losses and Waste: Extent, Causes and Prevention*; Food and Agriculture Organization (FAO) of the United Nations: Rome, Italy, 2011. http://www.fao.org/3/i2697e/i2697e.pdf (accessed April 27, 2021).
32. Hala-Ahmed, H. Removal of Heavy Metals from Wastewater Using Agricultural and Industrial Wastes as Adsorbents. *HBRC J.* **2013,** *9*, 276–282.
33. Hamdaoui, O.; Saoudi, F.; Chiha, M. Utilization of an Agricultural Waste Material, Melon (Cucumis melo L.) Peel, as a Sorbent for the Removal of Cadmium from Aqueous Phase. *Desalination Water Treatment* **2010,** *21*, 228–237.
34. Horsfall M.; Spiff, A. I.; Abia, A. A. Studies on the Influence of Mercapto Acetic Acid (MAA) Modification of Cassava Waste Biomass on the Adsorption of Cu^{2+} and Cd^{2+} from Aqueous Solution. *Bull. Korean Chem. Soc.* **2004,** *25*, 969–976.
35. Hossain, M. A.; Ngo, H. H.; Guo, W. S.; Nguyen, T. V.; Vigneswaran, S. Performance of Cabbage and Cauliflower Wastes for Heavy Metals Removal. *Desalination Water Treatment* **2013,** *52* (4–6), 844–860.
36. Huang, K.; Zhu, H. Removal of Pb^{2+} from Aqueous Solution by Adsorption on Chemically Modified Muskmelon Peel. *Environ. Sci. Pollut.. Res.* **2013,** *20*, 4424–4434.
37. Husein, D. Z. Adsorption and Removal of Mercury Ions from Aqueous Solution Using Raw and Chemically Modified Egyptian Mandarin Peel. *Desalination Water Treatment* **2013,** *51*, 6761–6769.
38. Iqbal, M.; Saeed, A.; Kalim, I. Characterization of Adsorptive Capacity and Investigation of Mechanism of Cu^{2+}, Ni^{2+} and Zn^{2+} Adsorption on Mango Peel Waste from Constituted Metal Solution and Genuine Electroplating Effluent. *Sep. Sci. Technol.* **2009,** *44*, 3770–3791.
39. Joshi, V. K.; Kumar, A.; Kumar, V. Antimicrobial, Antioxidant and Phyto-Chemicals from Fruit and Vegetable Wastes: A Review. *Int. Food Farm Technol.* **2012,** *2*, 123–136.
40. Kader, A. A. Increasing Food Availability by Reducing Postharvest Losses of Fresh Produce. *Acta Hort.* **2005,** *682*, 2169–2175.
41. Kamsonlain, S.; Balomajumder, C.; Chand, S. Studies on Surface Characterization and Isotherm Modelling: Biosorption of Arsenic (III) Onto Low-Cost Biosorbent Derived from Orange Peel. *J. Sci. Ind. Res.* **2012,** *71*, 810–816.
42. Klump, S. Food Manufacturer Perspective on Food Packaging Safety. Unpublished Paper Presentation at Quality & Safety Food Packaging Forum Workshop, Zurich; 8 October 2015; p 13.
43. Kummu, M.; de Moel, H.; Porkka, M.; Siebert, S.; Varis, O.; Ward, P. J. Lost Food, Wasted Resources: Global Food Supply Chain Losses and Their Impacts on Freshwater, Cropland, and Fertilizer Use. *Sci. Total Environ.* **2012,** *438*, 477–489.
44. Kuswandi, B.; Wicaksono, Y.; Abdullah, A.; Heng, L. Y.; Ahmad, M. Smart Packaging: Sensors for Monitoring of Food Quality and Safety. *Sens. Instr. Food Qual. Safet.* **2011,** *5* (3–4), 137–146.
45. Liang, S.; Guo, X.; Tian, Q. Adsorption of Pb2+, Cu2+ and Ni2+ from Aqueous Solutions by Novel Garlic Peel Adsorbent. *Desalination Water Treatment*, **2013,** *51*, 7166–7171.

46. Liu, J. Y.; Huang, G. H.; Deng, J. Q.; Liu, K.; Xie, Y. B. Adsorbent Prepared from Waste Pomelo Peel and Its Adsorption of Pb^{2+} in Wastewater. *J. Ecol. Rural Environ.* **2012,** *28*, 187–191.
47. Liu, W.; Liu, Y.; Tao, Y.; Yu, Y.; Jiang, H.; Lian, H. Comparative Study of Adsorption of Pb (II) on Native Garlic Peel and Mercerized Garlic Peel. *Environ. Sci. Pollut. Res.* **2014,** *21*, 2054–2063.
48. Mallampati, R.; Valiyaveettil, S. Apple Peels: A Versatile Biomass for Water Purification. *ACS Apple Mater. Interf.* **2013,** *5*, 4443–4449.
49. Mateo-Sagasta, J.; Raschid-Sally, L.; Thebo, A. Global Wastewater and Sludge Production, Treatment and Use. *Wastewater* **2015,** *2015*, 15–38.
50. Mitra, S. K.; Pathak, P. K.; Devi, H. L.; Chakraborty, Part I: Utilization of Seed and Peel of Mango. *Acta Hort.* **2013,** *992*, 593–596.
51. Moghadam, M.; Nasirizadeh, N.; Dashti, Z.; Babanezhad, E. Removal of Fe (II) from Aqueous Solution Using Pomegranate Peel Carbon: Equilibrium and Kinetic Studies. *Int. J. Ind. Chem.* **2013,** *4*, 1–6.
52. National Academy of Agricultural Science (NAAS). *Saving the Harvest: Reducing the Food Loss and Waste*; National Academy of Agricultural Sciences; New Delhi, India, 2019; p *5*. http://naasindia.org/documents/Saving%20the%20Harvest.pdf (accessed March 19, 2021).
53. Okafor, P. C.; Okon, P. U.; Daniel, E. F.; Ebenso, E. E. Adsorption Capacity of Coconut Shell for Lead, Copper, Cadmium and Arsenic from Aqueous Solutions. *Int. J. Electrochem. Sci.* **2012,** *7*, 12354–12369.
54. Orunima, N.; Brij, B.; Vinod, K. G.; Laura, R. T. Development of a Green and Sustainable Clean Up System from Grape Pomace for Heavy Metal Remediation. *J. Environ. Chem. Eng.* **2016,** *4*, 4342–4353.
55. Panda, S. K.; Mishra, S. S.; Kayitesi, E.; Ray, R. C. Microbial-Processing of Fruit and Vegetable Wastes for Production of Vital Enzymes and Organic Acids: Biotechnology and Scopes. *Environ. Res.* **2016,** *146*, 161–172.
56. Pandey, R.; Ansari, N. G.; Prasad, R. L.; Murthy, R. C. Removal of Cd (II) Ions from Simulated Wastewater by HCl Modified *Cucumis sativus* Peel: Equilibrium and Kinetic Study. *Air Soil Water Res.* **2014,** *7*, 93–101.
57. Panouille, M.; Ralet, M. C.; Bonnin, E.; Thibault, J. F. Recovery and Reuse of Trimmings and Pulps from Fruit and Vegetable Processing. In *Waste Management and Co-product Recovery in Food Processing*; Waldron, K. W., Ed.; Woodhead Publishing Limited: Cambridge, 2007; pp 417–447.
58. Park, D.; Yun, Y. S.; Park, J. M. The Past, Present, and Future Trends of Biosorption. *Biotechnol. Bioprocess Eng.* **2010,** *15*, 86–102.
59. Patel, S. Potential of Fruit and Vegetable Wastes as Novel Biosorbents: Summarizing the Recent Studies. *Rev. Environ. Sci. Biotechnol.* **2012,** *11*, 365–380.
60. Pehlivan, E.; Altun, T.; Parlayici, S. Modified Barley Straw as a Potential Biosorbent for Removal of Copper Ions from Aqueous Solution. *Food Chem.* **2012,** *135*, 2229–2234.
61. Plazzotta, S.; Manzocco, L.; Nicoli, M. C. Fruit and Vegetable Waste Management and the Challenge of Fresh-Cut Salad. *Trends Food Sci. Technol.* **2017,** *63*, 51–59.
62. Pranav, D. P. Characterizing Fruit and Vegetable Peels as Bioadsorbents. *Science*, **2016,** *110* (11), Online; E-article 18520; DOI: 10.18520/cs/v110/i11/2114–2123

63. Pranav, D. P.; Sachin, A.; Mandavgane, B.; Kulkarni, D. Fruit Peel Waste as a Novel Low-Cost Bioadsorbent. *Rev. Chem. Eng.* **2014,** *31* (4), 361–381.
64. Rahaman, M. S.; Basu, A.; Islam, M. R. The Removal of As (III) and As (V) from Aqueous Solutions by Waste Materials. *Bioresour. Technol.* **2008,** *99*, 2815–2823.
65. Rahul, N.; Gouri, S.; Yogesh, K. T.; Rajinder, K. G. Biosorption of Heavy Metals by Utilizing Onion and Garlic Wastes. *Int. J. Environ. Pollut.* **2012,** *49* (3–4).
66. Ramakrishna, M.; Suresh, V. Application of Tomato Peel as an Efficient Adsorbent for Water Purification-Alternative Biotechnology. *R. Soc. Chem.* **2012,** *2*, 9914–9920.
67. Rao, R. A. K.; Ikram, S. Sorption Studies of Cu (II) on Gooseberry Fruit and Its Removal from Electroplating Wastewater. *Desalination* **2011,** *277*, 390–398.
68. Rao, R. A. K.; Rehman, F.; Kashif U. M. Removal of Cr (VI) from Electroplating Wastewater Using Fruit Peel of Litchi. *Desalination Water Treatment* **2012,** *49*, 136–146.
69. Risha, J. S.; Candace, E. M. Immobilized Apple Peel Bead Biosorbent for the Simultaneous Removal of Heavy Metals from Cocktail Solution. *Cogent Environ. Sci.* **2019,** *5* (1), E-article 1673116; p 9; https://doi.org/10.1080/23311843.2019.1673116
70. Risha, J. S.; Candace, E. M.; Dave, B.; Rhonda, J. R. Cucumber Peel Bead Biosorbent for Multi-Ion Decontamination of Drinking Water Collected from a Mine Region in New Zealand. *Environ. Technol.* **2019,** *42* (16), 2461–2477
71. Sabarna R. *Wastewater Generation and Treatment – Present Status in India*; 2020. http://indianinfrastructure.com (accessed April 24, 2021).
72. Sagar, N. A.; Pareek, S.; Sharma, S.; Yahia, E. M.; Lobo, M. G. Fruit and Vegetable Waste: Bioactive Compounds, Their Extraction, and Possible Utilization. *Compreh. Rev. Food Sci. Food Safet.* **2018,** *17*, 512–531.
73. Saxena, A.; Bawa, A. S.; Raju, P. S. Jackfruit (*Artocarpus heterophyllus* Lam.) In *Postharvest Biology and Technology of Tropical and Subtropical Fruits*; Yahia, E. M., Ed.; Woodhead Publishing Ltd.: Cambridge, 2011; pp 275–298.
74. Schieber, A.; Stintzing, F. C.; Carle, R. By-Products of Plant Food Processing as a Source of Functional Compounds - Recent Developments. *Trends Food Sci. Technol.* **2001,** *12*, 401–413.
75. Seema, P. Potential of Fruit and Vegetable Wastes as Novel Biosorbents: Summarizing the Recent Studies. *Rev. Environ. Sci. Biotechnol.* **2012,** *11*, 365–380.
76. Seema, P.; Goyal, A. Recent Developments in Mushrooms as Anti-Cancer Therapeutics: A Review. *Biotech* **2012,** *2*, 1–15. https://doi.org/10.1007/s13205–011–0036–2
77. Silva, C. L.; Gomes, T. F.; G. C. R. M.; Andrade, S. H.; Monteiro, A. C. R.; Dias, E. A. G.; Zagatto, V. L. Banana Peel as Bioadsorbent for Removing Atrazine and Ametryne from Waters. *J. Agric. Food Chem.* **2013,** *61*, 2358–2363.
78. Singh, R.; Dwivedi, U. N. Effect of Ethrel and 1-Methylcyclopropene (1-MCP) on Antioxidants in Mango (*Mangifera indica*) During Fruit Ripening. *Food Chem.* **2008,** *111*, 951−956.
79. Sivakumar, D.; Jiang, Y. Maintaining Mango (*Mangifera indica* L.) Fruit Quality During the Export Chain. *Fruits Res. Int.* **2011,** *44* (5), 1254–1263.
80. Sivapalasingam, S.; Barrett, E.; Kimura, A.; Van Duyne, S.; De, U. I. T.; Ying, M. Salmonella Enteric Serotype Newport Infection Linked to Mango Consumption: Impact of Water-Dip Disinfection Technology. *Clin. Infect. Dis.* **2003,** *37*, 1585−1590.
81. Tasaso, P. Adsorption of Copper Using Pomelo Peel. *J. Clean Energy Technol.* **2014,** *2*, 154–157.

82. Torab-Mostaedi, M.; Asadollahzadeh, M.; Hemmati, A.; Khosravi, A. Equilibrium, Kinetic, and Thermodynamic Studies for Biosorption of Cadmium and Nickel on Grapefruit Peel. *J. Taiwan Inst. Chem. Eng.* **2013,** *44*, 295–302.
83. Vilarino, M. V.; Franco, C.; Quarrington, C. Food Loss and Waste Reduction as an Integral Part of a Circular Economy. *Front. Environ. Sci.* **2017,** *5*, 1–5.
84. Yasar, N.; Emine, M. Thermodynamic and Kinetic Studies for Environmentally Friendly Ni (II) Biosorption Using Waste Pomace of Olive Oil Factory. *Bioresour. Technol.* **2009,** *100* (8), 2375–2380.
85. Zou, W.; Zhao, L.; Zhu, L. Adsorption of Uranium (VI) by Grapefruit Peel in a Fixed-Bed Column: Experiments and Prediction of Breakthrough Curves. *J. Radioanalyt. Nucl. Chem.* **2013,** *295*, 717–727.

INDEX

Q